Polymers for Fibers and Elastomers

ACS SYMPOSIUM SERIES 260

Polymers for Fibers and Elastomers

Jett C. Arthur, Jr., EDITOR
Technical Consultant

ASSOCIATE EDITORS
R. J. Diefendorf, T. F. Yen,
H. L. Needles, J. R. Schaefgen,
M. Jaffe, A. L. Logothetis

Based on a symposium sponsored by
the Macromolecular Secretariat
at the 186th Meeting
of the American Chemical Society,
Washington, D.C.,
August 28–September 2, 1983

American Chemical Society, Washington, D.C. 1984

Library of Congress Cataloging in Publication Data

Polymers for fibers and elastomers.
(ACS symposium series, ISSN 0097-6156; 260)

"Symposium on Polymers for Fibers and Elastomers"—Pref.

Includes bibliographies and indexes.

1. Textile fibers, Synthetic—Congresses.
2. Elastomers—Congresses. 3. Polymers and polymerization—Congresses.

I. Arthur, Jett C. II. American Chemical Society. Macromolecular Secretariat. III. American Chemical Society. Meeting (186th: 1983: Washington, D.C.) IV. Symposium on Polymers for Fibers and Elastomers (1983: Washington, D.C.) V. Series.

TS1548.5.P635 1984 677'.4 84-14635
ISBN 0-8412-0859-X

PRINTED IN THE UNITED STATES OF AMERICA

ACS Symposium Series

M. Joan Comstock, *Series Editor*

FOREWORD

The ACS SYMPOSIUM SERIES was founded in 1974 to provide a medium for publishing symposia quickly in book form. The format of the Series parallels that of the continuing ADVANCES IN CHEMISTRY SERIES except that in order to save time the papers are not typeset but are reproduced as they are submitted by the authors in camera-ready form. Papers are reviewed under the supervision of the Editors with the assistance of the Series Advisory Board and are selected to maintain the integrity of the symposia; however, verbatim reproductions of previously published papers are not accepted. Both reviews and reports of research are acceptable since symposia may embrace both types of presentation.

CONTENTS

CARBON FIBERS IN POLYMERS

FORMATION AND MODIFICATION OF FIBERS

STATE OF THE ART IN FIBER TECHNOLOGY

PREFACE

THIS IMPORTANT VOLUME is concerned with giant molecules, polymers, that have been indispensable and abundant since the beginning of human existence. Naturally occurring polymers, such as cellulose, starch, and proteins, sustained human beings and other living things for many centuries. Now we depend upon both natural and man-made polymers.

The timing of this volume is appropriate. Last year was the 50th anniversary of Gibson and Fawcett's polymerization of ethylene, an accidental discovery that led eventually to the commercial manufacture of polyethylene by England's Imperial Chemical Industries. Parenthetically, Carl S. ("Speed") Marvel made linear polyethylene with a metal catalyst 54 years ago. However, the opportunity presented by his discovery was not exploited by industry.

Several decades ago our knowledge of polymer science was so limited that we could not manufacture even one synthetic polymer fiber. The situation is entirely different today. Tremendous progress has been made in polymer and textile science; we now can make polymers and fibers from almost any raw material, including petroleum, natural gas, corncobs, sand, and pitch. Today, approximately one dozen synthetic fibers are commercially important. A recent addition to the commercial fibers is poly(benzimidazole), a high-temperature and chemical-resistant fiber intended for use in protective apparel. In addition, an incredibly simple method for making carbon fibers, recently discovered, consists in passing natural gas through a hot stainless steel tube.

The need to develop synthetic fibers was recorded as early as the 17th century (Hooke's "Micrographia," 1665): "There might be a way found out to make an artificial glutinous composition much resembling, if not fully as good, nay better, than that excrement or whatever other substance it be out of which the silk-worm wire-draws his clew." This theme was resurrected in 1734 by the French naturalist, de Reaumur, but another 150 years passed before Swann in England and de Chardonet in France produced the first synthetic fibers in which nitrocellulose served as the "artificial glutinous composition" resembling natural silk.

Total production of synthetic polymer fibers is high. In 1978, 3.53 billion kg and 11 billion kg of noncellulosic organic fibers were produced in the United States and world, respectively. It has been predicted that production of the truly synthetic fibers will double to about 24.5 billion kg during the next decade.

A similar story can be told for synthetic rubbers. Several decades ago we could not even make them. Now, several types of synthetic elastomers, each having special properties and uses, are manufactured in large quantities from various raw materials.

The picture is even more exciting when we contemplate the entire synthetic polymer scene. Although polymer science as a distinct field is barely 60 years old, synthetic polymers are ubiquitous in modern life. A recent study sponsored by the National Research Council estimates that at least one-third of all industrial chemists deal with polymers of one kind or another, and that polymer processing accounts for nearly $100 billion of American manufacturing.

If the natural fiber proponents were complacent before the advent of the man-made fibers, they quickly lost their complacency when faced with the new competition. Research on the natural fibers was expanded, with special emphasis on efforts to minimize market losses to the man-made fibers. The natural fiber scientists also enjoyed substantial success in their research. As one result, modified natural fibers were given new properties, for example, wrinkle resistance, flame resistance, enhanced susceptibility to dyeing, and increased resistance to deterioration caused by heat and microorganisms.

The natural fibers, particularly those based on cellulose, do have some advantages. For example, cellulose is abundant, annually renewable, inexpensive, and gets its carbon from the air instead of petroleum. In addition, cellulose is amenable to many types of chemical and mechanical treatments. Moreover, cellulose is the most abundant organic material on earth, with some 5×10^{14} kg being formed each year by plants.

The Macromolecular Secretariat Symposium upon which this book is based had as its objective interdisciplinary discussions of the topic from the points of view of five divisions of the American Chemical Society: Cellulose, Paper, and Textile; Colloid and Surface; Polymer; Polymeric Materials Science and Engineering; and Rubber Divisions. The international interest in this topic is evident from the fact that contributions were received from 10 countries.

This volume will facilitate further progress in polymers as it will disseminate valuable information on a wide variety of topics. The information that this volume provides gives support to the optimistic view of the future of polymer science and technology.

CHARLES H. FISHER
Roanoke College
Salem, VA 24153

CHARACTERIZATION OF FIBERS BY INSTRUMENTAL METHODS

1

Instrumental Analyses of Modified Cottons

RUTH R. BENERITO, TRUMAN L. WARD, and OSCAR HINOJOSA

Southern Regional Research Center, Agricultural Research Service, U.S. Department of Agriculture, New Orleans, LA 70179

Chemically modified celluloses have been analyzed by conventional wet methods and by various instrumental methods designed to differentiate bulk and surface properties. Electron emission spectroscopy for chemical analyses (ESCA) used alone and in combination with radiofrequency cold plasmas yielded elemental analyses, oxidative states of the element, and distribution of the element. Techniques of electron paramagnetic resonance (EPR), chemiluminescence, reflectance infrared spectroscopy, electron microscopy, and energy dispersive X-ray analyses were also used to detect species on surfaces and to obtain depth profiles of a given reagent in chemically modified cottons.

Tetraalkylammonium salts, first synthesized by Hofmann[1] in 1851 by the reaction of a tertiary amine with an alkyl halide, are soluble in various polar solvents. These salts form crystalline hydrates that contain large numbers of water of hydration molecules. The unusual physical properties of tetraalkylammonium salts and their effects on the structure of bulk water have been reported.[2,3,4,5] Currently, several theories exist as to the effects of the tetraalkylammonium cations on the structure or entropy of bulk water. Yet to be understood are the effects of the anions associated with the tetraalkylammonium ions on the overall structure of water. Not only is the theory of interactions of such salts with water of interest to those engaged in basic chemistry, but the salts also are used in various applications. The ability of quaternary ammonium cations to attract a variety of dyes and both hydrophilic and hydrophobic microorganisms and particles of soil has been used to advantage by textile chemists. Of interest to the cellulose chemist are surface-active quaternary ammonium salts that have at least one long chain alkyl group and the chemistry of anion exchange resins that contain quaternary ammonium groups. Boyd and co-workers[3] have reported on such anion exchange resins, and we have reported the preparations of various quaternary ammonium cellulose anion exchangers. Our interest has been in the preparation of quaternary cellulose anion exchangers from cotton. Details of preparations of tertiary amine and quaternary ammonium anion exchange celluloses in aqueous[6,7] and in non-aqueous[8,9] media have been described.

Use of non-aqueous solvents such as methanol for the preparation of the sodium cellulosates, tertiary butanol for the preparation of the tertiary amine, diethylaminoethylcellulose and absolute ethanol for the preparation of quaternary ammonium compounds from the tertiary amine celluloses has made it possible to form cellulose anion exchangers in fabric form that contain up to 4.0% nitrogen. Usually, the maximum amount of nitrogen obtained with aqueous preparations is approximately 0.7%.

Recently, we have characterized tertiary amine and quaternary ammonium celluloses made either aqueously or non-aqueously by a variety of instrumental techniques. These techniques have also been used to follow effects of a given anion on the cellulose matrix of a quaternary ammonium cellulose exchanger when the exchanger prepared non-aqueously comes in contact with water. This report illustrates the importance of using several instrumental methods for the characterization of chemically modified cottons with the majority of specific examples being the anion exchange celluloses prepared non-aqueously.

Experimental

Reagents

Alkyl halides, diahaloalkanes, and tertiary butanol were reagent-grade chemicals from Eastman Organic Chemicals; the absolute methanol was from Mallinckrodt Chemical Works*, and β-chloroethyldiethylamine hydrochloride and dimethylformamide were reagent grade chemicals from Matheson, Coleman, and Bell. Argon and nitrogen were commercial-grade cylinder gases, purchased from Union Carbide Corporation.

Fabrics

The nonaqueous preparations of sodium cellulosates in the fabric form (4.4 oz/yd^2 cotton sheeting) and their subsequent reactions at room temperature with β-chloroethyldiethylamine in _t_-butanol to form the free amine forms of diethylaminoethyl (DEAE)-cottons of varying nitrogen contents have been reported.[10,11] To prepare a DEAE-cotton of approximately 0.6% N, cotton was reacted with 0.2_M_ sodium methoxide, and the resultant sodium cellulosate was reacted for 5 hr with a 3% by weight solution of β-chloroethyldiethylamine in _t_-butanol. The fabric of 3.0% N required a 1.25_M_ sodium methoxide solution to prepare the sodium cellulosate and reaction for 24 hr with a solution of β-chloroethyldiethylamine.

Preparation of Monoquaternary and Diquaternary Ammonium Celluloses

The DEAE-cottons containing from about 0.6% to 3.0% nitrogen in free amine form were washed free of *t*-butanol with absolute methanol and then refluxed for about 5 hr in an ethanolic solution of about 10-20% of the alkyl halide or dihaloalkane. Treated fabrics were then washed in ethanol and dried. Storage was in *t*-butanol.

Exchange of Anions

Replacement of the halide ion furnished by the original alkyl halide or dihaloalkane used in preparation of the quaternary-ammonium exchangers was accomplished in non-aqueous media. In some instances the original halide ion (such as iodide) had to be exchanged for another halide (chloride) before introduction or exchange with an oxidizing anion (such as super oxide anion). In general, exchange of anions was in absolute alcohol or dried dimethylformamide (DMF).

Chemical Analyses

Nitrogen contents of all modified cottons were determined by the Kjeldahl method and reported as milliequivalents per gram of fabric (meq N/g fabric).

All anion exchange fabrics were analyzed by conventional analytical procedures for halogen and halide ion and reported as meq X or $\bar{X}$ per gram of fabric before and after their exchanges with other anions. After each exchange with $\bar{BH_4}$, $\bar{SCN}$, or $Cr_2\bar{O_7}$, the fabric was analyzed by conventional wet analyses for the elements B, S, or Cr respectively. Percent of conversion with superoxide anions ($\bar{O_2}$) was calculated from the amount of $\bar{X}$ ions ions removed during the exchange process.

Titrations

Titrimetric methods reported previously[12] were also used to determine capacities of the anion exchangers. For potentiometric and conductometric titrations, the anion exchange fabric in salt form was cut in a Wiley mill to pass a 20-mesh screen and then regenerated to the base form in excess 0.05*M* NaOH. An accurately weighed sample sufficient to furnish approximately one meq of replaceable hydroxide anions was added to specially designed titration flasks. In potentiometric titrations, the flasks contained 50.00 ml of 1 M NaCl solution prepared in conductivity water and kept in a nitrogen atmosphere, and the titration was a standardized dilute HCl solution. All pH measurements were made to 0.01 pH with a Beckman research pH meter.

In conductometric titrations, the flasks contained 250 ml of conductivity water to which a known amount (approximately 1.5 meq) of HCl was added. The titrant was a standardized NaOH solution of approximately 1 M. Measurements were made on a Jones Conductivity bridge under a nitrogen atmosphere.

Electron Spectroscopy for Chemical Analyses (ESCA)

ESCA spectra were obtained on a Varian spectrometer Model VIEE-15 with a Mg_{K_α} X-ray source. Spectra were obtained on fabrics and on those fabrics ground in a Wiley mill to pass a 20-mesh screen. All binding energies for the nitrogen 1_s, halogens 2p, and metal 2p electrons are reported with reference to carbon 1_s electrons of binding energy of 285.6 eV. All spectra were analyzed for relative amounts of nitrogen of low (tertiary amine) and of high (quaternized ammonium) binding energies (BE).

Electron Paramagnetic Resonance Spectra (EPR)

Selected samples of anion exchange fabrics containing $\bar{O}_2$ anions or metallic ions capable of forming paramagnetic complexes with tertiary amine groups were subjected to EPR analyses. Samples were ground to pass a 40 mesh screen, packed in quartz EPR tubes, dried in vacuum oven P_2O_5, sealed in a nitrogen atmosphere and then analyzed. Spectra were obtained on a Varian 4502-15 EPR spectrometer system. In a few instances spectra were obtained before and after irradiations with UV light.

Chemiluminescence

Chemiluminescence (CL) was monitored with a Packard 3255 liquid scintillation spectrometer equipped with low dark-noise photomultiplier tubes (RCA 4501/4V) and a Packard model 585 linear recorder. Experimental details have been described previously[13] in a statistical report that shows CL values must differ by 2500 counts per minute to be significantly different.

Infrared Spectra

Multiple internal reflectance spectra of fabrics were obtained with a Wilks spectrophotometer Model 8C as described previously.[14]

Cold Plasma Treatments

The glass reactor with curved external electrodes and with inlets for introduction of gas either downstream from the electrodes or between electrodes has been described.[15] The system utilizes a radio frequency (rf) generator capable of producing up to 100 W of power at a frequency of 13.56 MHz. It can be operated in either a continuous or pulse mode. An impedance matching network, a vacuum pump with a capacity of 150 l/min and appropriate meters and gauges for measuring power output, gas flows, and pressures complete the system. In this investigation, samples were located downstream from the electrodes, but within the glow regions, on glass holders that allowed for suspension of fabrics. In some instances, the plasma generator was attached directly to the ESCA spectrometer cavity. In other instances, samples that were subjected to plasma irradiations were immediately removed and stored in inert atmosphere before being tested by other techniques.

Microscopical Examinations

Fibers of the control and selected chemically modified cottons were examined by techniques of optical microscopy described previously.[16] Ultra thin cross sections of the fibers were subjected to layer expansion by polymerization of methyl methacrylate and to solubility tests in 0.5 M cupriethylenediamine (cuene) and were examined by the techniques of transmission electron microscopy as previously reported.[17] Scanning electron micrographs of fibers of selected samples before and after subjection to various solvents were also obtained.[18]

Cross-sections of a few fibers were examined via energy dispersive x-ray analysis (EDAX) techniques[18] to determine distribution of reagent in the chemically modified cottons.

Results and Discussion

Surface analyses of unreacted and chemically modified cottons can be used to advantage in textile problem solving, provided there is an understanding of the principles and limits of surface analytical techniques. Surface analyses in combination with other techniques can be used to characterize surfaces, detect changes in composition with depth from surfaces, and even to determine differences in surface and bulk compositions of textile samples.

Diethylaminoethylcellulose (DEAE) has been made aqueously and nonaqueously. In an aqueous preparation, even with mercerized fabric, the maximum amount of N is about .8% (0.57 meq/g) and the product is Cell-O-C_2H_4 - $N(C_2H_5)_2{}^+H\ \bar{X}$, where $\bar{X}$ is the anion used in the neutralizing wash water. In the non-aqueous preparation of DEAE cellulose, (equations 1 and 2 of Figure 1) the product is a base, Cell-O-C_2H_4-N $(C_2H_5)_2$. In nonaqueous preparations, various amounts of N (up to 4%) can be added to cotton. The nonaqueously prepared DEAE cottons were reacted with alkyl halides or dihaloalkanes to form mono- and di-quaternary ammonium cellulose anion exchangers according to equations 3 and 4 in Figure 1, where I represents mono- and II the di-quaternary compounds. The DEAE celluloses were also reacted with BI_3 and with salts of transition

$$\text{Cell OH} + \text{NaOCH}_3 \xrightarrow{\text{CH}_3\text{OH}} \text{Cell-}\bar{\text{O}}\,\overset{+}{\text{Na}} \qquad (1)$$

$$\text{Cell-}\bar{\text{O}}\,\overset{+}{\text{Na}} + (\text{C}_2\text{H}_5)_2\,\text{N}\,\text{C}_2\text{H}_4\text{Cl} \xrightarrow[\underline{\text{t}}\text{-butanol}]{} \text{Cell-O-C}_2\text{H}_4\text{-}\underset{\text{C}_2\text{H}_5}{\overset{\text{C}_2\text{H}_5}{\text{N}}}\,: \qquad (2)$$

DEAE

$$\text{DEAE} + \text{RX} \xrightarrow{\text{CH}_3\text{OH}} \text{Cell-OC}_2\text{H}_4\,\underset{\text{C}_2\text{H}_5}{\overset{\text{C}_2\text{H}_5}{\text{N}}}\text{-R}^{+}\quad \bar{\text{X}} \qquad (3)$$

(I)

$$\text{DEAE} + \text{X H}_2\text{C -(CH}_2)_n\text{-CH}_2\text{X} \rightarrow \text{Cell-OC}_2\text{H}_4 - \underset{\text{C}_2\text{H}_5}{\overset{\text{C}_2\text{H}_5}{\text{N}}} - (\text{CH}_2)_{n+2} - \underset{\text{C}_2\text{H}_5}{\overset{\text{C}_2\text{H}_5}{\text{N}}} - \text{C}_2\text{H}_4\text{O-Cell} \qquad (4)$$

(II)

$$\text{DEAE} + \text{BI}_3 \xrightarrow{\text{DMF}} \text{Cell-OC}_2\text{H}_4\,\underset{\text{C}_2\text{H}_5}{\overset{\text{C}_2\text{H}_5}{\text{N}}}\,: \longrightarrow \text{BI}_3 \qquad (5)$$

$$\text{DEAE} + \text{Me}^{+2} \xrightarrow{\text{DMF}} \text{Cell-O-C}_2\,\text{H}_4 - \underset{\text{C}_2\text{H}_5}{\overset{\text{C}_2\text{H}_5}{\text{N}}}\,: \longrightarrow \text{Me}^{+2} \qquad (6)$$

Figure 1. Equations for non-aqueous preparations of sodium cellulosate (1), diethylaminoethylcellulose (DEAE) (2), mono-quaternary ammonium celluloses (3), where R varies from CH_3 to $C_{10}H_{22}$, diquaternary ammonium celluloses (4), where n varies from 2 to 10, complex formation between DEAE and Lewis base, BI_3 (5), and coordination of tertiary amines of DEAE with a transition metal ion (6).

metal ions that were capable of acting as Lewis acids in reactions with the tertiary amine groups of DEAE cellulose as indicated in equation 5 and 6 of Figure 1.

In Figure 2 are shown the ESCA spectra for a DEAE cotton before and after its conversion to the mono-quaternary salt with CH_3I that gave a 90% conversion. The N_{1s} electron binding energy of 402 eV is characteristic of the more positive quaternary nitrogen and the BE value of 399 eV is characteristic of the more electronegative tertiary amine in DEAE cotton. Reaction of DEAE with longer chain alkyl halides resulted in less conversion to the quaternary. However, a 50% conversion to the mono-quaternary could be realized with $C_{18}H_{37}I$. Poor conversion was obtained with all alkyl chlorides and with some dihaloalkanes as shown in Fig. 3.

The presence of both tertiary and quaternary groups can be detected in titration curves as well. In Fig. 4 is a titration curve of a cellulose anion exchanger having both types of groups. The relative amounts of tertiary amino and quaternary were in agreement with those amounts indicated by ESCA.

The titration curve for a mono-quaternary ammonium cellulose exchanger that has been totally converted with CH_3I is shown in Fig. 5.

Confirmation of the type of titration curve typical of a quaternary ammonium cellulose exchanger was obtained by reacting cotton with trimethylglycidylammonium chloride and then titrating the product:

$$[\text{Cell-O-}CH_2CHOHCH_2N(CH_3)_3{}^{+}\overline{Cl}]$$

Its titration curve is shown in Figure 6.

There was good agreement between the relative amounts of tertiary and quaternary groups in the cellulose anion exchangers as determined by ESCA and by potentiometric titrations (Table I).

A typical conductometric titration curve for a DEAE cellulose anion exchanger of 2.58% N is shown in (Fig. 7).

Distribution of some substituents in chemically modified cottons can be obtained through depth profiles. For example, conventionally applied fluorochemical finishes are concentrated on the surfaces. As a result, the intensity of the F_{1s} electron peak decreases as the fabric is etched with an argon cold plasma within the spectrometer cavity. In our analyses of anion exchange celluloses, we were not able to use the rf cold plasma etching method because nitrogen, that we wanted to measure, is chemically bound to cotton as it is subjected to cold plasmas of argon as

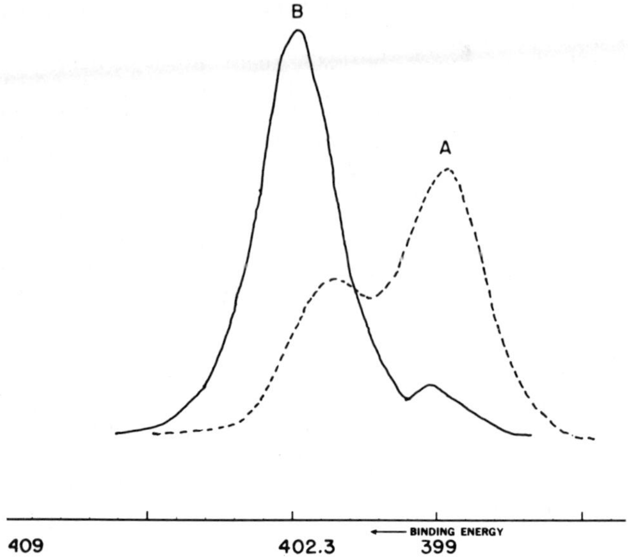

Figure 2. ESCA spectra of N_{1s} electrons from a DEAE cotton of 1.7% N before (A) and after (B) its conversion (90%) to a monoquaternary ammonium cellulose with CH_3I.

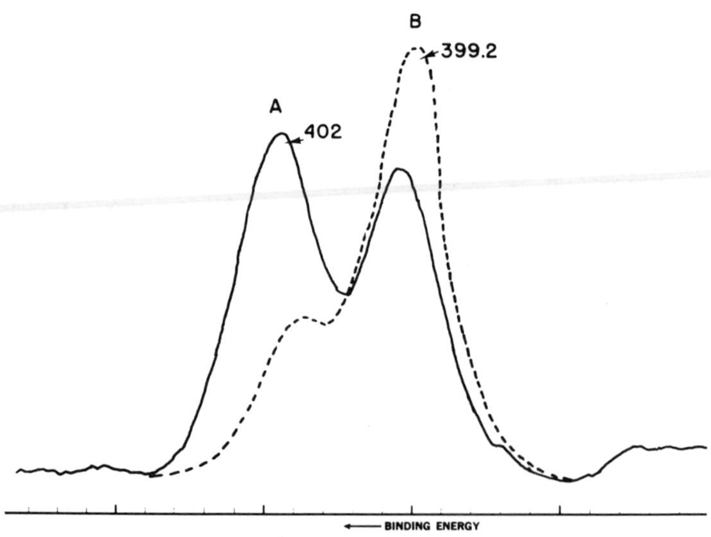

Figure 3. ESCA spectra of N_{1s} electrons from a DEAE cotton of 1.7% N before (A) and after (B) its partial conversion to a diquaternary ammonium cellulose with $ICH_2(CH_2)_mCH_2I$.

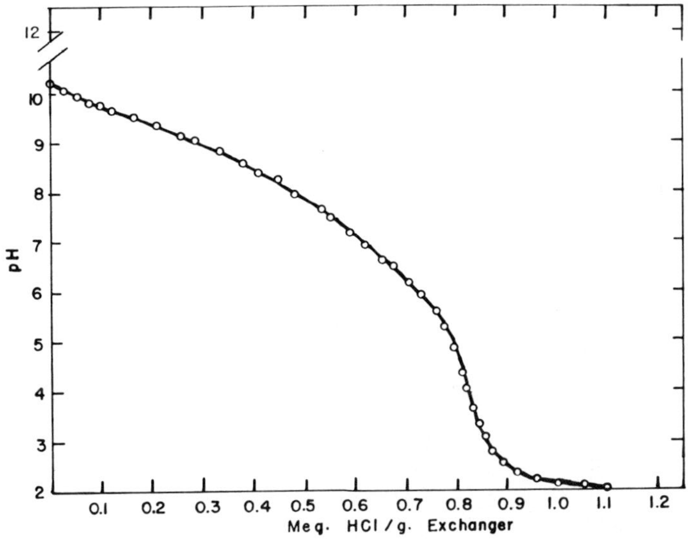

Figure 4. Potentiometric titration curve of a product containing both tertiary amine and quaternary ammonium groups formed by incomplete conversion of a DEAE cellulose of 1.57% N (1.12 meq/g) by its reaction with C_2H_5Br. (ESCA showed 33% conversion.)

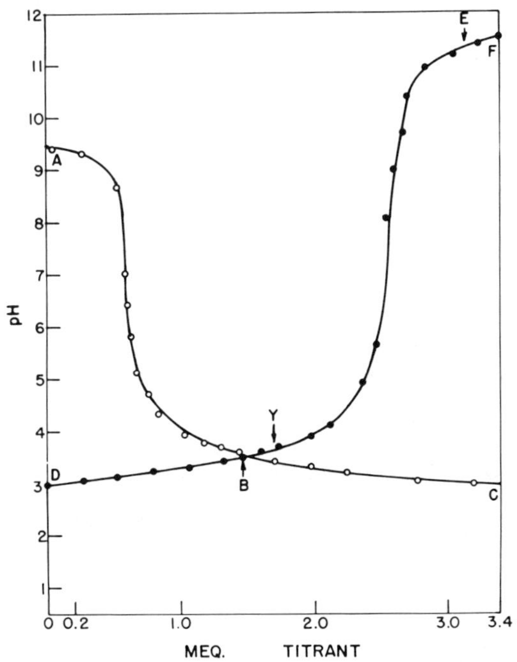

Figure 5. Potentiometric titration curves of a quaternary ammonium cellulose formed by complete conversion of a cellulose of 2.58% N with CH_3I. Curve ABC is for a 0.8043 g sample in base form (1.48 meq) vs 0.1 N HCl. Curve DYEF represents titration of the chloride form in excess HCl (1.72 meq) vs 0.1 N NaOH.

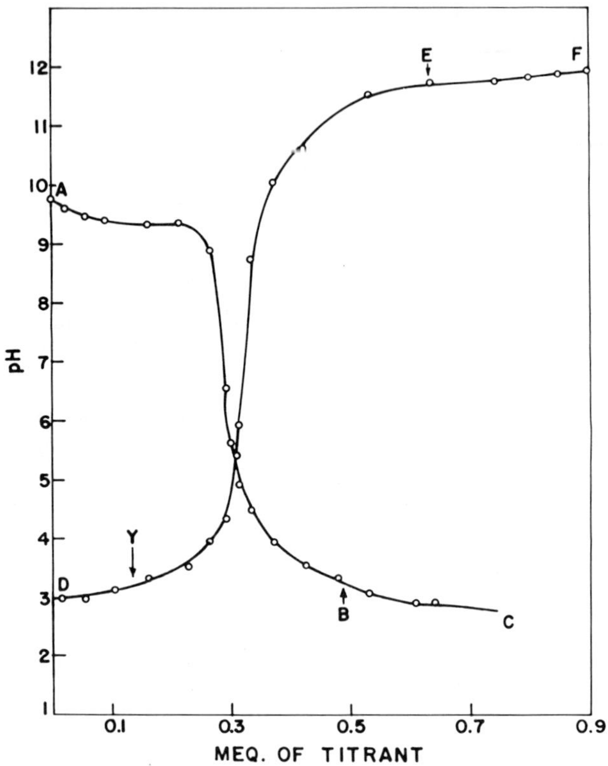

Figure 6. Potentiometric titration curve of a 1.6400 g sample of the quaternary ammonium cellulose, Cell-O-$CH_2CHOHCH_2\overset{+}{N}$-$(CH_3)_2$ $\overline{Cl}$, (0.5001 meq N). Curve ABC is for basic form in 50 ml of 1 M NaCl vs 0.1 M HCl, Curve DYEF is for the chloride form in presence of excess HCl (.15 meq) vs 0.1 M NaOH. AB and YE represent titrations of exchangers.

Table I. Relative Amounts of Quaternary and Tertiary Amines

Exchanger	ESCA Analyses: N_{1s}, eV: $\overset{+}{NR_4}$	ESCA Analyses: N_{1s}, eV: $R_3N:$	ESCA Analyses: $\overset{+}{NR_4}$ / $R_3N:$	Titrimetric Analyses: meq/g: End Pt 1	Titrimetric Analyses: meq/g: End Pt 2	Titrimetric Analyses: $\overset{+}{NR_4}(2)$ / $R_3N:(1)$
DEAE	402	399	0.9	0.8	0.6	0.9
DEAE + CH_3I	402	399	3.4	0.3	0.9	3.0
DEAE + BrC_2H_4Br	402	399	1.6	0.6	0.9	1.5

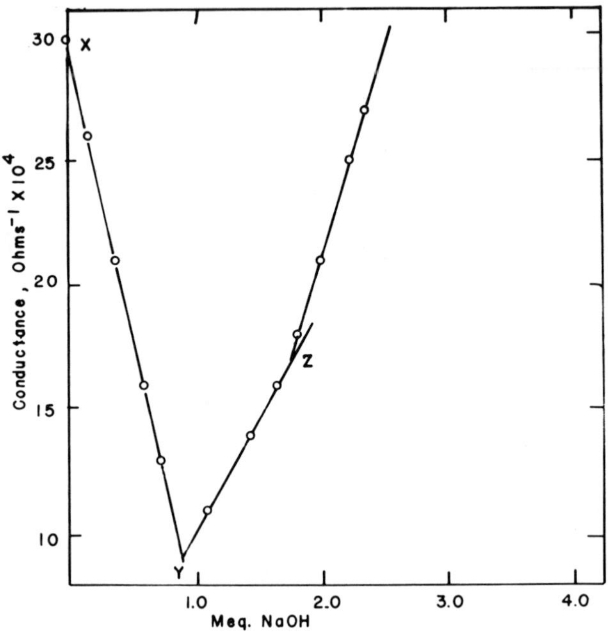

Figure 7. Conductometric titration curve of a 0.5001 g sample of DEAE cellulose of 2.58% N (0.92 meq) in base form in presence of 1.802 meq HCl vs. standardized 1.0 M NaOH. XY is neutralization of excess HCl; YZ represents exchange of Cl^- ions of exchanger for OH^- ions.

well as of nitrogen. Thus far, we have been unable to completely eliminate the source of nitrogen in argon plasma.

In Figure 8 are ESCA spectra of a purified cellulose before (bottom curves) and after (middle curves) its irradiation for only 20 minutes in an argon plasma within the ESCA spectrometer. Included in this figure are the difference spectra (irradiated - unirradiated) for the C_{1s}, O_{1s} and N_{1s} electrons. In addition to showing the increase in bound N after irradiation, these spectra indicate formation of C having 1_s electrons of higher BE and of O having 1_s electrons of lower BE than respective control values. Formation of carbon free radicals and carbonyl groups during plasma irradiations account for these new spectra. These differences have been observed in all polysaccharides treated in rf cold plasmas and have been discussed in an earlier report.[15]

Another method to detect differences in surface and bulk analyses is to compare ESCA spectra of modified cottons before and after they have been ground in a Wiley Mill. For example, such spectra of a plasma irradiated cotton show that the nitrogen added to cotton during its irradiation is concentrated at the surface. In Figure 9 are ESCA spectra of the C_{1s}, O_{1s} and N_{1s} electrons of a cotton fabric that had been irradiated for 90 min in an argon plasma and then analyzed via ESCA in fabric and pulverized forms. These difference spectra show that the N was predominantly on the surface, as were the carbons having highest BE (free radicals) and the oxygens of lowest BE (carbonyl oxygens). Therefore, to see if N was uniformly distributed in non-aqueously prepared cellulose anion exchangers, we obtained differences in N_{1s} peaks for samples before and after they were ground in a Wiley Mill. Not expected was the fact that the nitrogen was uniformily distributed.

Uniformity of distribution was also indicated by SEM and EDAX data. A mono-quaternary ammonium cellulose anion exchanger (3.46% N, 31.3% I) was exchanged in non-aqueous media with NaSCN. Wet analyses (4.86% N, 0% I, and 5.70% S) indicated it was 100% converted to the thiocyanate form. Ultra-thin cross sections of the fibers in the thiocyanate form examined by EDAX revealed that sulfur was uniformly distributed throughout the fibers. The cross-section of the fiber shown in (Fig. 10) (upper) and EDAX scan (lower) are illustrative of the uniform distribution of sulfur in a diquaternary ammonium cellulose thiocyanate. This is typical of a quaternary or diquaternary product in the SCN^- form.

The physical properties of the quaternary ammonium cellulose exchanger varied with the associated anions. All fabrics maintained original fabric structure regardless of anion when in non-aqueous media. Usually, anions were exchanged in DMF. Of particular interest was the fact that exchange of chloride for large SCN^- or BH_4^- anions resulted in fabrics that were easily dis-dispersed or almost soluble in water. The effects of the anion on

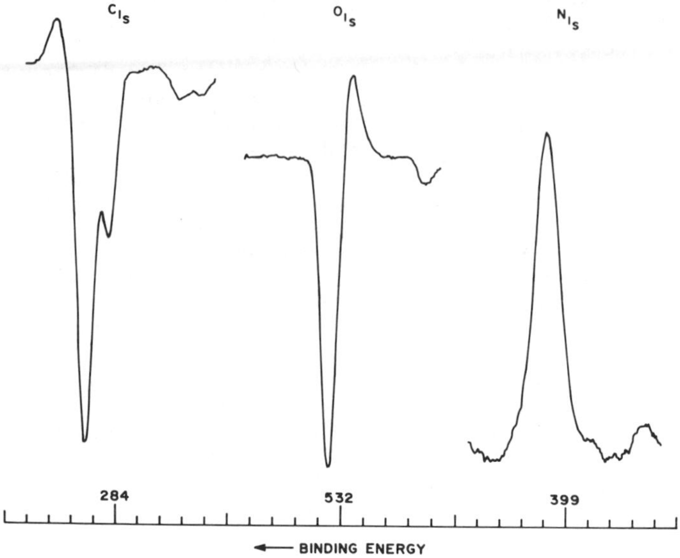

Figure 8. ESCA spectra of a purified cellulose before (bottom curves) and after irradiation in argon plasma for 20 min (middle curves) and difference spectra (irradiated-unirradiated) for C_{1s}, O_{1s}, and N_{1s} electrons.

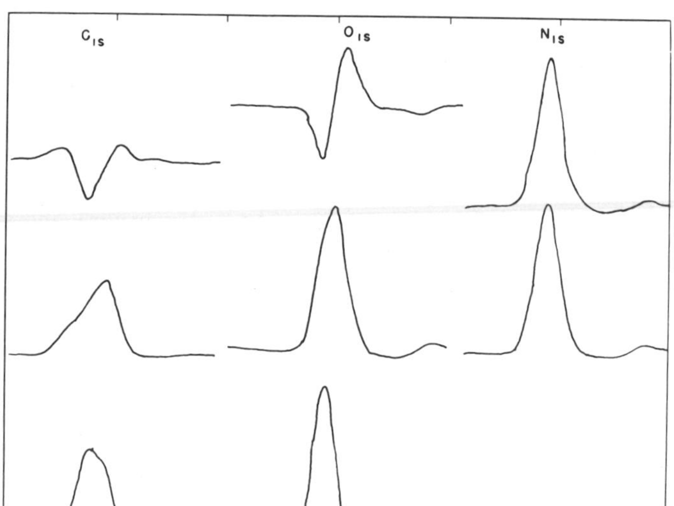

Figure 9. Difference spectra for C_{1s}, O_{1s}, and N_{1s} electrons from cotton. Fabric irradiated for 90 min in argon plasma and then analyzed via ESCA as fabric and pulverized samples. Difference = (fabric - pulverized).

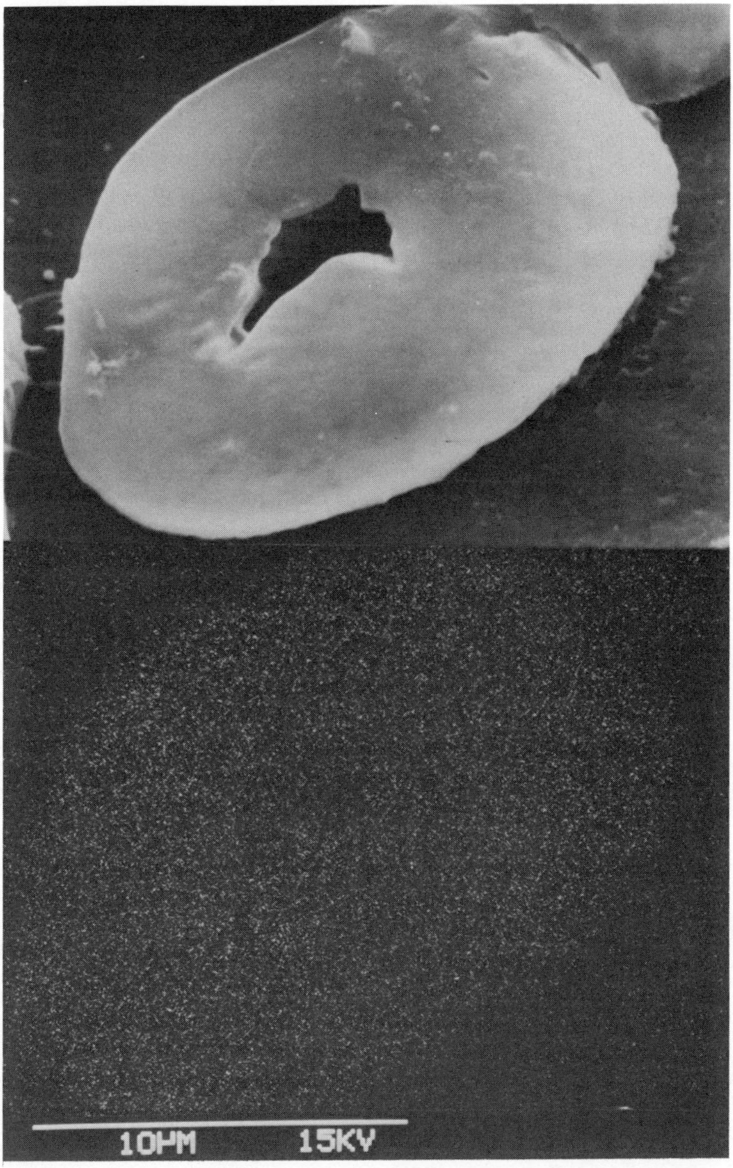

Figure 10. Scanning electron micrograph of a cross-section (top) of a diquaternary ammonium cellulose prepared by reaction of a DEAE cellulose (2.57% N) with diiodopentane and exchanged to thiocyanate form (100% conversion) and EDAX (bottom) showing distribution of S throughout fiber.

the swelling of fibers in water have been followed with the techniques of light and electron microscopy and will be the subject of another report.

For a mono-quaternary anion exchanger of a given N content, those most swollen or most easily dispersed in water were those having $\bar{O}_2$, BH_4^- or $SC\bar{N}$ as anions. The light micrographs of whole fibers of a mono-quaternary of 3.20% N in the $\bar{I}$, BH_4^- and $SC\bar{N}$ forms before and after being contacted with water are shown in (Fig. 11). These micrographs illustrate the swelling in water of fibers from all three exchanges and the partial dissolution in water of the $SC\bar{N}$ and BH_4^- forms of the exchangers.

Formation of diquaternary ammonium cellulose anion exchangers prevented the dispersion in water of the fibers even when they were in the BH_4^- or $SC\bar{N}$ form. Micrographs of whole fibers of diquaternary exchangers of 3.46% N before and after contact with water are shown in Figures 12, 13, and 14 for the I^-, BH_4^- and $SC\bar{N}$ forms, respectively. Also included in each figure is a micrograph of fibers after immersion in 0.5M cuene, a solvent that dissolves cotton, DEAE cottons and mono-quaternary ammonium celluloses. Insolubility in cuene indicates crosslinking of cellulose chains by the dihaloalkanes.

Scanning electron micrographs of cross-sections of the mono- and diquaternary exchangers in the BH_4^- or $SC\bar{N}$ form also show that crosslinking decreased the amount of swelling and dispersion or dissolution of fibers when BH_4^- or $SC\bar{N}$ is the associated anion.

Use of nonaqueously prepared DEAE celluloses as Lewis bases offers potential for formation of metal ion complexes between transition metal ions such as Cu^{+2} and Ni^{+2}, formation of charge-transfer complexes with electron acceptor compounds such as ortho- and para-chloranils or bromanils and with other Lewis acid compounds such as BI_3.

Electron spin resonance spectra of those products that were paramagnetic were used in conjunction with ESCA spectra of the products to gain information on types of bondings. With copper and nickel complexes, the products were paramagnetic and the BE of the N_{1s} electrons shifted to higher BE values as the BE of the Cu_{2p} and Ni_{2p} electrons shifted to lower binding energies. Both the Ni^{+2} and Cu^{+2} complexes with DEAE were paramagnetic. Complexing of DEAE with BI_3 resulted in an increase of BE of the N_{1s} electrons of DEAE from 399 eV to approximately 402 eV. With charge transfer complexes, formed with ortho- or para-chloranils, the BE of N_{1s} electrons shifted to 401.4 eV (Fig. 15) and the Cl_{2p} electrons had BE values of 197 for the Cl ion and 200.4 for the organic chlorine. These complexes were paramagnetic, as illustrated by the EPR spectrum for the DEAE-o-chloranil complex shown in Figure 16. The singlet of g-value of 2.0059, indicates

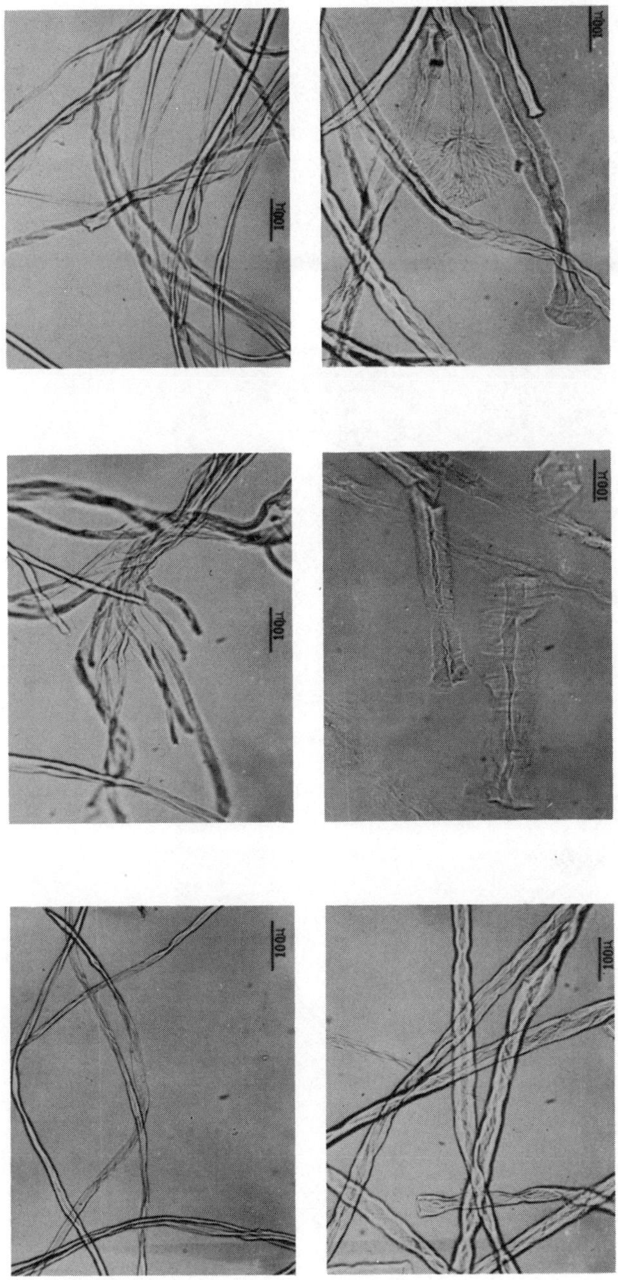

Figure 11. Light micrographs of fibers from mono-quaternary ammonium celluloses (3.27% N) prepared nonaqueously from DEAE and CH_3I in the I^-, BH_4^-, and SCN^- forms (left to right) before (upper) and after (lower) immersion in water.

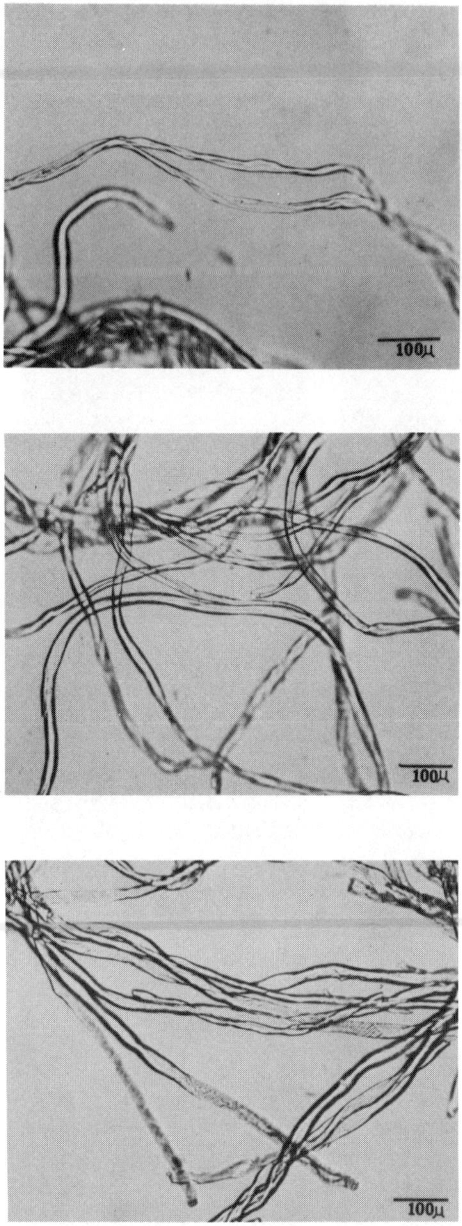

Figure 12. Light micrographs of fibers from diquaternary ammonium cellulose (3.46% N) prepared non-aqueously from DEAE and diiodopentane. Fibers immersed in mineral oil, water, and 0.5 M cupriethylenediamine are shown in top, middle, and bottom micrographs, respectively.

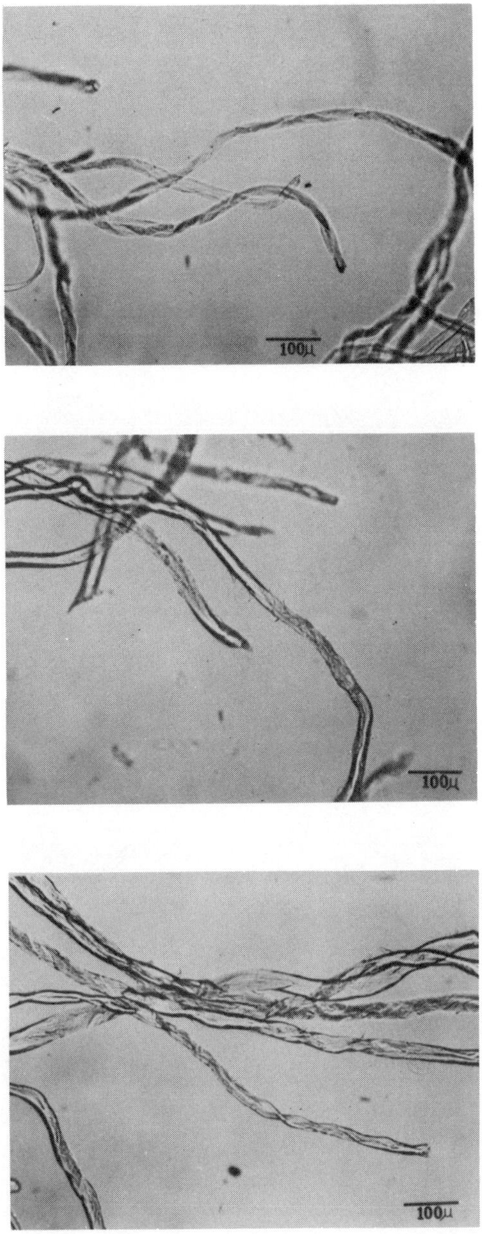

Figure 13. Light micrographs of the diquaternary iodide of Figure 12 after its conversion to the borohydride ($B\overline{H}_4$) form and immersion in mineral oil (top), water (middle), or 0.5 M cupriethylenediamine (bottom).

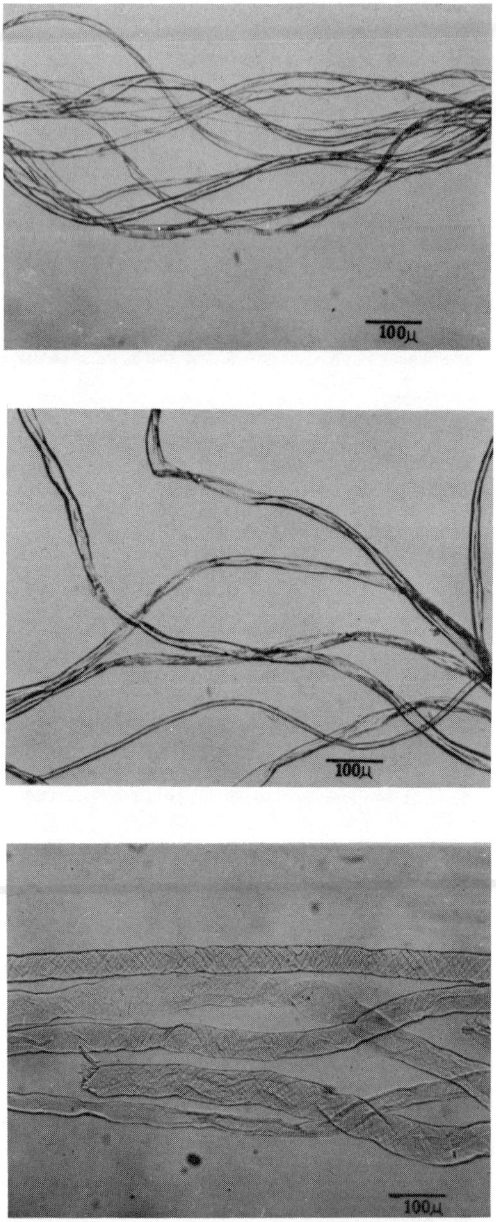

Figure 14. Light micrographs of the diquaternary of Figure 12 after its conversion to the $SC\overline{N}$ form and immersed in mineral oil (top), water (middle) or 0.5 M cupriethylenediamine (bottom).

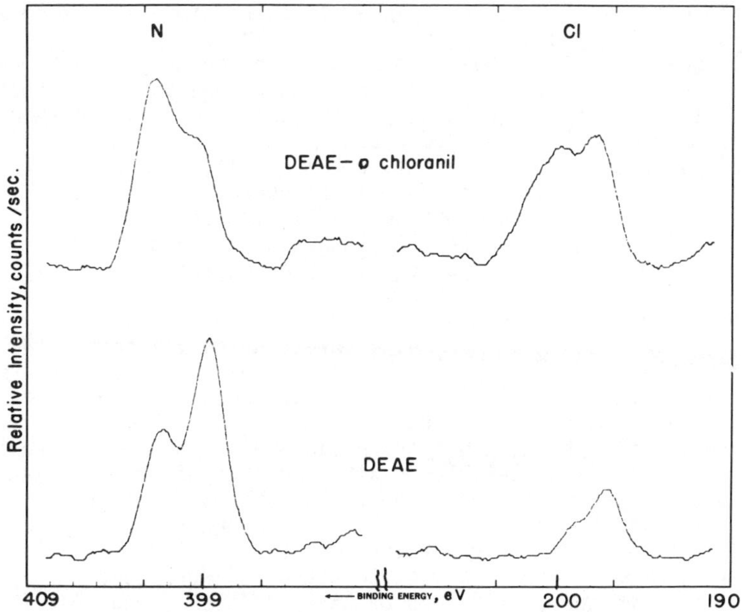

Figure 15. ESCA spectra for the N_{1s} and Cl_{2p} electrons from DEAE cotton before and after its reaction with o-chloranil.

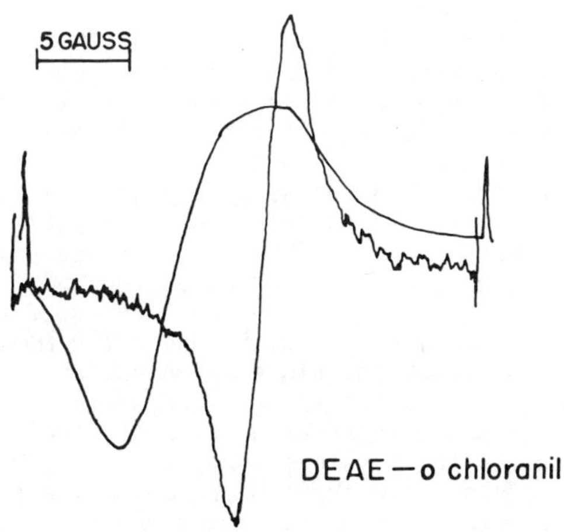

Figure 16. EPR spectrum obtained at room temperature of a DEAE-cotton of 2.58% N after its reaction with 2,3,5,6-tetrachloro-p-benzoquinone (o-chloranil) in chloroform. (g-values = 2.0059.)

that the complex does not contain a radical cation, as a single electron on nitrogen would generate a triplet signal.

Another technique which can be used in conjunction with EPR spectral changes is chemiluminescence (CL). We have found that certain types of free radicals, particularly those formed when cottons or chemically modified cottons are subjected to UV radiations or to rf cold plasmas, react readily with oxygen to produce CL. There is an inverse linear correlation between CL and EPR intensities. The intensities of EPR signals decrease as CL increases.[19] Frequently, as EPR signals decrease or as CL increases, the intensities of C-O bonds as determined by multiple internal reflectance IR increase.

The electro-kinetic properties of DEAE and quaternary ammonium celluloses are also being investigated via techniques of streaming electrode potentials. However, some difficulties are encountered, particularly in those products of relatively high D.S. that form gels when ground and immersed in water.

The drastic changes in hydrophilic and hydrophobic properties of the anion exchange celluloses with exchange of the associated anion are not yet fully understood. The attraction of the quaternary ammonium cellulose exchanger for neutral, cationic and anionic substances is of great interest to textile chemists. Quaternary groups have been used to attract fluorescent whiteners[20], to scavenge anions of heavy metals[21], and most recently to scavenge dyes during launderings.[22] More investigations using a variety of techniques are needed to fully elucidate the effects of anions as well as the quaternary ammonium cations on the structures of both water and the overall hydrogen bonded network with the cellulose matrix.

Summary

The materials selected in this report for surface analyses and bulk analyses were nitrogen containing cellulose anion exchangers of the weak-base and strong-base types. The surface analysis technique used (ESCA) has an intrinsic surface sensitivity to within a range of 50 A. Unfortunately, our Center does not have instrumentation for secondary ion mass spectroscopy (SIMS) which is the most surface sensitive method. In order to probe deeper than 50 A with ESCA, it is necessary to apply ion etching techniques via such methods as radio frequency cold plasmas of argon. This technique has been used to obtain depth profiles of various elements with chemically modified cellulosic fibers. However, with cellulose anion exchangers used in this discussion, consideration has to be taken of the fact that nitrogen is bound to cellulose in an amide type of bonding whenever cellulose is subjected to rf cold plasmas of argon, air, or nitrogen. The exact BE of core electrons being investigated depends on the nature of the chemical bonding of the element. In some instances, the chemical bonding affects the spectral line shape and the appearance of satellites. With experience and the use of spectra of known compounds as fingerprints, it is possible to distinguish among oxidation states of a given element in chemically modified cottons. Surface analyses used in conjunction with EPR and CL give some insight into the types of active species formed in chemically modified cottons. Data from all of these techniques can be correlated with that from conventional chemical analyses. Once it is established that analyses of surfaces correlate with bulk analyses, the solid state analyses are very useful and give more detailed information than conventional chemical analyses. Even in those instances where composition varies with depth from surfaces, the methods of surface analyses can be useful to the cellulose chemist.

Acknowledgement

The authors wish to thank Jarrel H. Carra, Wilton R. Goynes, Jr. and Bruce F. Ingber for the microscopic analyses.

Mention of a company and/or product does not imply approval or recommendation by the U.S. Department of Agriculture, to the exclusion of others which may also be suitable.

References

1. Hofmann, A. W. Ann. Chem. (1851), 78, 253-258.
2. Frank, H. S. and Evans, M. W. J. Chem. Phys. (1945), 13, 507-532.
3. Boyd, G. E., Schwarz, A., and Lindenbaum, S. J. Phys. Chem. (1966), 70, 821-825.
4. Lindenbaum, S. J. Phys. Chem. (1966), 70, 814-821.
5. Diamond, R. M. J. Phys. Chem. (1964), 67, 2513-2517.
6. Soignet, D. M., Berni, R. J., and Benerito, R. R. Textile Res. J. (1966), 36, 978-989.
7. Soignet, D. M. and Benerito, R. R. Textile Res. J. (1967), 37, 1001-1003.
8. Perrier, D. M. and Benerito, R. R. J. Appl. Polym. Sci., (1975), 19, 3211-3220.
9. Perrier, D. M. and Benerito, R. R. Appl. Polym. Sci. (1976), 29, 213-221.
10. Berni, R. J., Soignet, D. M. and Benerito, R. R. Textile Res. J. (1970), 40, 999-1006.
11. Soignet, D. M., Berni, R. J. and Benerito, R. R. J. Appl. Polym. Sci. (1971), 15, 155-168.
12. Benerito, R. R., Woodward, B. B. and Guthrie, J., Anal. Chem. (1965), 1693-1699.
13. Jung, H. Z., Ward T. L. and Benerito, R. R., J. Macromol. Sci. Chem. (1979), A13, 1117-1133.

14. O'Connor, R. T. Instrumental Analysis of Cotton Cellulose and Modified Cotton Cellulose, Marcel Dekker, Inc., New York, (1972), pp 10-11, 441-443.
15. Ward, T. L., Jung, H. Z., Hinojosa, O. and Benerito, R. R. Surface Sci. (1978), 76, 257-273.
16. Tripp, V. W., Moore, A. T. and Rollins, M. L. Textile Research J. (1966), 36, 295.
17. Rollins, M. L. and Tripp, V. W. Methods of Carbohydrate Chemistry (1963) R. W. Whistler, Ed., Academic Press, New York, (1963), p.. 356.
18. Goynes, W. R., Jr. and Carra, J. H. A.C.S. Division of Organic Coatings and Plastics Chemistry, (1976), 36, 9-13.
19. Ward, T. L. and Benerito, R. R. J. Polymer Photochemistry (1983) 3 (4),267-277.
20. Berni, R. J., Benerito, R. R., and Pilkington, M. W. U. S. Patent 3,914,108 (Oct. 21, 1975).
21. Perrier, D. M., Benerito, R. R., and Steele, R. H. U.S. Patent 4,029,533 (June 14, 1977).
22. Claiborne, J. L. (to Dixie Yarns, Inc.) U.S. Patent 4,380,453 (April 19, 1983).

RECEIVED May 3, 1984

2

Cross-Polarization/Magic Angle Spinning ^{13}C-NMR Study

Molecular Chain Conformations of Native and Regenerated Cellulose

F. HORII, A. HIRAI, and R. KITAMARU

Institute for Chemical Research, Kyoto University, Uji, Kyoto 611, Japan

Quantitative analysis of CP/MAS ^{13}C NMR spectra has been carried out to characterize molecular chain conformations of crystalline and noncrystalline components of native cellulose such as cotton, ramie, bacterial, and valonia cellulose, and regenerated cellulose. The resonance lines of C1 and C4 carbons in the β-1,4-glycosidic linkage and C6 carbon of the CH_2OH side group are deconvoluted into the crystalline and noncrystalline components. The chemical shifts of the two components of the C1, C4, and C6 carbons as determined by the line deconvolution method are correlated to the torsion angles ϕ and ψ about the β-1,4-glycosidic linkage and χ about the exo-cyclic C5-C6 bond, respectively. On the basis of these results, the characteristic chain conformations of the crystalline and noncrystalline components of native and regenerated cellulose are discussed.

Cellulose, which is one of the most abundant organic substances found in nature, has been extensively studied by various techniques such as x-ray scattering, electron microscopy, IR and Raman spectroscopy, NMR spectroscopy etc. However, the crystal structure and noncrystalline state are not yet solved for cotton, ramie, bacterial and valonia celluloses which can be easily obtained in pure form. Cross-polarization/magic angle spinning(CP/MAS) ^{13}C NMR spectroscopy is a promising new method to study these unsolved problems of cellulose, because this method is very sensitive to local molecular conformations and dynamics.

Our recent CP/MAS ^{13}C NMR work(1,2) on native and regenerated celluloses has shown that the resonance lines can be analyzed in terms of the contribution from the crystalline and noncrystalline components. The chemical shifts of C1, C4, and C6 carbons thus ascribed to the respective components could be correlated to the torsion angles ϕ and ψ, which define the conformation about the β-1,4-glycosidic linkage, and χ about the exo-cyclic C5-C6 bond,

0097-6156/84/0260-0027$06.00/0

respectively(3,4). In this paper we review these results and discuss the chain conformations of the crystalline and noncrystalline components for both native and regenerated cellulose samples.

Experimental

Sample Preparation. Valonia cellulose was obtained from the cell walls of *valonia macrophysa*. The valonia cell walls were purified as follows. They were washed with water thoroughly, boiled in a 1% NaOH aqueous solution for 4 hr, and washed well with water. Then they were immersed in 0.1% acetic acid to neutralize the trace of the alkali, washed with water, and dried.

Acetobacter xylinum, American Type Culture Collection strain number 10245, which was kindly supplied by Dr. H. Takai of Hokkaido University, was grown in a sterile liquid medium. This consisted of(in g/liter): anhydrous glucose 20, yeast extract 5, polypeptons 5, disodium hydrogen phosphate(anhydrous) 2.7, and citric acid 1.15. The pH of the mixture was 7.0. 100 ml Erlenmeyer flasks filled with 15 ml of the liquid medium were simultaneously inoculated with 0.5 ml of inoculum from 2-4 day inoculation culture. Cultures were grown at 30°C for 4 days with no agitation. Cellulose membranes which were produced on the surface of the liquid were washed with water, immersed in 2% NaOH solution for more than 7 days at room temperature, then washed again with water. To free the membranes from the alkali they were immersed in 1% acetic acid for 1 day, washed with water and dried.

Egyptian cotton, ramie and cupra rayon fibers were purified according to the methods reported previously(5,6). A highly crystalline cellulose II sample with low molecular weight was prepared using a method based on that of Bulleon and Chanzy(7). Five grams of cellulose acetate were mixed with 150 ml dichloromethane, 100 ml acetic anhydride, 45 ml glacial acetic acid, and 0.2 ml $HClO_4$(density 1.67 gcm^{-3}) and maintained for 6 hr at 55°C. The reaction was stopped by adding sodium acetate(2.5 g) suspended in a small amount of water. The reaction mixture was then poured into 3 *l* of boiling 0.1 M aqueous sodium acetate. The precipitate was filtered, washed with water, then stirred in methanol to remove oligomers of low molecular weight and the yellow color. The cellulose triacetate(1.5g) with low molecular weight thus produced was dissolved in 1 *l* of a 40% solution of methylamine in water. The solution was held at 90°C for 5 hr to effect deacetylation and crystallization of the cellulose.

Other cellulose and monosaccharides and disaccharides were the same as reported previously(1,3). The samples except for hydrates and β-methyl cellobioside·CH_3OH were well dried at 50°C under vaccum for 2-3 days before and after packing in a rotor for CP/MAS measurements. The hydrates and β-methyl cellobioside·CH_3OH were used without drying.

CP/MAS ^{13}C NMR Spectroscopy. CP/MAS ^{13}C NMR spectra were recorded on a JEOL JNM FX-100 and FX-200 spectrometers equipped with a CP/MAS unit operating at 25 and 50 MHz for ^{13}C, respectively. The matched field strengths ν_{1C} and ν_{1H} of 71 kHz for 25 MHz and 62.5 kHz for 50 MHz were applied to ^{13}C and ^{1}H for 2.0 ms; ν_{1H} remained on during the signal acquisition period. Magic angle spinning was carried out at a rate of about 3.2 kHz for 25 MHz and 3.5 kHz for 50 MHz by use of a bullet type rotor of poly(chlorotrifluoroethylene), whose volume was about 0.5 cm^3. Chemical shifts relative to tetramethylsilane(TMS) were determined using a narrow crystalline resonance line at 33.6 ppm for polyethylene inserted in all samples.

Results and Discussion

1. Assignments of the Resonance Lines of the Crystalline and Noncrystalline Components.

Figure 1 shows a 50 MHz CP/MAS ^{13}C NMR spectrum of ramie cellulose and a stick-type ^{13}C NMR spectrum of low molecular weight cellulose($\overline{DP}$∿10) in deuterated dimethyl sulfoxide solution(DMSO)(8)(The broken and solid lines in the CP/MAS spectrum will be explained below.). As already reported(9,10), the assignments for the C1, C4 and C6 carbons are relatively easy, based on analogies with the solution state spectrum. However, it should be noted that these resonance lines shift downfield by 2.3-9.6 ppm in the solid state compared to the solution state. The cause of such large downfield shifts(to be explained in the next section) is attributed to the different conformations about the β-1,4-glycosidic linkage and the exo-cyclic C5-C6 bond in which these carbons are involved.

Another remarkable feature is that the resonance lines of the C4 and C6 carbons split into two components; sharp downfield and broad upfield components. Similar splittings have been observed for the C4 and C6 resonance lines of cotton, bacterial and valonia cellulose and for the C1, C4 and C6 resonance lines of regenerated cellulose(1) such as cupra rayon fibers. These spectra and splittings are shown in Figure 2. In the case of regenerated cellulose(1), we have already concluded that the components causing the sharp downfield and the broad upfield shifts are the crystalline and noncrystalline components, respectively. Therefore, we have assumed that the splittings are due to the same components in native cellulose. However, the C1 resonance line does not seem explicitly to contain two such components.

In order to confirm this assumption, we deconvoluted the C4 and C6 resonance lines into two components by use of a curve resolver and by assuming a Lorentzian line shape as shown by broken and thin solid lines in Figures 1 and 2. In Figure 3 the integrated fraction f_{nmr} of the sharp downfield component of the C4 carbon is plotted against the degree of crystallinity f_x determined by x-ray analysis based on a modified Hermans' method(1,11,12). This figure shows also data of hydrocellulose, some regenerated cellulose samples(1), and mercerized cotton and ramie. A linear relationship exists between f_{nmr} and f_x for all the samples. A similar relationship was also obtained for the case of the C6 resonance line. Therefore, it is concluded that the sharp downfield and broad upfield components of the C4 and C6 carbons of native cellulose are also the crystalline and noncrystalline components, respectively.*)

Recent work(13) reported that the broad upfield components of C4 and C6 carbons were not attributed to the noncrystalline component but to anhydroglucoses on the surface of elementary fibrils because the spectra of cotton, hydrocellulose, and ramie showed little differences in relative intensity and in appearance of the broad

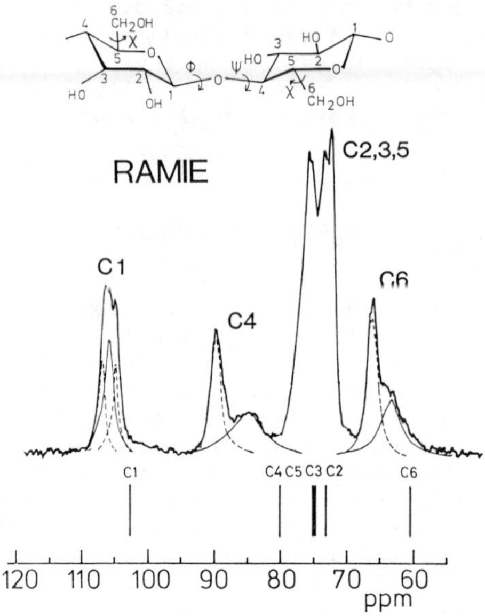

Figure 1. 50MHz CP/MAS ^{13}C NMR spectrum of ramie and stick-type scalar-decoupled ^{13}C NMR spectrum of low molecular weight cellulose in DMSO-d_6 solution (8). Broken and thin solid lines in the CP/MAS spectrum are for the crystalline and noncrystalline components, respectively.

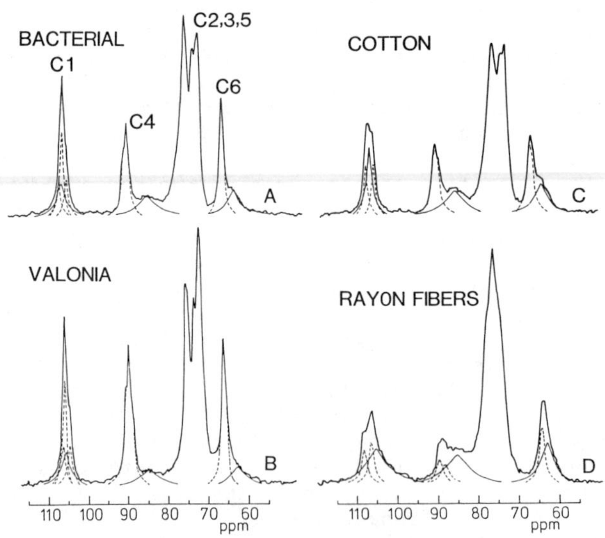

Figure 2. 50 MHz CP/MAS ^{13}C NMR spectra of bacterial cellulose, valonia cellulose, cotton, and cupra rayon fibers. Broken and thin solid lines in the CP/MAS spectra are for the crystalline and noncrystalline components, respectively.

component. However, it is very plausible that such differences could not be found among these samples, because their crystallinities are almost the same. More importantly, they believed that the crystallinities of these samples were 80-96%. We think that these values are evidently overestimated. According to our quantitative analysis described above it is not necessary to consider the contribution of the surface of elementary fibrils to the broad components.

On the other hand, the resonance line of the C1 carbon is also assumed to contain crystalline and noncrystalline components. In order to examine this problem, we measured spin-lattice relaxation times T_1 using a pulse sequence as developed by Torchia(16). As a result, it has been found that each resonance line of ramie contains two components with T_1 values of 4.6-16s and 65-130s which correspond to the noncrystalline and crystalline components, respectively(15). On the basis of these different T_1 values, we were able to record the spectra of the crystalline and noncrystalline components of native cellulose separately by appropriate pulse techniques.

Figure 4(A) shows the spectrum of the crystalline component of ramie which was obtained under the MAS condition by a pulse sequence for T_1 measurements developed by Torchia(16). The time between the two $\pi/2$ pulses was about 30s. As is clearly seen, the broad upfield components of the C4 and C6 resonance lines disappear and another fine splitting appears in the C1 resonance line which originates only from the crystalline component. Figure 4(B) shows the spectrum of the noncrystalline component of ramie which was obtained by a conventional $\pi/2$ single pulse sequence with a short waiting time of 4s employing the DD/MAS technique. As the difference in T_1 between the crystalline and noncrystalline components is not very large, a small amount of the downfield crystalline lines also appears for the C4 and C6 carbons. Although a small amount of the crystalline component is also involved in the C1 resonance line, the characteristics of the noncrystalline component of this line can be well observed: the chemical shift of this component is about 106 ppm from TMS and the line width is about 90 Hz.

By considering the features of the C1 resonance line for both components, we deconvoluted the C1 resonance line into the crystalline and noncrystalline components as shown in Figure 1. Although two Lorentzian lines were used for the crystalline components based on the result in Figure 4 (A), the C1 line could be deconvoluted into the two components. The integrated fraction of the crystalline component was also in good accord with the f_{nmr}. Similar deconvolutions were also possible for the C1 resonance lines of cotton, bacterial and valonia celluloses as shown in Figure 2. It shoud be noted that three Lorentzian lines were used for the crystalline component of the latter two samples.

---

*) Garroway et al.(14) have recently pointed out that the relative intensities of resonance lines of CP/MAS ^{13}C NMR spectra are proportional to the number of carbons, when the time constant T_{CH} for ^1H-^{13}C cross polarization is much shorter than the spin-lattice relaxation times $T_{1\rho}^H$ and $T_{1\rho}^C$ in the rotation frames of ^1H and ^{13}C nuclei, respectively. Since this condition has been found to be fulfilled under our experimental condition(15), such a quantitative analysis as shown in the text can be reasonably carried out for cellulose samples.

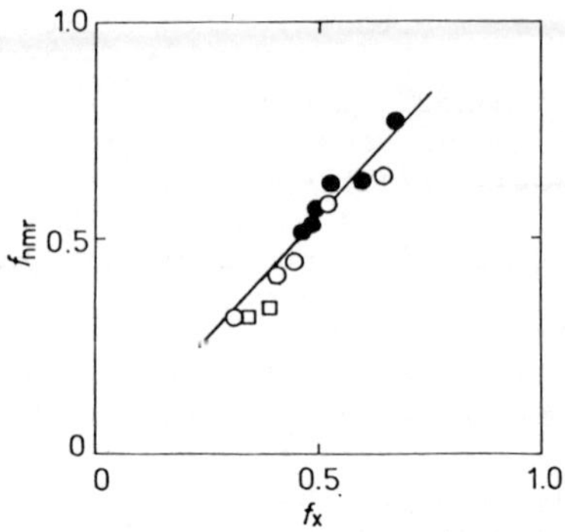

Figure 3 Integrated fraction f_{nmr} of the downfield component of the C4 resonance line vs. degree of crystallinity f_x determined by x-ray analysis. ●: native cellulose, o: regenerated cellulose, □: mercerized cotton and ramie. (The data of regenerated cellulose samples except for cupra rayon fibers were reproduced from Ref. 1.)

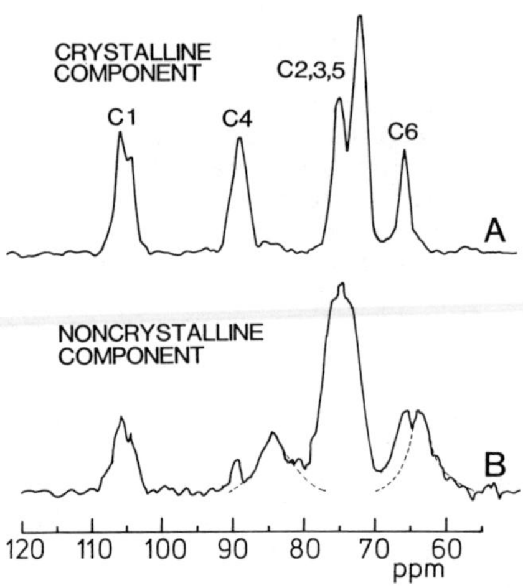

Figure 4 Partially relaxed 25 MHz ^{13}C NMR spectra of the crystalline and noncrystalline components of ramie which were separately measured by a pulse sequence for CP T_1 measurements developed by Torchia(16) and a single $\pi/2$ pulse sequence, respectively.

The pulse techniques employed for the selective observation of the crystalline and noncrystalline components for native cellulose are also useful for regenerated cellulose. Figure 5 shows the spectrum of the crystalline component of cupra rayon fibers together with that of a regenerated cellulose sample(1) of high crystallinity. The good agreement of the two spectra confirms that very sharp lines of the crystalline component are involved in the respective lines of the whole spectrum of the rayon fibers which is relatively broad as shown in Figure 2.

The chemical shifts of C1, C4, and C6 carbons of the crystalline and noncrystalline components of native and regenerated celluloses are tabulated in Tables I and II, respectively. Table II shows also the line width $\Delta\nu$ of each resonance line. Before we discuss these results, let us consider the meanings of the solid-state ^{13}C chemical shifts in the next section.

2. Relationships between ^{13}C Chemical Shifts and the Corresponding Torsion Angles.

The ^{13}C isotropic chemical shifts obtained by CP/MAS technique are also determined by the magnetic shielding of electrons surrounding concerned carbons just as in solution. Nevertheless, MAS can produce high-resolution lines without any aid of molecular motions. Therefore, the respective states of electron shielding fixed in the solid state are reflected on the respective chemical shifts of a given carbon, whereas such states are usually averaged out in solution by rapid inner molecular motions. This suggests that the solid-state ^{13}C chemical shifts provide important information of the local conformations and alignments of molecules in the solid state which possibly determine the extent of the electron shielding.

However, the solid-state chemical shifts are not well understood at present. The main effects which have been pointed out hitherto are the effects of packing, hydrogen bonding, and torsion angles about chemical bonds involving concerned carbons. In relation to the packing effect, VanderHart(17) has found that the resonance line of CH_2 carbons of triclinic crystals of n-$C_{20}H_{42}$ is shifted 1.3 ± 0.4 ppm downfield in comparison to that of n-paraffins of other crystal forms. As for the hydrogen bonding effects, Terao et al.(18-20) have reported that large downfield shifts appear for the carbons bonded chemically to hydroxyl groups which form strong hydrogen bonds. First we have examined these two effects for the crystals of monosaccharides and disaccharides which are compounds related to cellulose.

Table III compares the ^{13}C CP/MAS chemical shifts in the solid state and the corresponding shifts in D_2O. The comparison is made only for carbons not associated with exo-cyclic C-C bonds and β-1,4-glycosidic linkages. The respective resonance lines of the solid-state spectra of α-glucose and β-glucose and of the solution spectra of all the samples were assigned according to the results of Pfeffer et al.(21,22) In the table are also shown the O···O distances of the hydrogen bonds in the solid state(23-26). The difference $\Delta\delta$ in chemical shift between the two states ranges from -1.51 ppm to 2.43 ppm and are independent on the O···O distances. Therefore, the hydrogen bonding seems not to significantly affect the chemical shifts in these carbohydrates. However, the somewhat large

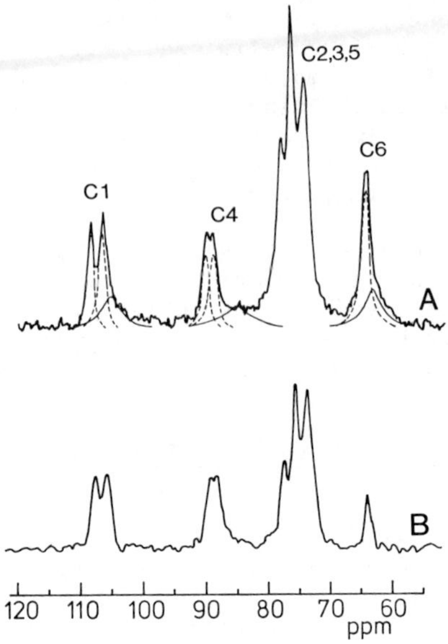

Figure 5 50 MHz CP/MAS ^{13}C NMR spectrum (A) of highly crystalline cellulose II sample and partially relaxed 25 MHz ^{13}C NMR spectrum (B) of the crystalline component of cupra rayon fibers which was separately measured by the pulse sequence for CP T_1 measurements.

Table I

^{13}C Chemical Shifts of the Crystalline Components of Native and Regenerated Cellulose

sample	chemical shift/ppm					
	C1			C4		C6
<u>native cellulose</u>						
valonia	107.0	106.4	105.3	91.3	90.1	66.5
bacterial	107.3	106.3	105.3	91.0	90.1	66.4
cotton	107.1		105.4	89.9		66.5
ramie	107.2		105.0	89.7		66.0
<u>regenerated cellulose</u>						
cupra rayon	108.0		106.3	89.5	88.2	64.1
high-crystalline cellulose II	108.1		106.3	89.6	88.4	63.7

Table II

^{13}C Chemical Shifts and Line-Widths of the Noncrystalline Components of Native and Regenerated Cellulose

	C1		C4		C6	
sample	δ/ppm	Δν/Hz	δ/ppm	Δν/Hz	δ/ppm	Δν/Hz
native cellulose						
valonia	106.2	114	85.2	249	63.4	185
bacterial	106.6	120	85.0	242	63.4	163
cotton	106.0	143	85.1	256	63.7	185
ramie	105.8	128	84.8	284	63.0	171
regenerated cellulose						
cupra rayon	105.3	284	85.1	327	62.8	185
amorphous	105.2	284	81.6	284	62.8	284

Table III

Comparison of ^{13}C Chemical Shifts of Monosaccharides and Disaccharides in the Solid State and in Solution

Sample	Carbon	δ/ppm		Δδ/ppm	O···O distance/Å
		Solid	D_2O solution		
α-D-glucose	1	93.54	93.30	0.24	2.847[a]
	2	71.22	72.73	-1.51	2.776[a]
	3	73.75	74.05	-0.30	2.707[a]
	4	73.36	70.93	2.43	2.773[a]
β-D-glucose	1	97.34	97.15	0.19	2.666[b]
	2	76.39	75.46	0.93	2.686, 2.708[b]
	3	76.39	77.17	-0.78	2.772[b]
	4	70.34	70.93	-0.59	no[b]
β-D-cellobiose	1'	98.02	97.05	0.97	2.776[b]
α-D-lactose monohydrate	1'	93.50	93.15	0.35	2.790[c]
β-D-lactose	1'	99.00	97.10	1.90	2.772[d]

a. Ref. 23 b. Ref. 24 c. Ref. 25 d. Ref. 26

difference $\Delta\delta$ may originate from total intermolecular effects such as packing and hydrogen bonding effects.

In Table IV the solid-state chemical shifts of the carbons of ramie which are involved in the β-1,4-glycosidic linkages and in the exo-cyclic C-C bonds are compared with the corresponding values of low molecular weight cellulose in DMSO solution(8). In this case the $\Delta\delta$ values are as large as 2.3-9.8 ppm, suggesting that such larger differences may stem from the difference in the torsion angles ϕ, ψ, χ about the corresponding chemical bonds. Therefore we next examined the effects of the torsion angles.

Figure 6 shows the chemical shifts of the C1 carbon against the torsion angles ϕ, determined from x-ray analyses(24-28), for different disaccharides of the β-1,4-glycosidic linkages(4). The chemical shifts of the crystalline components of different cellulose samples which are shown in Table I are also plotted against the corresponding ϕ values(29-32). In the latter case two or three different shift values are plotted against the same ϕ values except for the case of ramie where two different ϕ values have been determined by x-ray analysis(31). Although the data are somewhat scattered, the chemical shift of C1 carbon seems to be correlated to the torsion angle ϕ.

Figure 7 shows the similar plot of the chemical shifts of C4 carbon against the torsion angles ψ determined by x-ray analyses. As a linear relationship also exists between them, the chemical shift of the C4 carbon seems to be dependent on the torsion angles ψ.

In Figure 8, the ^{13}C chemical shifts of C6 carbon are plotted against the torsion angles χ about C5-C6 bonds for glucose and glucose residues of different disaccharides and cellulose(3). In this case there exists also a linear relationship between the two parameters. According to a recent survey(33) of the rotations about the C-C bonds in low molecular weight glucosides, *gauche-gauche* and *gauche-trans* are the two favored conformers. In agreement with this fact, no data of monosaccharides and disaccharides are included in this figure which correspond to the χ values around 300°. On the other hand, the conformation of the CH_2OH groups of the crystalline component of native cellulose is found to be *trans-gauche*(29,31) against the expectation from the data of low molecular weight glucosides. Figure 8 shows that these three preferred conformations are well correlated to the chemical shifts of the C6 carbon; *gauche-gauche*, *gauche-trans*, and *trans-gauche* correspond to the chemical shifts of about 62 ppm, 62.7-64.5 ppm, and about 66 ppm, respectively.

It is concluded that the solid-state chemical shifts of the C1, C4, and C6 carbons are mainly correlated to the torsion angles ϕ, ψ, and χ, respectively. However, the intermolecular effects such as packing and hydrogen bonding produce totally the downfield or upfield shift of 0.2-2.4 ppm for the carbohydrates studied here. Therefore, when the downfield or upfield shift of such an order appears even in the solid-state ^{13}C NMR spectra, the intermolecular effects must be also considered.

3. Chain Conformation of the Crystalline and Noncrystalline Components of Cellulose.

First we discuss the chemical shifts of the crystalline components shown in Table I. The C1 and C4 resonance lines of valonia and

Table IV

Comparison of ^{13}C Chemical Shifts of C1, C4 and C6 Carbons of Ramie in the Solid State with Those of Low Molecular Weight Cellulose in Solution

Carbon	δ/ppm			Δδ/ppm	
	Crystalline	Non-crystalline	Solution[a]	Crystalline	Non-crystalline
1	107.2, 105.0	105.8	102.7	4.5, 2.3	3.1
4	89.9	84.8	80.1	9.8	4.7
6	66.0	63.0	60.6	5.4	2.4

a. in deuterated dimethyl sulfoxide(8).

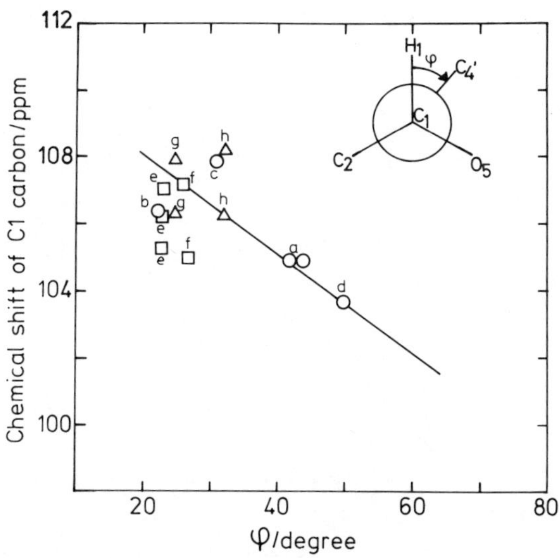

Figure 6 ^{13}C chemical shifts of the C1 carbons vs. torsion angles φ. a: β-D-cellobiose, b: β-D-methyl cellobioside·CH_3OH, c: α-D-lactose monohydrate, d: β-D-lactose, e: cellulose I (29), f: cellulose I (31), g: cellulose II (30), h: cellulose II (32). (Data a-d, g-h; Ref. 4.)

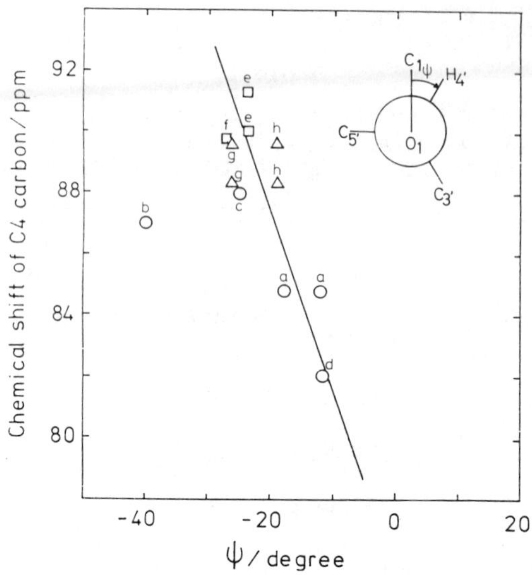

Figure 7. ^{13}C chemical shifts of the C4 carbons vs. torsion angles ψ. a: β-D-cellobiose, b: β-D-methyl cellobioside·CH_3OH, c: α-D-lactose monohydrate, d: β-D-lactose, e: cellulose I (29), f: cellulose I (31), g: cellulose II (30), h: cellulose II (32). (Data a-e, g-h; Ref. 4.)

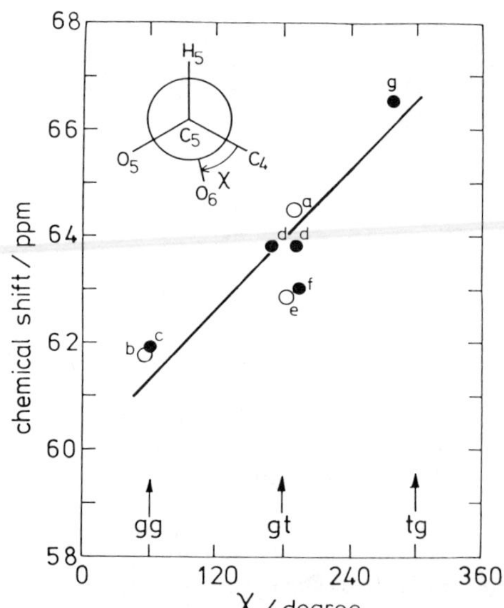

Figure 8. ^{13}C chemical shifts of the C6 carbons vs. torsion angles χ. a: α-D-glucose, b: α-D-glucose monohydrate, c: β-D-glucose, d: β-D-cellobiose, e: α-D-lactose monohydrate, f: β-D-lactose, g: cellulose I. o: α-glucose and α-glucose residues, •: β-glucose and β-glucose residues. (Data a-f; Ref. 1.)

bacterial celluloses split, respectively, into three and two lines in good accord with the results reported by others(13,34). On the other hand, in cotton and ramie the splitting into two lines appears only in the C1 line. This is possibly caused by the disappearance of the centered line of the C1 carbon and the downfield line of the C4 carbon of valonia and bacterial celluloses. If these splittings are assumed to be associated with the different torsion angles ϕ and ψ in the β-1,4-glycosidic linkage, the differences in ϕ and ψ which correspond to the splittings of 0.6-2.2 ppm in the C1 carbon and 0.9-1.2 ppm in the C4 carbon can be estimated to be 4.0-14.7° and 1.5-2.0° from the slopes of the straight lines in Figures 6 and 7, respectively. The validity of the assumption and the estimation will be examined for ramie by an x-ray crystal refinement because in this sample a relatively large splitting appears only for the C1 carbon.

An alternative cause of the splittings may be the intermolecular effects because the extent of splittings are in the order of these effects as described in the preceding section. In that case molecular chains must be packed in such a manner that at least two different intermolecular environments surrounding the C1 and C4 carbons are defined. The effects of hydrogen bonding associated with O5 and O1 oxygens might be somewhat significant. These effects are very difficult to be precisely estimated at present. Nevertheless, it can be concluded here that the crystal structure of valonia and bacterial celluloses are different from that of cotton and ramie, although the same crystal structure of cellulose I has been assumed for all these samples in the x-ray crystal analyses.

On the other hand, the results for regenerated cellulose are different from both types of native cellulose; both C1 and C4 carbons split into two lines. The chemical shift of the downfield line of the C1 carbon is the highest value of the C1 resonance lines and, on the contrary, that of the upfield line of the C4 carbon is the lowest value of the C1 resonance lines. These facts may be related to the crystal structure of cellulose II but the detailed structure must be also re-examined by considering two kinds of values for both ϕ and ψ or two different kinds of intermolecular environments for C1 and C4 carbons. In addition, the chemical shifts of the C6 carbon are 63.7-64.1 ppm for the regenerated cellulose, which correspond to the *gauche-trans* conformation as is seen in Figure 8. This conflicts with the result of x-ray crystal analyses(30,32), where both *gauche-trans* and *trans-gauche* are assumed. In an x-ray crystal refinement this fact must be also considered.

Table II shows that the noncrystalline components of native cellulose have no significant difference in chemical shift and line width from each other. The chemical shifts of the regenerated cellulose fibers, except for the amorphous cellulose, are also almost the same as those of native cellulose. This may suggest that the most probable conformation of the regenerated fibers is very similar to that of native cellulose. However, there is a marked difference in line width $\Delta\nu$ of the C1 carbon between native and regenerated celluloses; the line width of the C1 carbon of native cellulose is about half that of the regenerated cellulose. Although the cause of the line broadening in CP/MAS spectra is not clear as yet, it is most likely that the line width of the C1 carbon is primarily dependent on the distribution of the torsion angle ϕ. Thus, it is suggested that the distribution in the angle ϕ is relatively narrow for the noncrystalline component of native cellulose compared with that of regenerated cellulose. This means that the conformation of the noncrystalline component is rather hindered for native cellulose, whereas in the regenerated cellulose fibers it is relatively relaxed. Figure 9 shows the schematic models of the noncrystalline components of both samples. The result obtained here is in good accord with our previous ^{1}H broad-line NMR analyses(5,6) which indicate that the noncrystalline component of regenerated cellulose fibers is much enhanced in molecular mobility by a swelling agent such as DMSO, whereas no significant effect appears for native cellulose.

In the amorphous cellulose sample the chemical shift of the C4 carbon is as low as 81.6 ppm and very close to the corresponding value of the low molecular weight cellulose in DMSO solution(see Table IV). This sample was prepared by dissolution of Watman cellulose powder CF-1 in DMSO-paraformaldehyde followed by precipitation in ethanol. Therefore, the molecular chains of this sample must be fully disordered in comparison with those of the regenerated cellulose fibers and native cellulose. The more detailed structure of the noncrystalline components of different cellulose samples will be discussed elsewhere.

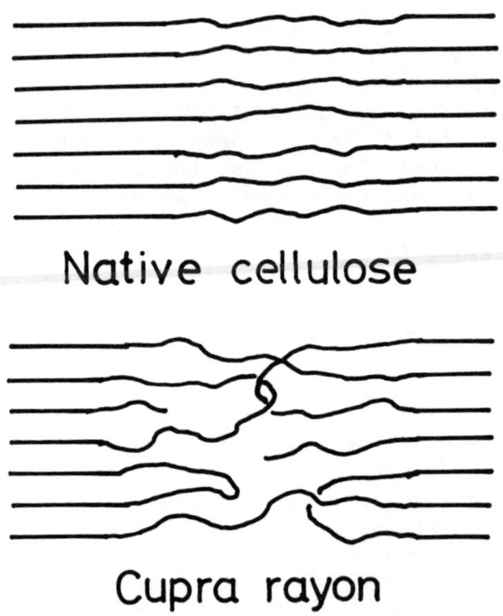

Figure 9 Schematic structure models of native cellulose and cupra rayon fibers.

Conclusions

1. The contributions of the crystalline and noncrystalline components to the resonance lines of C1, C4 and C6 carbons of different native and regenerated cellulose samples can be analyzed by using the respective line shapes. The chemical shifts of both components thus obtained were precisely determined.
2. The chemical shifts of C1, C4 and C6 carbons are correlated to the torsion angles ϕ, ψ and χ, respectively.
3. The crystalline component of ramie and cotton differs in chemical shift and line splittings from that of valonia and bacterial cellulose. This suggests that there is some difference in crystal structure among these samples, although the crystalline form is generally assumed to be cellulose I for all these materials.
4. The linewidth of C1 noncrystalline resonance of the native cellulose fibers is about half that of the regenerated cellulose fibers. This suggests that the conformational freedom of the noncrystalline molecular chain in native cellulose is strongly limited in comparison with that of regenerated cellulose.

Acknowledgments

We wish to thank Dr. M. Takai of Hokkaido University for providing the bacterial strain and for his helpful suggestions for preparing the bacterial cellulose. We are also grateful to Mr. S. Nagata, Laboratory of Microbial Biochemistry, Institute for Chemical Research, Kyoto University, for assistance in the preparation of the bacterial cellulose.

Literature Cited

1. Horii, F.; Hirai, A.; Kitamaru, R. Polym. Bull. 1982, 8, 163.
2. Hirai, A.; Horii, F.; Kitamaru, R. Bull. Magn. Reson. 1983, 5, 233.
3. Horii, F.; Hirai, A.; Kitamaru, R. Polym. Bull. 1983, 10, 357.
4. Horii, F.; Hirai, A.; Kitamaru, R. Bull. Magn. Reson. 1983, 5, 190.
5. Hirai, A.; Horii, F.; Kitamaru, R. J. Polym. Sci. Polym. Phys. Ed. 1980, 18, 1801.
6. Hirai, A.; Kitamaru, R.; Horii, F.; Sakurada, I. Cellulose Chem. Technol. 1980, 14, 611.
7. Buleon, A.; Chanzy, H. J. Polym. Sci. Polym. Phys. Ed. 1978, 16, 833.
8. Gast, J. C.; Atalla, R. H.; McKelvey, R. D. Carbohydr. Res. 1980, 84, 137.
9. Atalla, R. H.; Gast, J. C.; Sindorf, D. W.; Bartuska, V. J.; Maciel, G. E. J. Am. Chem. Soc. 1980, 102, 3249.
10. Earl, W. L.; VanderHart, D. L. J. Am. Chem. Soc. 1980, 102, 3251.
11. Hermans, P. H.; Weidinger, A. J. Appl. Phys. 1948, 19, 491.
12. Hermans, P. H.; Weidinger, A. J. Polym. Sci. 1949, 4, 135.
13. Earl, W. L.; VanderHart, D. L. Macromolecules 1981, 14, 570.
14. Garroway, A. N.; Moniz, W. B.; Resing, H. A. In "Carbon-13 NMR in Polymer Science"; ACS SYMPOSIUM SERIES No. 103, American Chemical Society: Washington D.C., 1979; p. 67.

15. Horii, F.; Hirai, A.; Kitamaru, R., to be published.
16. Torchia, D. A. J. Magn. Reson. 1978, 30, 613.
17. VanderHart, D. L. J. Magn. Reson. 1981, 44, 117.
18. Imashiro, F.; Maeda, S.; Takegoshi, K.; Terao, T.; Saika, A. Chem. Phys. Lett. 1982, 92, 642.
19. Imashiro, F.; Maeda, S.; Takegoshi, K.; Terao, T.; Saika, A. Chem. Phys. Lett. 1983, 99, 189.
20. Terao, T.; Maeda, S.; Saika, S. Macromolecules 1983, 16, 1535.
21. Pfeffer, P. E.; Valentine, K. M.; Parrish, F. W. J. Am. Chem. Soc. 1979, 101, 1265.
22. Pfeffer, P. E.; Hicks, K. B.; Frey, M. H.; Opella, S. J. J. Carbohydr. Chem., in press.
23. Brown, G. M.; Levy, H. A. Science 1965, 147, 1038.
24. Chu, S. S. C.; Jeffrey, G. A. Acta Cryst. 1968, B24, 830.
25. Fries, D. C.; Rao, S. T.; Sundaralingam, M. Acta Cryst. 1971, B27, 994.
26. Hirotsu, K.; Shimada, A. Bull. Chem. Soc. Japan 1974, 47, 1872.
27. Brown, C. J. J. Chem. Soc. (A) 1966, 927.
28. Ham, J. T.; Williams, D. G. Acta Cryst. 1970, B26, 1373.
29. Gardner, K. H.; Blackwell, J. Biopolymers 1974, 13, 1975.
30. Kolpak, K. J.; Blackwell, J. Macromolecules 1976, 9, 273.
31. Woodcock, C.; Sarko, A. Macromolecules 1980, 13, 1183.
32. Stipanovic, A. J.; Sarko, A. Macromolecules 1976, 9, 851.
33. Perez, S.; S.-Pierre, J.; Marchessault, R. H. Can. J. Chem. 1978, 56, 2866.
34. Kwoh, D. L. W.; Bhattacharjee, S. S.; Cael, J. J.; Patt, S. L. TAPPI Proceedings 1982, p. 114.

RECEIVED January 11, 1984

3

Conformational Accessibility of Some Simple Polyglucosides

ALFRED D. FRENCH

Southern Regional Research Center, Agricultural Research Service, U.S. Department of Agriculture, New Orleans, LA 70179

Conformational studies were made of the homopolymers of 1,2-, 1,3-, 1,4-, and 1,6- linked alpha- and beta-D-glucose. Flexiblity within the monomeric units and in the value of the glycosidic bond-angle was permitted, and the modeling method was able to depict the conformations of each of the 1,6-glucans on a single map. On the basis of the accessible conformations, it is predicted that the properties of alpha-1,3-glucan would be similar to those of cellulose, the beta-1,4-glucan, and therefore suited for use as a fiber. It is proposed that, when possible, polymers tend to crystallize in extended shapes with the fewest possible number of monomeric units per repeat. Other likely shapes include collapsed helices; shapes with large numbers of monomeric units per repeat and/or intermediate degree of extension are not likely to be found in crystalline regions.

D-Glucose can, at least theoretically, form a total of 8 homopolymers, ie. the alpha- and beta-1,2-, 1,3-, 1,4- and 1,6-glucans. Only one of these, the beta 1,4-glucan (cellulose), is of primary importance as a textile fiber. It is of interest to compare the structural properties of cellulose with those of other glucans in order to obtain greater insight to its unique qualities, as well as to better understand the overall relationship of polymer structure to polymer properties.

The flexibility of a polymeric material is an important property. In a fabric, for example, it can be assessed at a number of levels. Variations in structure of the finished assembly, the yarn, the fiber, the microfibril and the molecule all affect the flexibility. The flexibility at each of these levels depends in part on the structure at the next lower level. In turn, the molecular flexibility depends on the joints and on the constituent monomeric units of the polymer.

It is proposed in this paper that the molecular flexibility, or in the case of cellulose, the lack thereof, is one of the fundamental factors that determines the suitability of a polymer for use as a fiber. Another factor is the set of interactions between molecules in the crystalline material. Those interactions, which depend on the accessible molecular shapes, diminish solubility, especially if both hydrogen and hydrophobic bonds are formed, as in the case for the common conformation of cellulose. The relatively limited range of shapes (a definition of stiffness) not only keeps the cellulose chains in a conformation that retains the interchain attractions but also minimizes increases in entropy in solution.

As an approximation, the shape of a local region in a polymeric material can be described in terms of the number **n**, of monomeric units per helix pitch, P, and the rise per monomeric unit along the helix axis, **h**. A measure of flexibility proposed in this paper is simply the number of **n** and **h** combinations that is stereochemically possible. Reflection on the allowed **n-h** combinations permits the proposal of rules that determine the ranges of helical shapes which are likely to be found in crystalline regions. However, many of the interesting properties of polymeric materials depend substantially on non-crystalline components. Modeling studies are one approach to understanding those structures.

Flexiblity Within the Glucopyranose Ring

The glucans discussed herein are all based on the pyranose form. Shown in Figure 1 is the alpha form with only the carbon and oxygen atoms. The beta form has a bond between C1 and O1 that is equatorial to the ring. Almost all prior conformational analyses of polyglucosides have assumed that the monomers possess a rigid chair form. This assumption is very limiting, although it may appear unimportant if the first choice of monomeric shape gives an expected result. Workers such as Goebel et al (1) have pointed out the importance of monomeric flexibility but no other studies have explicitly incorporated flexibility in a fairly systematic way.

Evidence for the flexibility of the monomer comes from single crystal studies of compounds such as glucose and the di- and tri-saccharides. The structures of more than 200 glucose rings have been determined, including those in the special structures such as cycloamyloses. The glucose ring is always found in the $^4C_1$ shape but there are differences in the exact values of the molecular parameters.

Variations in the positions of the hydrogen atoms in hydroxyl groups are not very important because they adjust to allow a given molecular shape. The largest remaining variation lies in the position of the primary alcohol group. This group is free to rotate about the C5-C6 bond, and is found in a 60° range about one of two staggered conformations. The third staggered conformation is almost never found, to the extent that it is termed a "forbidden" conformation. It has been reported in single crystals, however, and is often proposed for polysaccharides.

The most important variations within the glucose ring itself are the ranges of the dihedral angles such as C1-C2-C3-C4. These angles each are seen to have ranges of about 20° when the results from

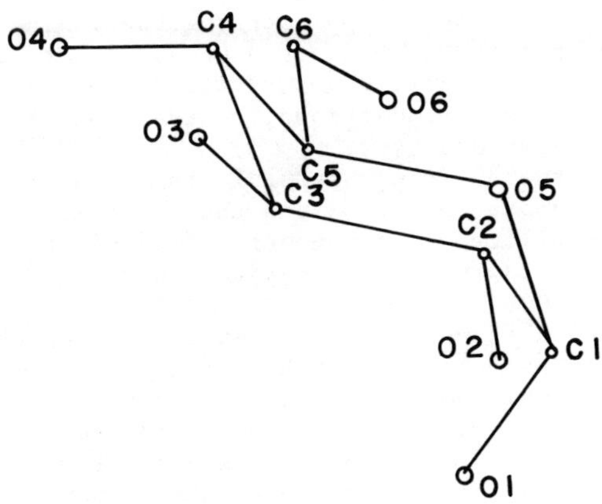

Figure 1.

Alpha-D-Glucopyranose, showing only the carbon and oxygen atoms for clarity.

currently available crystal structure determinations are compiled (2). Other molecular parameters, such as individual bond lengths and angles, have much less variation. Thus, variations in structure over the full observed range of those parameters have little effect on the shape of a polymer based on that monomer.

Work by Rees (3) shows the various geometries of the monomer linkages for the polypyranosides. When glucose polymerizes, C1 is always involved, and the linkage is to the O2, O3, O4 or O6 of the adjacent monomer. The relative angles of the C1-O1 bond with the Cn-On bond in the same monomeric unit determine a catagory of shapes that the polymer can acquire.

The intra-ring torsion angles affect the shape of the polymer in a cummulative way. The variations are summed along the collection of bonds involved in linking the polymer together. The effect of linkage type pointed out by Rees is a gross effect, and the effect from variations in monomeric geometry for a given type of linkage is more subtle. It is well understood, however, at least for the alpha-1,4-glucan, amylose (4).

In amylose, the 30° range possible in the angle between C1-O1 and C4-O4 of a glucose residue gives the alpha-1,4-linked glucan a substantial flexibility. The monomeric flexibility allows the formation of helices with the same pitch but with values of **n** ranging from 5 to 9.

That variation in monomeric geometry has been checked by the energy calculated by molecular mechanics for changes in the O4-O1 distance (5). A small amount of energy, about 2 Kjoules, is required to deform the glucose monomer from one extreme of the observed range to the other extreme. This amount of energy is available from formation of hydrogen bonds and vander Waals attractions. The population distribution calculated from the energy correlates with a histogram of O4-O1 distances observed in the monomer.

Flexiblity at the Linkage Between Monomers

While the flexiblity of a polysaccharide depends on the flexibilities of both the monomers and the linkages, the only variables usually considered in modeling studies have been the rotations about the bonds directly involved with the linkage oxygen atom. The primary factor which determines the degree of rotation about the linkage bonds is steric encounters. Another factor is the exo-anomeric effect, which tends to limit the rotation about the glycosidic bond to about a 60° range. The glycosidic angle usually is within $\pm 3^{\circ}$ of 117°, so relatively small differences in overall conformation are caused by its variation. Typically, larger glycosidic angles slightly increase the number of the allowed conformations.

We have developed a new representation of conformational space and a computer program to prepare it (4). This representation, the **n-h** map, allows many combinations of monomeric geometry and glycosidic linkage angle to be considered. A bonus in our method is that representations for the various polymers are quite different, whereas the Phi-Psi maps (without the iso-**n** and iso-**h** contours) are all very similar (3). Some polymers are characterized by **n-h** maps that have large, continuous allowed areas, while others have small or multiple, disconnected areas.

The **n-h** Map

The **n-h** map is a simple grid (Figure 2), with axes of **n** and **h**. Negative values of **n** indicate left-handed helices. The allowed combinations of **n** and **h** are enclosed by contours. At each gridpoint, the question was asked, "Is a stereochemically reasonable helix with these **n-h** values possible?" If a conformation was not allowed but had only 1 short contact, it might be represented by a cross-hatched region.

Twelve different coordinate sets from the single-crystal literature were used to make the maps of the alpha-1,2-, 1,3-, and 1,4-linked polymers in the present work. Another 12 geometries were used for the beta-linked molecules. The minimally allowed atomic radii were:

Carbon:	0.15 nm
Oxygen:	0.13 nm
Hydrogen:	0.09 nm

Minimally allowed interatomic distances between non-bonded atoms corresponded to the sum of the radii for the two atoms in question. Acceptable models had glycosidic bond angles within a range of 110 to 122°.

The computer program systematically checks the variation space of interest. The algorithms were designed to execute as quickly as possible for FORTRAN. Program NHMAP has run on IBM-370, Heathkit H-89 and Hewlett-Packard 3000 computers.

Results from **n-h** Map Studies

The alpha- and beta-1,2-glucans do not occur in large amounts. Only right-handed, 3- and 4-fold helices are permitted for alpha-1,2-glucans (Figure 2). The 3-fold helices are fully extended and the 4-fold ones are half-extended. There is a sub-marginal allowance of left-handed 4- and 5-fold helices of moderate extension, but there is no path from the fully allowed right-handed helices.

A beta-1,2-glucan is called sophoran. Our studies show that only one helical conformation is allowed, a left-handed helix with **n** = 3 and **h** = 2.5. This linkage occurs in a cyclic molecule consisting of 17 to 24 residues (6). According to our conformational analysis, the cyclic molecule must itself be composed of a 3-fold "helix" that curves back on itself to form the ring.

The alpha-1,3-glucans arise from fungi and other primitive plants. A number of different crystalline forms, or allomorphs, have been reported (7,8). In every case of which this writer is aware, the shape of crystalline alpha-1,3-glucan is a fully extended 2-fold helix. The differences in crystal structure arise from different packings of similar chains and depend on the history and the amount of water in the structure. From the **n-h** map (Figure 3), it is easy to see why only one shape is found. The extended conformation is isolated from all other conformations on the **n-h** map. Therefore, if the molecule is formed in this shape, it is likely that it will keep that conformation.

The drawings (Figure 4) of possible models of the polysaccharide in question are tied to the **n-h** maps. For the alpha-1,3-glucans,

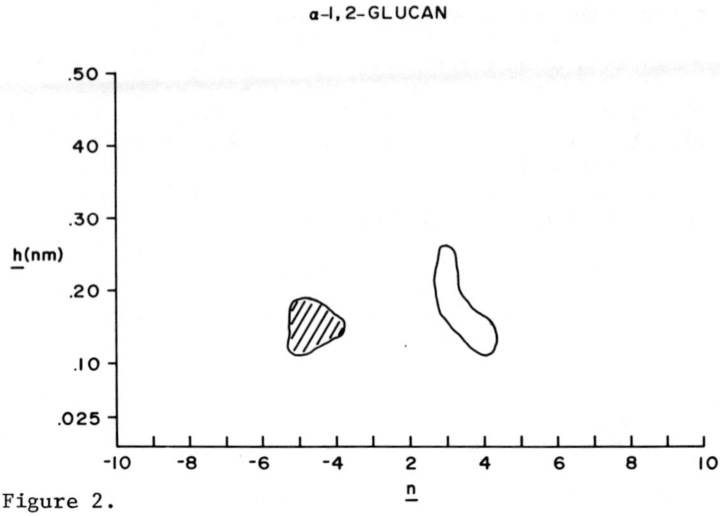

Figure 2.
An **n-h** map for alpha-1,2-glucan. Allowed conformations are within contours, disallowed conformations with only 1 short contact are cross-hatched. The negative values of **n** represent left-handed helices.

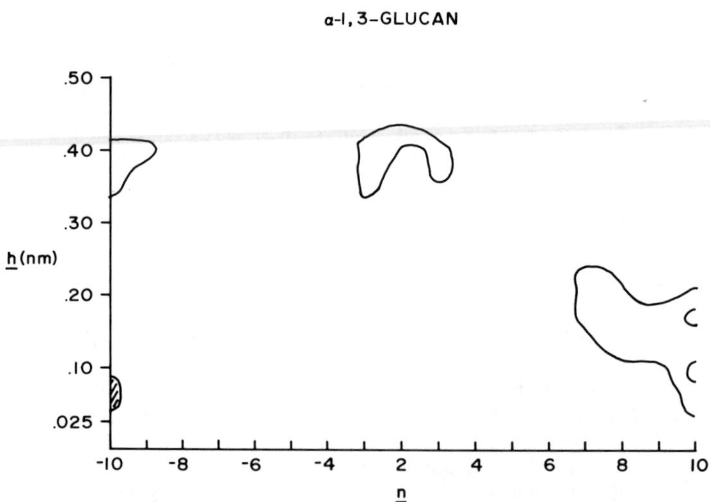

Figure 3. **n-h** map for alpha-1,3-glucan.

the main, 2-fold conformation is shown in the center. Another set of models on the map were collapsed right-handed helices with **n** = 7 or greater. An extended left-handed helix was also allowed, with **n** = 9. However, there is no path on the **n-h** map permitting conversion of the extended helices with low values of **n** to the larger values of **n**.

The beta-1,3-glucans are dramatically different from the glucans described so far. This polymer is extremely flexible, and occurs in many instances in nature. The polymer goes by the names of curdlan, pachyman, laricinan, schleroglucan, paramylon, lentinan, laminarin, callose and schizophylan. The prevalent form appears to be a triple helix, with **n** = 6 and **h** = 0.3 nm for each of the three strands (9). If triacetate derivatives are made, **n** = 6 and **h** = 0.31 to 0.36 nm (10). The acetate helices are single. Of some interest is the lack of allowed 2-fold conformations on the **n-h** map (Figure 5). The allowed zones close up when **n** = 20 or so, but some small adjustments in the monomer shape could allow an infinite number of monomeric units per helix repeat.

The drawing of the beta-1,3 helices (Figure 6) shows a few of the wide variety of possible shapes. Such flexiblity is probably necessary in order to form multiple helices.

The alpha-1,4-glucan is commonly called amylose. It takes on a large variety of different crystalline conformations. The **n** and **h** values found are all accounted for by Figure 7. The prevalent native forms are apparently double helices, with each strand having **n** = 6 and **h** of 0.35 nm. Right-handed models have been proposed (11). The other proposed shapes of amylose are all left-handed. Collapsed single helices with **n** = 6 have been indicated by diffraction studies to be left-handed (12). The KBr amylose complex has an **h** of 0.4 nm and **n** = 4. This conformation must be left-handed, as shown by the lack of 4-fold right-handed shapes. The KOH amylose complex is left-handed, with **n** = 6 and **h** = 0.377 (13). Note that there is no path of conversion between right- and left-handed helices. This observation has some implications for the validity of the right-handed proposal for native starch.

Figure 8 shows a few of the large variety of amylose helices.

The restricted area on the map (Figure 9) for beta-1,4-glucans (celluloses) is restricted similarly to the maps for the 1,2- and alpha-1,3-glucans. Only the extended 2- and 3-fold helices are fully allowed. However, there is an area with only one contact leading to collapsed, left-handed, 6- and 7-fold helices. There is some speculation that a blue stain resulting from phosphoric acid and iodine treatment of cellulose causes a 6-fold collapsed helix to be formed. According to the present work, if such a helix exists, it would probably be left-handed. There is no allowed path permitting conversion between the known extended helices and the collapsed, right-handed helices.

Both the 2- and 3-fold conformations are known to exist (Figure 10). The fiber properties are good for both conformations, but it may be that the native 2-fold shape was important to formation of good fibers. The 3-fold shapes are not stable; they depend on alkali for retention of conformation.

The alpha-1,6-glucan is called dextran. Because of rotation about the C5-C6 bond, in addition to the rotations about the C6-O and

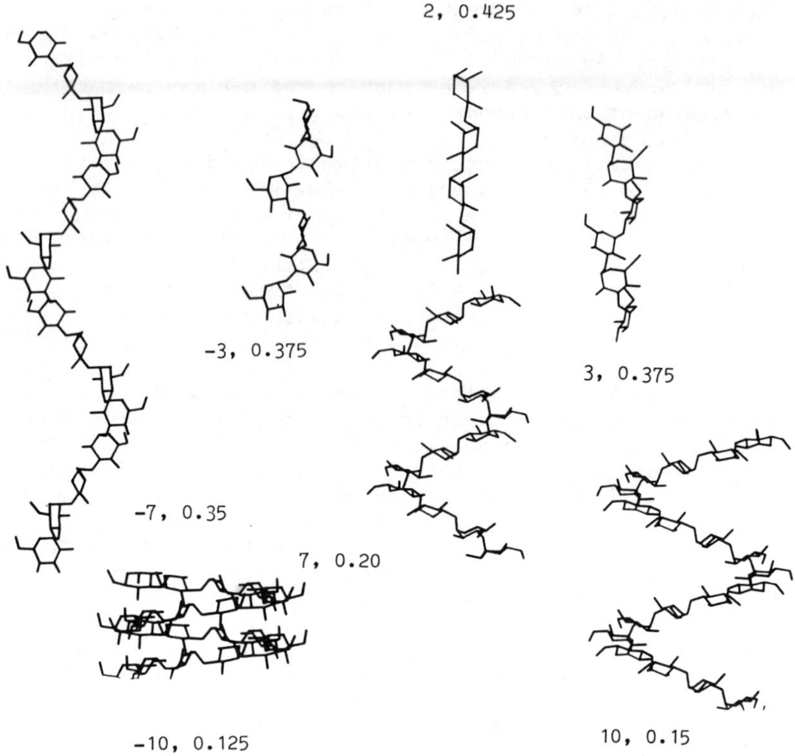

Figure 4.
Some of the allowed helical shapes for alpha-1,3-glucan. The **n** and **h** values are indicated.

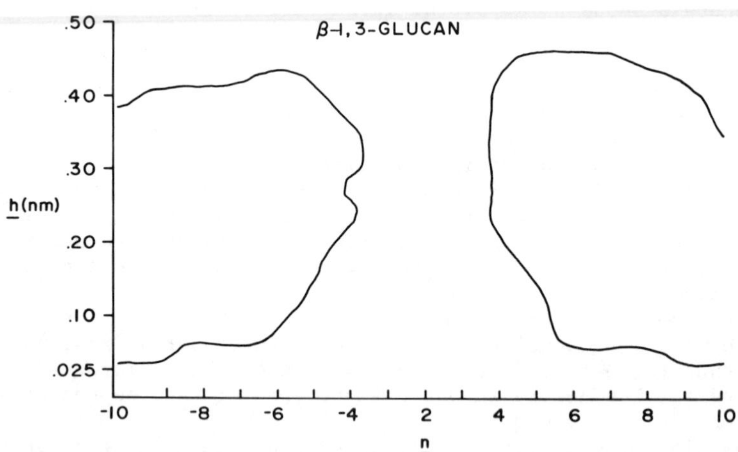

Figure 5. **n-h** map for beta-1,3-glucan (curdlan, etc.)

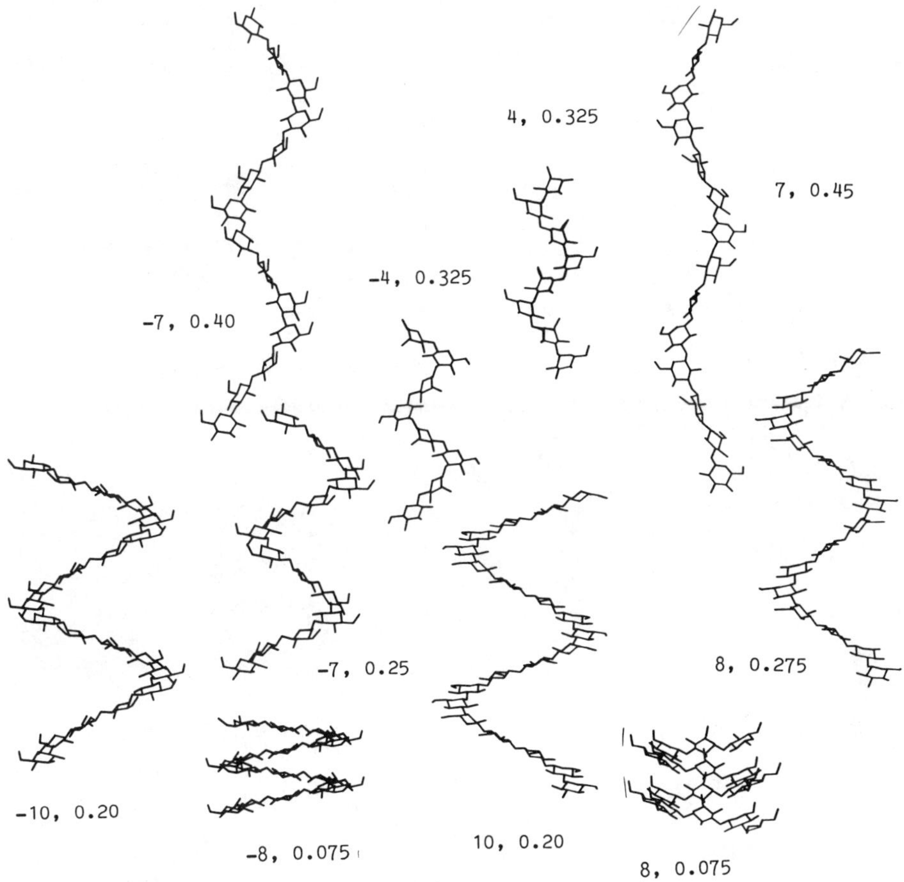

Figure 6.
Some of the allowed helical shapes for beta-1,3-glucan.

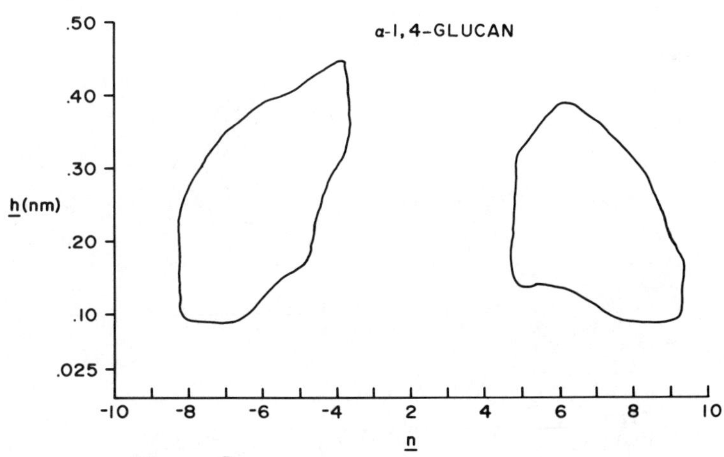

Figure 7.
n-h map for alpha-1,4-glucan (amylose).

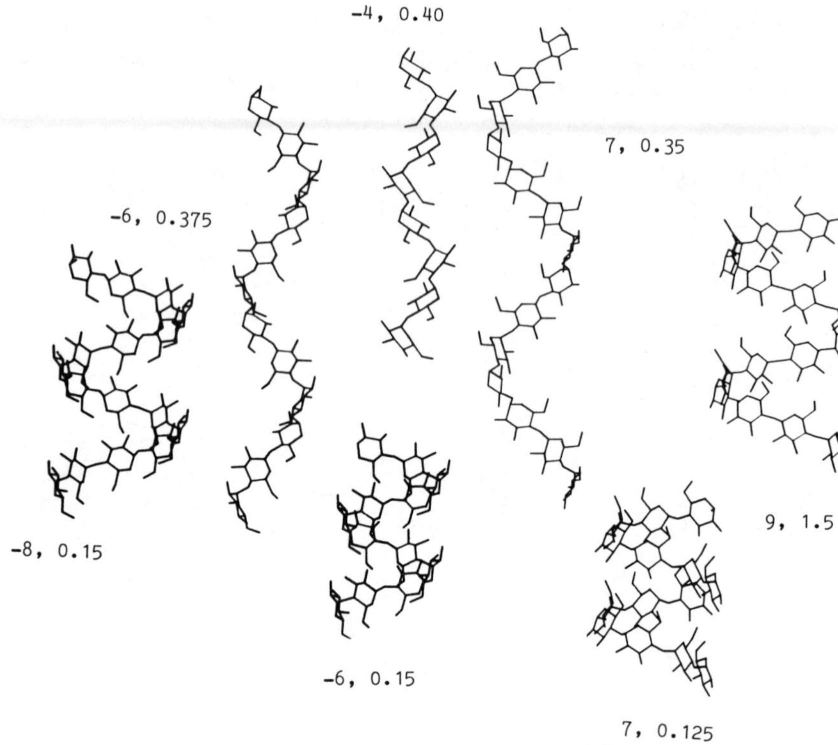

Figure 8.
Some of the allowed helical shapes for alpha-1,4-glucan.

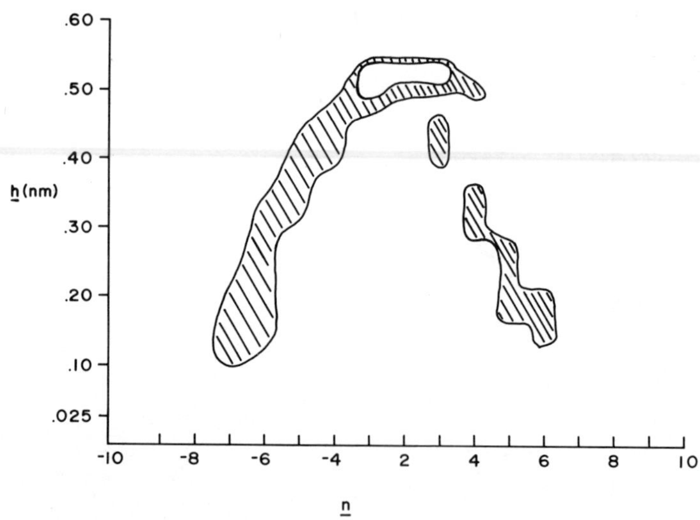

Figure 9.
n-h map for beta-1,4-glucan (cellulose).

O-C1 bonds, this molecule is more flexible than the other molecules. It is more flexible than the beta-1,3-glucan in that there is a path between right- and left-handed helices. Almost any shape is possible, and crystallization of this material is difficult. The **n-h** map in Figure 11 is a strength of our technique that allows large or small variations in monomeric geometry. Only one monomeric geometry (with O6 in 9 different positions) was used to make this map.

The wide open **n-h** map corresponds to a wide variety of helices (Figure 12). Dextran would probably be difficult to make into useful fibers because of the lack of reason to crystallize in any particular shape. It has been crystallized, however, in an extended 2-fold conformation (14).

A natural beta-1,6-glucan is called pustulan (15). It is also an extremely flexible molecule, as indicated by Figure 13. It is also very difficult to crystallize.

Figure 14 shows the wide variety of shapes possible for pustulan, because of the 1,6-linkage.

Discussion

The above helices are all related to a linear helix axis. If the molecules contain monomeric units with geometries that differ in a random manner, the polymer will no longer be a regular helix. Instead, it will have segments corresponding to the above helical segments, but the axis will deviate in a random manner from linearity.

The **n-h** mapping program can model molecules that have regular sequences of different geometries. This capablity has not been implemented yet, but is similar to the idea used in modeling the 1,6-linked glucans. This gives an additional flexiblity to the molecule. As an example of the departure from a linear helix axis brought about by using geometrically different monomeric units within the same molecule, the model shown in Figure 15 can be formed. Only 14 residues are required to form a ring. The glycosidic linkages and hydrogen bonds in this model are the same as proposed for native cellulose. The amount of distortion of the chain implied by this model ring shows the amount of bending of the molecule possible in non-crystalline or amorphous regions of cellulose.

Although the model in Figure 15 does not correspond to a known structure, the known, cyclic sophorans would be somewhat similar to the model. The difference would be that the 1,4-linked model is a 2-fold "helix" that curves around, whereas the cyclic sophorans (1,2-linked) would be 3-fold curved "helices" with several more monomeric units.

At least two generalizations can made from a study of these accessible conformations of glucosides. Although a large number of amylose conformations is observed, a great many more are possible. The factors involved in crystallization seem to center on dense, efficient packing. No examples of single helices with large values of **n** and intermediate values of **h** are found in crystalline samples. Either the helices are quite extended or they are collapsed (16).

Amylose complexes with foreign agents such as iodine, alcohol, or fatty acids, by wrapping around a linear array of these

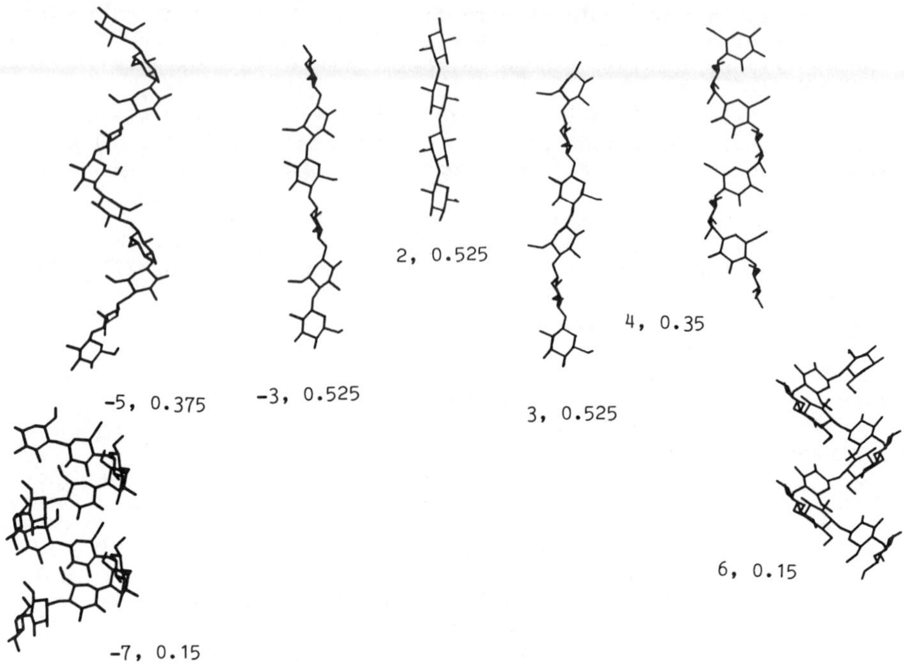

Figure 10.
Some of the helical shapes for beta-1,4-glucan. Only the shapes with **n** = 2 or + 3 are fully allowed. The other shapes have one short contact.

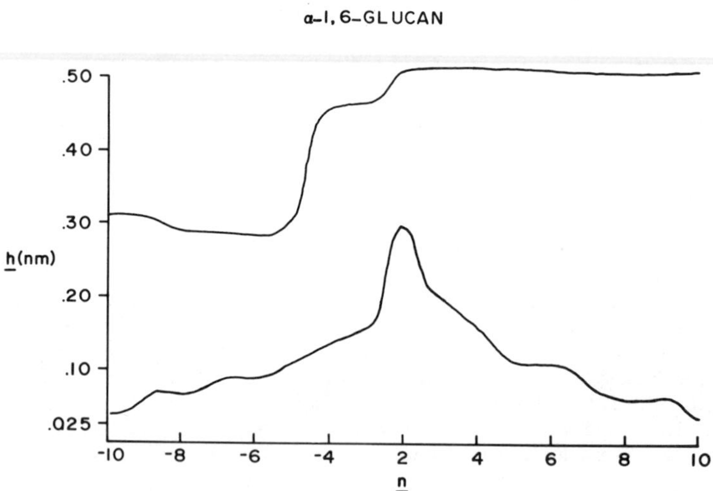

Figure 11.
n-h map for alpha-1,6-glucan (dextran).

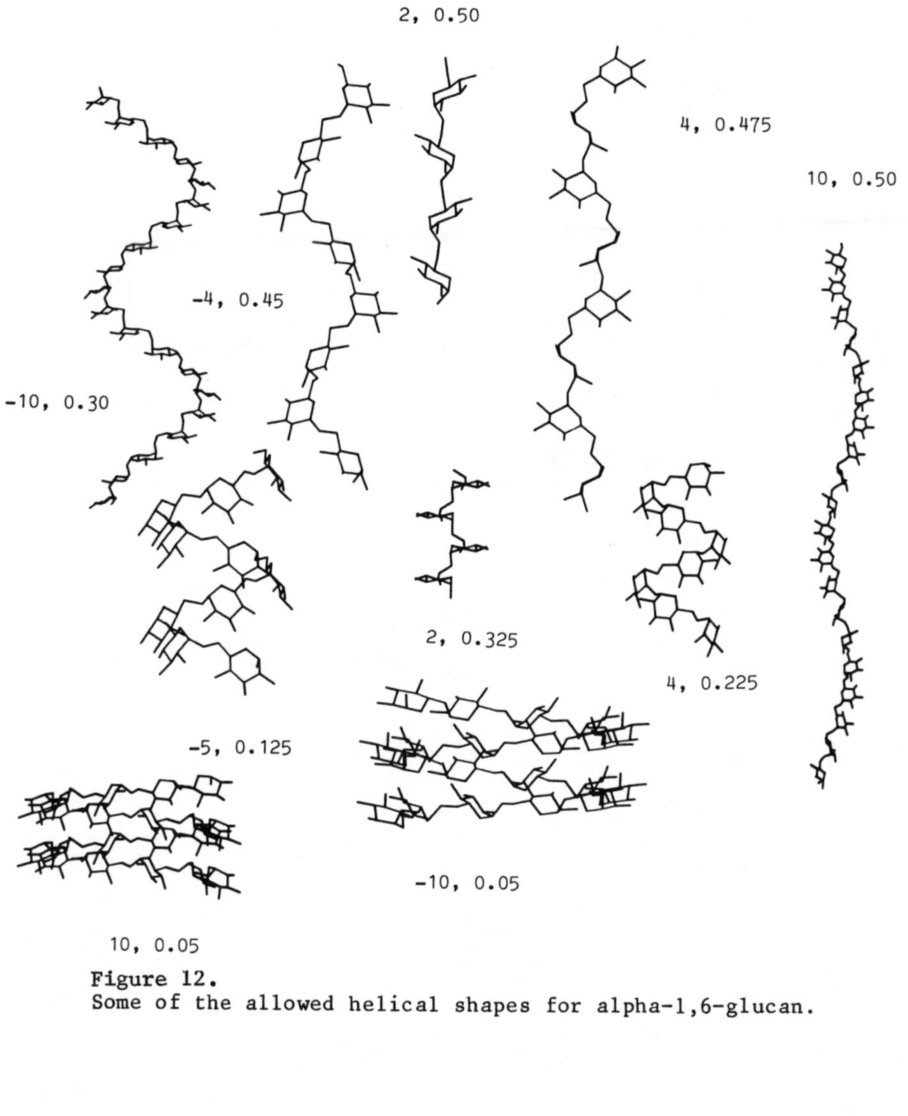

Figure 12.
Some of the allowed helical shapes for alpha-1,6-glucan.

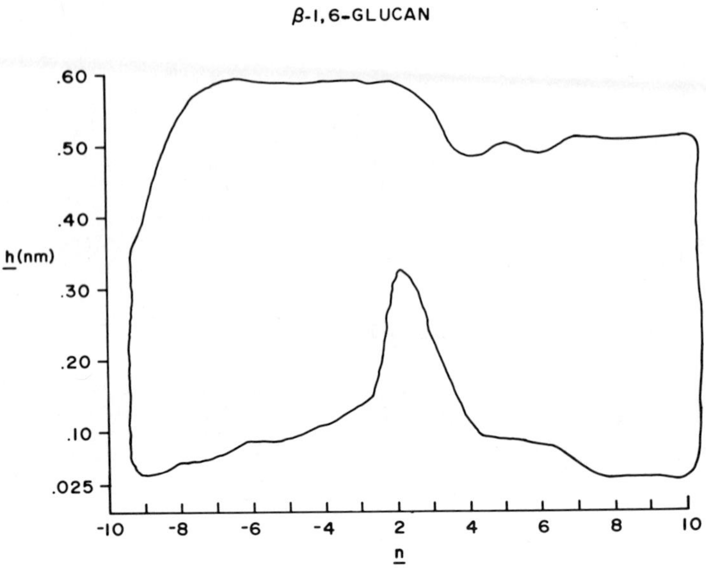

Figure 13.
n-h map for beta-1,6-glucan (pustulan).

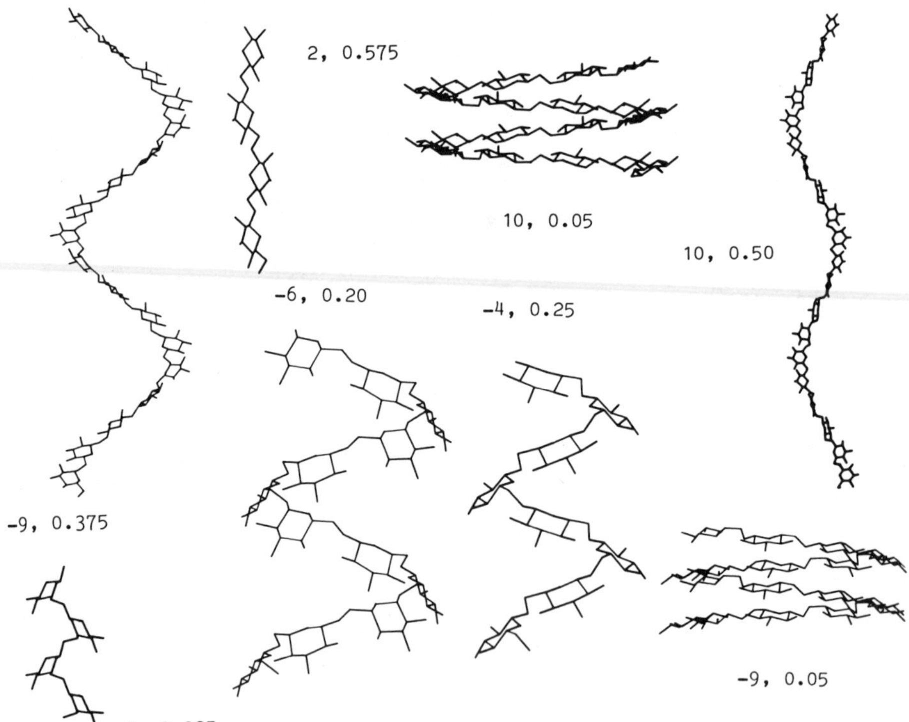

Figure 14.
Some of the allowed helical shapes for beta-1,6-glucan.

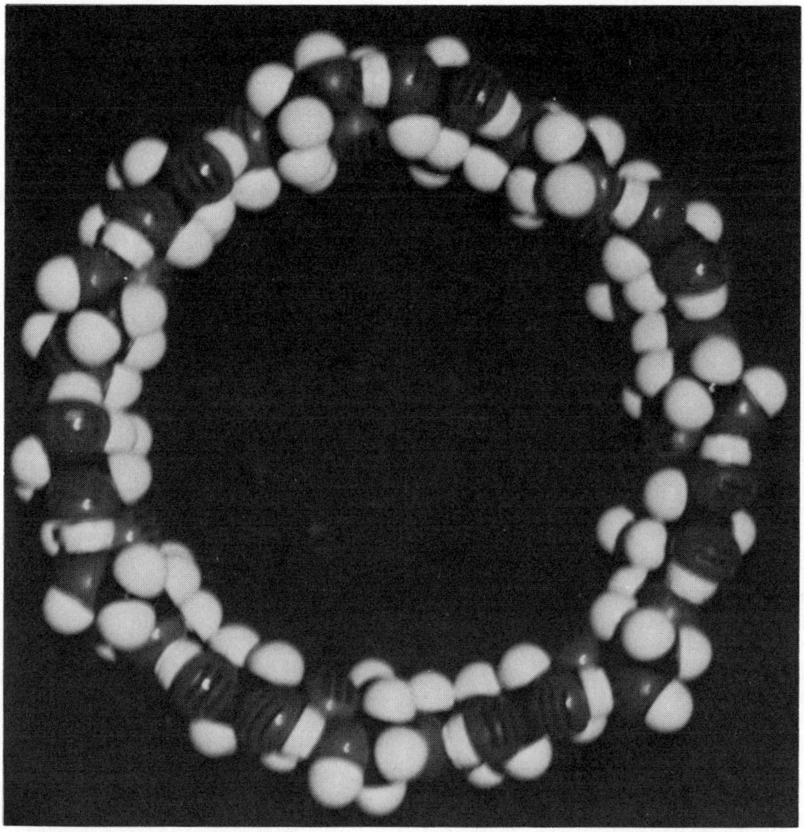

Figure 15.
Space-filling model of circular cellulose. This model does not represent an observed conformation, but does indicate the flexibility of the cellulose chain that arises from the monomeric flexibility. Only 14 glucopyranose residues are required to complete the ring.

substances. In these instances, the helix is collapsed. If the foreign substance is ionic, such as KBr or KOH, the helix is extended, with nests formed for the ionic material. If no other material is available at time of crystallization, a double helix is apparently formed. Surprisingly, crystalline symmetry is unimportant. Both 7-fold and 5-fold helical structures have been reported (17). No crystallographic symmetry applies to those shapes. In order to make reasonable models of all the amylose allomorphs, the monomeric geometry must be flexible. No one monomer geometry would be able to model all the observed forms.

It appears that one way to predict whether a given combination of **n** and **h** will be observed is to determine whether it is on the boundary of the **n–h** map. This will not apply to multi-stranded helical conformations. For the single stranded helices, however, it is a good guide. The lower bound of an **n–h** map is a place where atoms are likely to be brought into contact with others in the same helix, strengthening vander Waals forces or possibly hydrogen bonds. The fully extended helices will have small diameters and therefore be able to pack with a high density of close contacts with neighboring helices.

Conclusions

The accessible conformations for a complete series of simple polyglucosides have been presented. They are all composed of the same building block, yet their conformational and physical properties are different. Two glucans, the beta-1,4 and the alpha-1,3, are inflexible and can form extended 2-fold helices. They are both found in cell walls. Two flexible glucans, with 1,6-linkages, can also form the extended 2-fold conformations but do not take a structural role in cell walls. The highly flexible alpha-1,4- and beta-1,3-glucans, which cannot form extended 2-fold helices, form multiple helices. The 1,2-glucans have little flexibility and cannot form the extended 2-fold conformations. Therefore, except for the alpha-1,3- and beta-1,4-polymers, the simple glucans are not suited for use as fibers.

In addition to the better understanding of what the molecular requirements are for use as fibers, this work provides a picture of the total flexibility of a molecule. The **n–h** map gives a good idea of the available structures in solutions and in amorphous regions. Those structures must be studied with modeling techniques.

Acknowledgments

Margaret Berg was of great assistance in implementing and operating the **n–h** mapping program, which was written by my brother, Walt French.

Literature Cited

1. Goebel, C. F.; Dimpfl, W. L.; Brant, D. A. Macromolecules 1970, **3**, 644.
2. French, A. D.; Murphy, V. G. Carbohydr. Res. 1973, **27**, 391.

3. Rees, D. A. "Polysaccharide Shapes"; Halsted Press: New York, 1977; p. 55.
4. French, A. D.; French, W. A. "Fiber Diffraction Methods", ACS SYMPOSIUM SERIES No. 141, American Chemical Society: Washington, D. C., 1980, p. 239.
5. Pensak, D. A.; French, A. D. Carbohydr. Res. 1980, **68**, 1.
6. Hisamatsu, M.; Amemura, A. Carbohydr. Res. 1983, **121**, 31.
7. Ogawa, K.; Misaki, A; Oka, S.; Okamura, K. Carbohydr. Res. 1979, **75**, c13.
8. Jelsma, J.; Kreger, D. R. Carbohydr. Res. 1979, **71**, 51.
9. Chuah, C. T.; Sarko, A.; Deslandes, Y.; Marchessault, R. H. Macromolecules 1977, **55**, 293.
10. Bluhm, T. L.; Sarko, A. Biopolymers 1977, **9**, 2067.
11. Wu, H. H.; Sarko, A. Carbohydr. Res. 1979, **61**, 7.
12. French, A. D.; Zaslow, B. J. C. S. Chem. Comm. 1972, 41.
13. Sarko, A.; Biloski, A. Carbohydr. Res. 1980,**79**, 11.
14. Sarko, A. Presented at the 186th National Meeting of the American Chemical Society, Washington, D. C. August 30, 1983.
15. Chanzy, H.; Guizard, C.; Vuong, R. J. Microsc. (Oxford) 1977, **111**, 143.
16. French, A. D. Bakers Digest 1979, **53**, 39.
17. French, A. D.; Murphy, V. G. Cereal Foods World 1977, 61.

RECEIVED January 10, 1984

4

Quantitative Analysis of Textiles by IR Spectroscopy

A Comparison of Fourier Transform and Dispersive Techniques

NANCY M. MORRIS and RALPH J. BERNI

Southern Regional Research Center, Agricultural Research Service, U.S. Department of Agriculture, New Orleans, LA 70179

Enhancement of the accuracy of quantitative infrared determinations by use of mathematical manipulation of the spectral data as performed by Fourier Transform (FTIR) and Computerized Dispersive (CDIS) Infrared Spectroscopy has been compared. Cotton-polyester blends and cotton treated with THPOH-NH_3 and with dimethyloldihydroxyethylene urea (DMDHEU) were analyzed by FTIR and CDIS. The mathematical techniques used included direct spectral subtraction and spectral subtraction combined with analysis of peak areas. Advantages and disadvantages of each method were found.

Determination of finish add-ons or blend ratios for textiles containing cotton are of prime interest in textile finishing laboratories. Many approaches have been taken to acquire this information including Kjeldahl nitrogen analysis, other wet chemical determinations, x-ray fluorescence or diffraction (1), and infrared spectroscopy (2-6).

The use of infrared spectroscopy required that a band due to the treatment or blend component be isolated from bands due to the cellulose so that peak height or peak area measurements could be made without interference from overlapping peaks. Peak area measurements involved counting squares on the graph paper, cutting out the peak and weighing the graph paper, or making planimeter measurements. With the advent of computerized spectroscopy, first with FTIR and then with CDIS, the question arose whether spectral subtraction routines and other mathematical manipulation of the spectra could provide better and/or easier quantitation of these blends or treatments.

Comparison of CDIS and FTIR

CDIS interfaces a standard grating spectrophotometer and a microcomputer. The spectrophotometer alone gives a usable infrared spectrum (a plot of intensity *versus* wave number or wavelength). Use of the computer adds the ability to:

1. Store the spectral data on a microfloppy disk for later manipulation and retrieval.
2. Mathematically manipulate the spectra (i.e., addition, subtraction, multiplication, baseline flattening, smoothing, etc.).
3. Replot the spectrum on a video display unit or on paper.
4. Perform repetitive runs.
5. Measure the peak heights or integrated areas with the computer.

FTIR takes a completely different approach. The spectral data are acquired as an interferogram (Figure 1) which must be transformed into a plot of intensity versus wavenumber or wavelength through the application of Fourier transform equations. Thus, the computer is an integral part of the system without which little useful information could be obtained. FTIR has the following advantages over computerized dispersive infrared spectroscopy:

1. The multiplexing, or Fellgett's, advantage which arises because all of the resolution elements are observed all of the time.
2. The throughput, or Jacquinot's, gain which is produced by the large sampling area and absence of narrow slits in the interferometer. These factors result in a large amount of energy reaching the detector and a significant improvement in the signal-to-noise ratio for FTIR.
3. The use of the He-Ne laser interferometer to reference the position of the moving mirror increases the accuracy of frequency determination (Connes' advantage) for FTIR.

In addition to the abilities provided dispersive infrared spectroscopy by interfacing a computer, the FTIR adds the ability to select the resolution through software parameters. This sets the mirror displacement which, depending on the instrument, can take on values from 0.125 cm^{-1} to 32 cm^{-1}. Good spectra are obtained with a resolution of 4 or 8 cm^{-1}. The farther the mirror travels, the greater the resolution; but, at greater resolutions, the time for collection and computation of the spectral data increases considerably.

Experimental

To study the quantitation of cellulosic textiles, potassium bromide disks were prepared by the method of O'Connor *et. al.* (3). The infrared spectra of the disks were obtained and stored for further calculation.

The Fourier transform IR spectra were obtained on a Digilab Model 15B (Company or trade names are given for information purposes only and their use does not imply that the USDA recommends those products over others that may also be suitable.) with a Nova 3 computer, a TGS (triglycine sulfate) detector, a globar source and a KBr beam splitter. The instrument was continuously purged with dry nitrogen. The following parameters were used to collect the

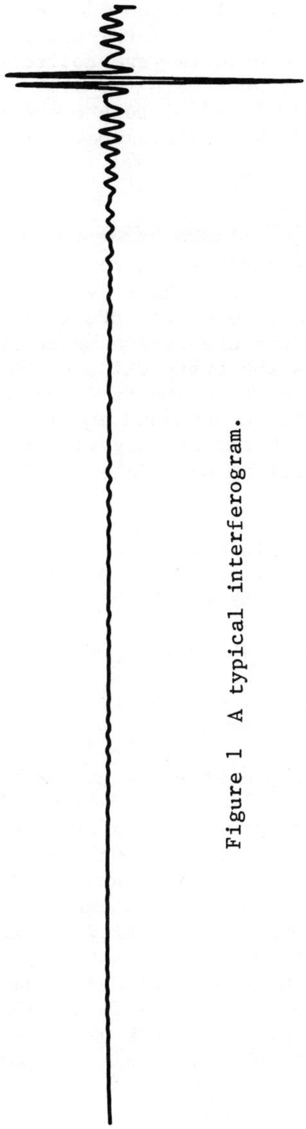

Figure 1 A typical interferogram.

spectral data: resolution, 8; number of scans, 200; triangular apodization; double precision.

The spectral data were weight normalized to a sample weight of 2.00 mg with the scaling function in the arithmetic command according to the equation (PFA) (SCA)=scaled spectrum. PFA is the contents of spectral file A and SCA is a scaling factor computed by multiplying the reciprocal of the sample weight by 2.

The dispersive spectra were collected on a Perkin-Elmer Model 283B infrared instrument with the 3500 data station.

The samples analyzed by both procedures were (1) THPOH-NH_3 treated cotton, (2) DMDHEU treated cotton, and (3) cotton-polyester blends.

Comparison of Subtraction Routines

FTIR. Because the subtraction routine on the FTIR is interactive, the operator has visual control over the scaling factor for the reference spectrum. To effect the subtraction, the operator selects a region in which the absorption bands are due to the reference material and gives the subtraction command while watching the screen. When a flat baseline is obtained in the region of interest, the operator breaks the command and the difference spectrum is displayed along with the scaling factor. The equation used to effect this subtraction was (PFA) (SCA)-(PFB) (SCB)= Difference Spectrum, where:

PFA = spectrum of treated cellulose.
PFB = spectrum of untreated cellulose.
SCB = scaling factor by which the spectrum of untreated cellulose is multiplied.
SCA is held constant at one.

A Fortran Program called "Ingrate" (7) was used to compute the area under the band of interest.

CDIS. A different procedure is used with the dispersive instrument. Again the operator selects the region over which the subtraction will take place and gives the subtraction command. With this software, the computer calculates the scaling factor and stores the difference spectrum which may then be displayed with a view command. The operator can change the scaling factor by entering a value for this parameter along with the difference command. This operation can be repeated as many times as the operator desires.

A quantitative package is available from the manufacturer, but we wrote our own OBEY routines to calculate (1) the ratios of two peaks, using one cellulose peak as an internal standard in each case and (2) the peak height of a selected band from the difference spectrum.

THPOH-NH_3 Treated Cotton

THPOH-NH_3 is a treatment used to impart durable flame retardancy to cotton fabric. The level of THPOH-NH_3 is very difficult to determine by conventional infrared spectroscopy due to the overlap of the reagent bands with those of cellulose.

A series of eight THPOH-NH_3 treated fabrics was examined. The spectra of the untreated cotton, the treated cotton, and the difference spectrum obtained by subtracting the untreated cellulose spectrum from that of the treated fabric as obtained on the FTIR are shown in Figure 2. It is very difficult to distinguish the 1655 cm^{-1} band which may be due to NH deformation (8) in the treated cellulose from that of the 1630 cm^{-1} band due to adsorbed water. For spectra obtained with FTIR, the area under the curve between 1740 and 1630 cm^{-1} was integrated with the Fortran "Ingrate" program. For the spectra obtained with the dispersive instrument, two methods were used. For the peak height method, the difference spectrum was obtained and the height of the peak at 1655 cm^{-1} was measured. With the second procedure, the ratio of the absorbance at 1655 cm^{-1} to that at 1163 cm^{-1}, which is an internal standard for cellulose, was measured. The linear regression lines for the data from the FTIR and CDIS are shown in Figure 3. Treatment levels calculated from both regression lines are given in Table I. The CDIS

Table I. Percent Phosphorus in THPOH-NH_3 Treated Fabrics

Wet Chemical	FTIR	CDIS peak heights	CDIS peak ratios
4.07	4.04	3.32	3.21
3.59	3.66	2.50	2.68
3.13	2.71	3.07	3.42
2.31	2.24	2.34	2.40
1.86	1.76	3.33	3.12
1.22	1.23	1.27	1.18
0.57	0.62	0.92	0.76

subtraction routine did not handle the peak overlap as well as the FTIR, and the reagent level as calculated from the CDIS data is not useful. The results obtained with FTIR are comparable to those obtained by wet chemical analysis and by x-ray diffraction (unreported).

DMDHEU Treated Cotton

Dimethyloldihydroxyethylene urea (DMDHEU) is used to impart durable press properties to cellulose. Spectra of the treated cellulose, untreated cellulose and the difference between the two are shown in Figure 4. In this case the area under the curve between 1740 and 1670 cm^{-1} was calculated for the FTIR spectra. For CDIS, the height of the C=O band in the difference spectrum of DMDHEU treated cotton minus untreated cotton or the ratio of the C=O band at 1711 cm^{-1} versus a band at 1370 cm^{-1} in cellulose was plotted versus percent nitrogen. The separation between the carbonyl band at 1715 and the 1630 cm^{-1} adsorbed water band was sufficient to provide good results regardless of the method used. The linear regression lines are shown in Figure 5. The treatment levels calculated from each regression line are given in Table II. There is essentially no difference in the values from FTIR and from CDIS by each procedure for these samples which had been prepared by a conventional pad,

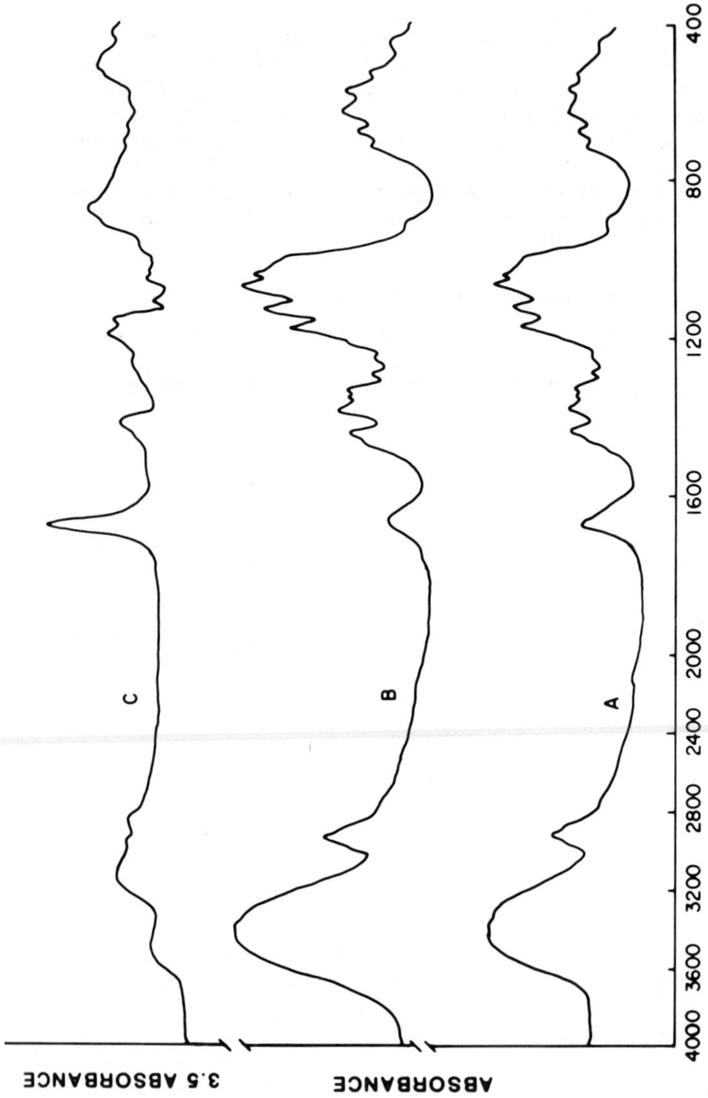

Figure 2 Spectra of: (A) THPOH-NH_3 treated cotton; (B) Untreated cotton; and (C) Difference (Untreated cotton subtracted from treated cotton and expanded 3.5 times).

Reproduced with permission from Ref. 6.

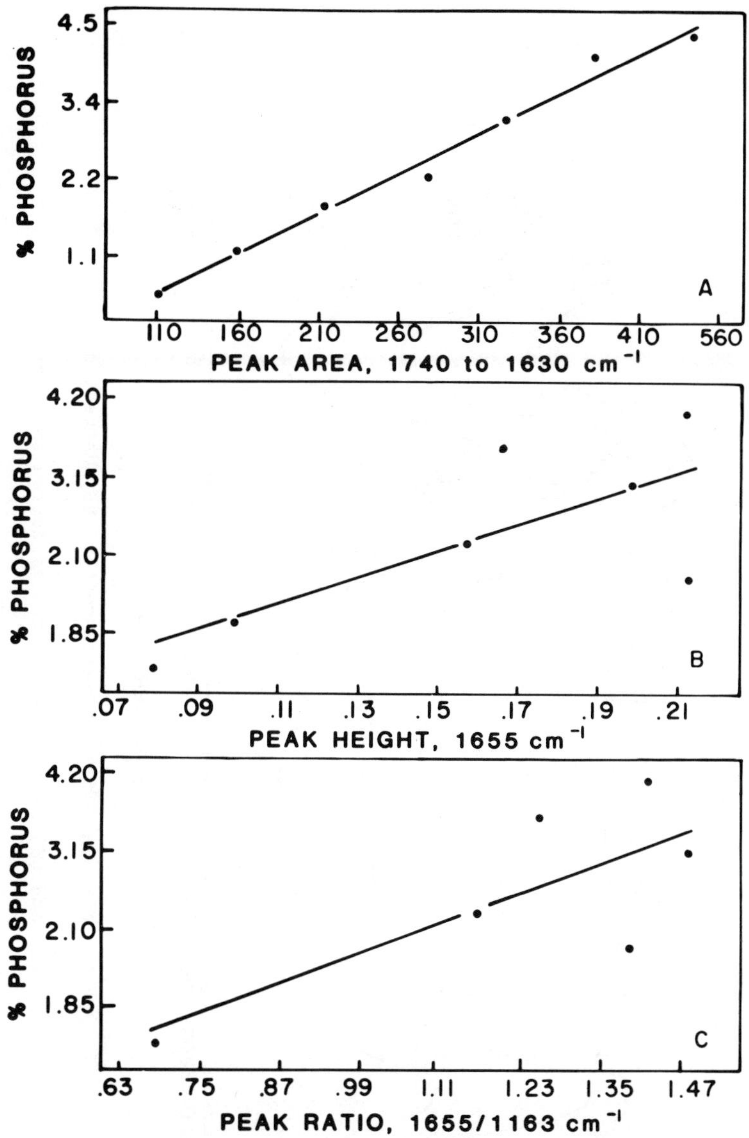

Figure 3 Linear regression lines for percent phosphorous <u>versus</u>: (A) Integrated area between 1740 and 1630 cm^{-1} determined by FTIR; (B) Peak height at 1655 cm^{-1} determined by CDIS; and (C) Ratio of peak at 1655 cm^{-1} to that at 1163 cm^{-1} determined by CDIS.

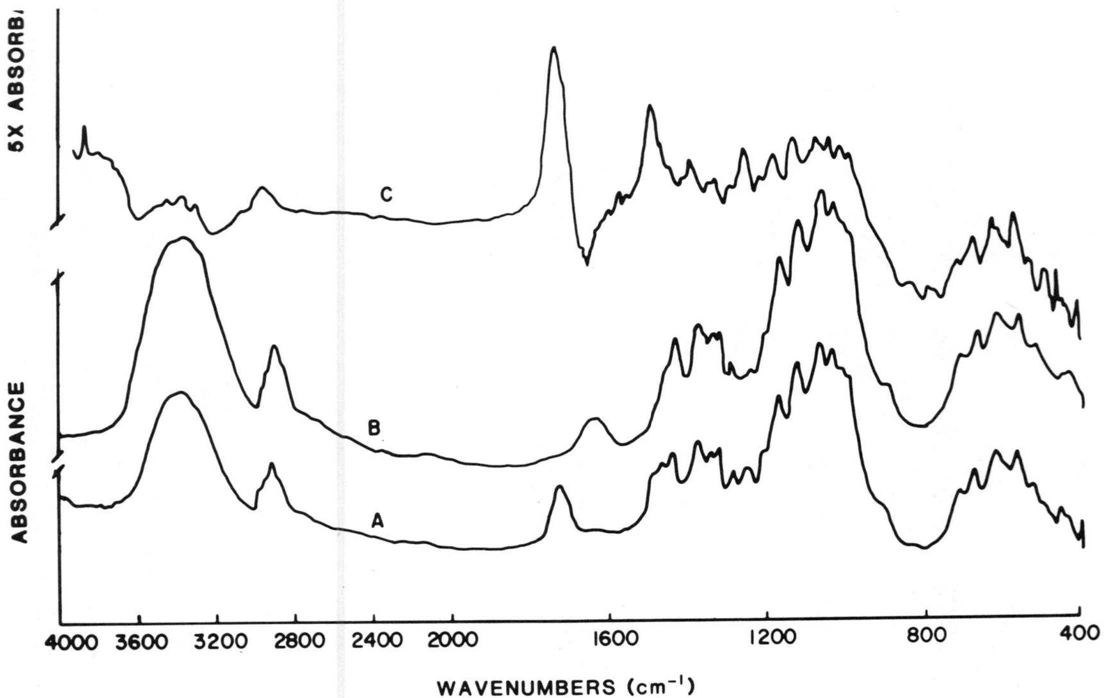

Figure 4 Spectra of: (A) DMDHEU treated cotton; (B) Untreated cotton; and (C) Difference (Untreated cotton subtracted from treated cotton and expanded 5 times). Taken by permission from reference 6.

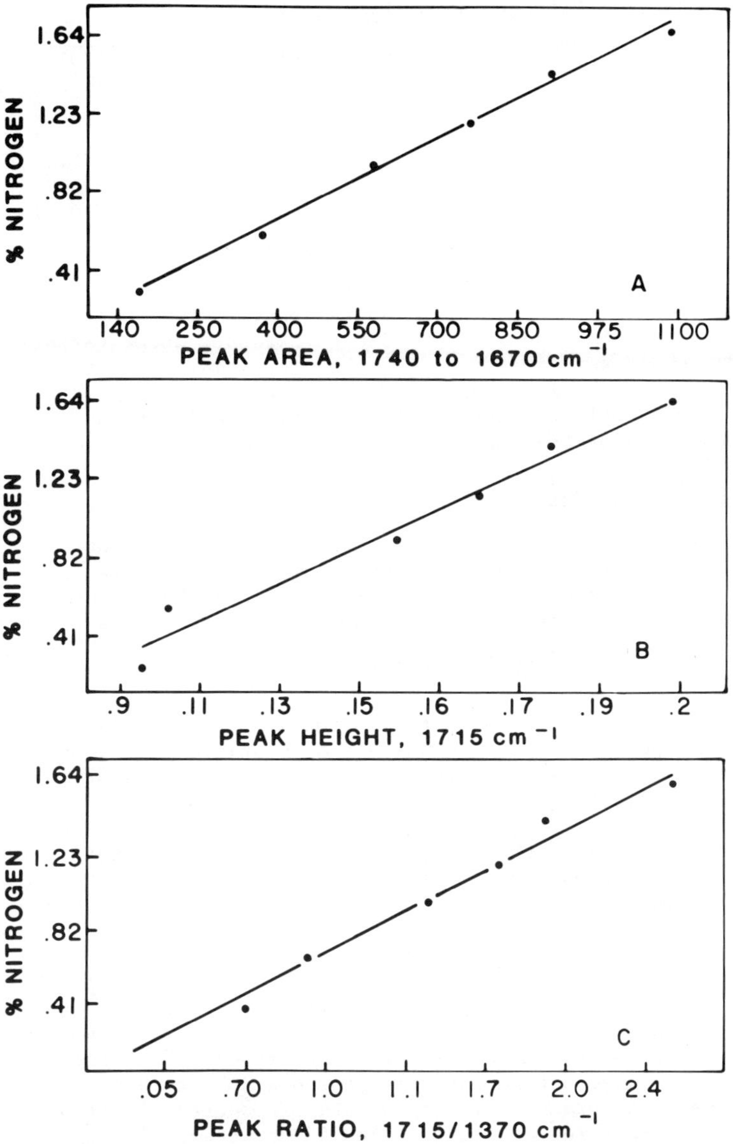

Figure 5 Linear Regression Lines for percent nitrogen <u>versus</u>: (A) Integrated area between 1740 and 1670 cm^{-1} determined by FTIR; (B) Peak height at 1715 cm^{-1} determined by CDIS; and (C) Ratio of peak at 1715 cm^{-1} to that at 1370 cm^{-1} determined by CDIS.

Table II. Percent Nitrogen in DMDHEU Treated Fabrics

Kjeldahl	FTIR	CDIS peak heights	CDIS peak ratios
1.66	1.71	1.62	1.69
1.44	1.39	1.32	1.27
1.18	1.18	1.15	1.12
0.96	0.92	0.96	0.91
0.59	0.62	0.41	0.52
0.29	0.30	0.35	0.32

dry, cure process. When this method was applied to a series of fabrics used to study resin migration some rather large variations in DMDHEU levels were observed (6). The very small sample size (2 mg) required for the infrared analysis can give a very good indication of sample homogeniety by analyzing replications from various portions of the fabric.

Cotton-Polyester Blends

Computerized infrared spectroscopy offers a relatively fast and simple procedure for the approximate determination of blend composition. The spectra of 100% polyester, a 50/50 cotton/polyester blend, and the difference spectrum in which the polyester spectrum has been subtracted from that of the blend are shown in Figure 6. The difference spectrum (C) is virtually identical to that of 100% cotton.

The FTIR determination of blend composition was effected by selecting a spectral region in which a band unique to the polyester appears and monitoring the screen until a flat line is obtained in this region indicating that all of the polyester has been subtracted. At this point the subtraction was stopped and the scaling factor (SCB) was displayed on the screen. This factor multiplied by 100 gave good correlation (6) with the nominal percent polyester in the blend. Two regions of the spectrum were used to estimate the blend composition: (1) the 1850 to 1650 cm^{-1} region and (2) the 850-600 cm^{-1} region.

The height of the carbonyl band in the cotton/polyester blend or the ratio of this carbonyl band at 1715 cm^{-1} to that of the hydroxyl region (at 3385 cm^{-1}) of cotton were used to obtain quantitative information from the dispersive instrument. The regression lines for the three determinations are shown in Figure 7. The percent polyester determined by each of the procedures is given in Table III. Except in three cases (the 65/35, the 55/45, and 40/60 blends), the values determined by infrared (either FTIR or CDIS peak height) were ± 3% of the nominal value. The samples for FTIR and CDIS were drawn at different times. It is interesting to note that, if the composition of the 65/35 and of the 40/60 blend as determined by FTIR is averaged with that as determined by CDIS (peak height), the average percent polyester becomes 65.3 and 41.0 percent, respectively. The value of the 55/45 blend is low regardless of the method used. These three values that fell outside this range (±3%) may have done so because of sampling and homogeneity problems. A blend is prepared by weighing the required amount of each component

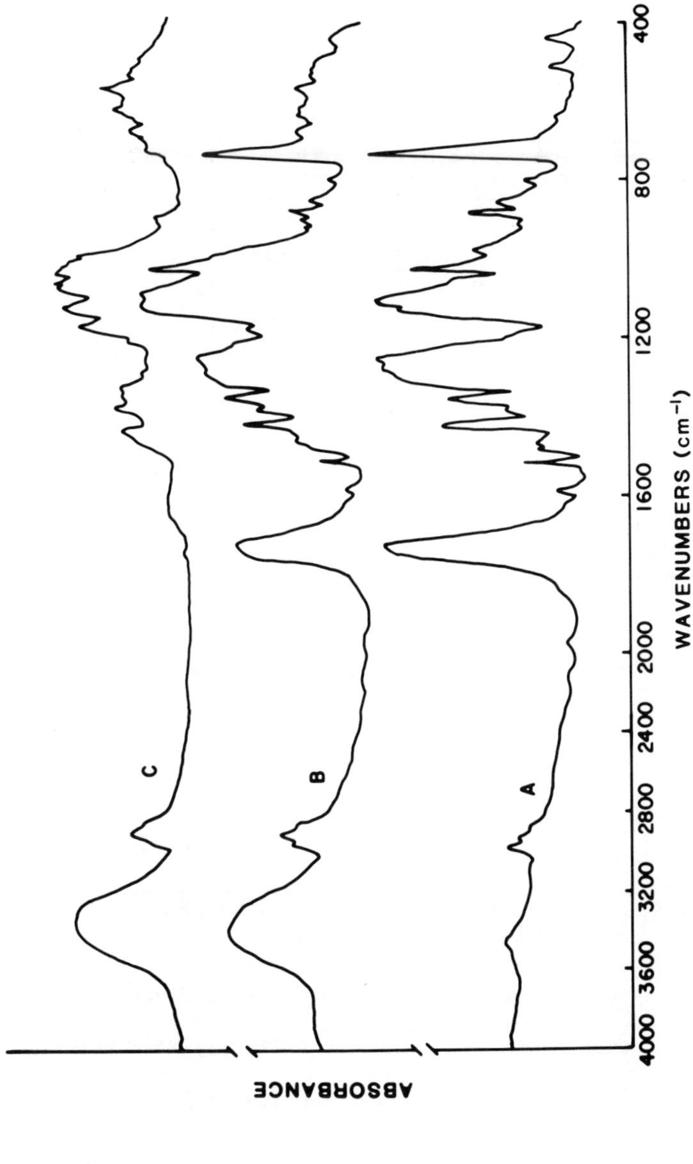

Figure 6 Spectra of: (A) 100% polyester; (B) 50/50 cotton/polyester blend; (C) Difference (100% polyester subtracted from blend, compare 6C with 4B). Taken by permission from reference 6.

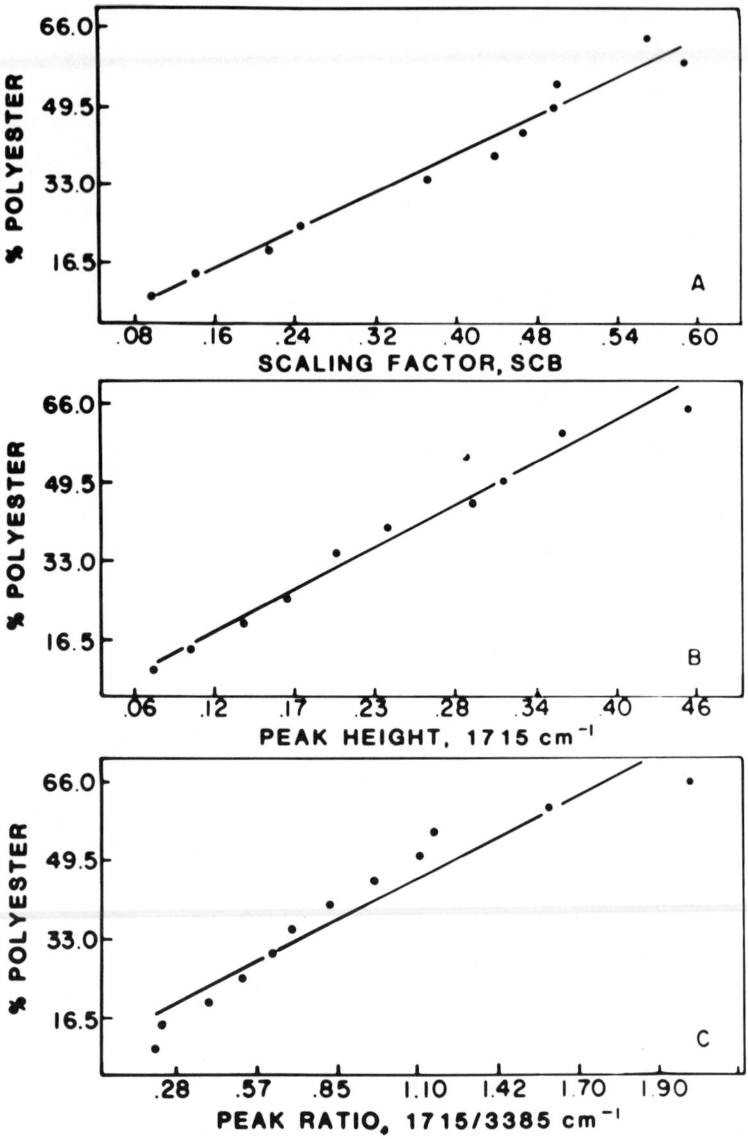

Figure 7 Linear Regression Lines for percent polyester _versus_: (A) Scaling factor (SCB) determined by FTIR; (B) Peak height at 1715 cm^{-1} determined by CDIS; and (C) Ratio of peak at 1715 cm^{-1} to that at 3385 cm^{-1} determined by CDIS.

Table III. Percent Polyester in Cotton/Polyester Blends

Nominal Value	FTIR	CDIS peak heights	CDIS peak ratios
65	59.4	71.3	74.9
60	63.0	57.2	59.3
55	50.2	46.6	47.1
50	49.8	50.7	45.7
45	46.9	47.3	40.8
40	44.0	37.9	36.1
35	37.2	32.3	32.2
30	*	29.7	29.8
25	24.7	27.1	26.4
20	21.4	22.0	22.8
15	14.2	16.1	17.8
10	9.8	12.0	16.9

* not determined by FTIR

and mechanically blending them to achieve the desired product. Because of the small sample size taken for these analyses, small differences in blend composition could become readily apparent. X-ray diffraction values for these samples were close to the nominal percent polyester for each, but the sample size for x-ray is over 100 times greater than that taken for infrared and small differences in composition would be averaged out.

Conclusions

Computerized infrared spectroscopy offers a relatively simple means of determining blend composition and level of treatment for cellulose. It is much more rapid than wet chemical procedures and avoids the hazards and expense of working with concentrated acids and other potentially harmful chemicals. The CDIS instrument and many models of FTIR are considerably less expensive than x-ray fluorescence and do not require monitoring for exposure to radiation. Infrared instruments are more common in colleges and universities as well as in industry so more people trained in their use are available.

The interactive subtraction routine for FTIR gives better results when the peak of interest lies close to another band. There is new software available for the dispersive instrument which includes an interactive subtraction routine which may provide results more nearly comparable to those obtained with FTIR. FTIR offers greater frequency accuracy, better signal-to-noise ratios, and is faster than CDIS. The FTIR instrument used in this work was much more expensive and more difficult to learn to use than the CDIS. Also, if there are problems with the computer, the entire system becomes useless until the situation is corrected.

Either type of instrument offers a reasonable solution to the problem of quantitative determinations in the textile laboratory. Choice of method would be dictated by needs and availability. The computer is an excellent tool to increase productivity and accuracy; however, computerized infrared instrumentation cannot replace an experienced and competent spectroscopist who is essential for interacting with the computerized instrument to obtain and interpret data.

Acknowledgments

The authors thank Glenn Johnson and Sandra Catchings for their assistance in preparing this manuscript.

References

1. Mitcham, D.; Berni, R. J.; Tripp, V. W. Textile Research Journal. 1977, 47, 386-388.
2. Morris, N. M.; Berni, R. J. In "Analytical Methods for a Textile Laboratory"; 3rd Edition; Weaver, J. W. Ed.; American Association of Textile Chemists and Colorists: Research Triangle Park, NC. In Press.
3. O'Connor, R. T.; DuPre', E. F.; McCall, E. R. Analytical Chemistry. 1957, 29, 998-1005.
4. McCall, E. R., Miles, S. H.; O'Connor, R. T. American Dyestuff Reporter. 1966, 55, issue of May 23, 31-35.
5. McCall, E. R.; Morris, N. M. In "Developments in Applied Spectroscopy"; Grove, E. L.; Perkins, A. J., Eds.; Plenum Press: New York, 1970; Vol. 7B, pp. 228-251.
6. Morris, N. M.; Pittman, R. A.; Berni, R. J. "Fourier Transform Infrared Analysis of Textiles", Textile Chemist and Colorist. In press.
7. Digilab, Inc., "Fortran Applications Programs User's Manual, M091-0121." 1979. Cambridge, Massachusetts 3/125 - 3/127.
8. Kasene, M. A., Richards, H. R.; Walker, C. C. J. Appl. Polym. Sci.. 1971, 15, 2237-2243.

RECEIVED January 10, 1984

Electron Spin Resonance Studies on Oxygen Diffusion

In γ-Irradiated Polypropylene Fibers Around the Glassy Transition Temperature

M. I. CHIPARA, V. LUPEI, L. GEORGESCU, D. CATANA, S. BERCEA, and M. VELTER-STEFANESCU

National Center for Physics, Magurele, Bucharest, P.O. Box Mg-6, R-76900, Romania

D. STAICU

Central Institute for Chemistry, 204 Splaiul Independentei Av., Bucharest, Romania

Radicals, generated in polypropylene fibers during irradiation as well as the transition from these radicals (R•) to peroxy radicals (RO_2•) are monitored by electron spin resonance spectroscopy. Experimental data exhibit an anomaly in the temperature dependence of RO_2• concentration, around the glassy transition temperature (T_G). The dependence of RO_2• concentration on temperature, around T_G, is described by a William-Landel-Ferry equation rather than by an Arrhenius one. Consequently, the transition consists of two steps; the first is associated with diffusion processes and the second with the proper chemical reaction. Around T_G, the conversion of R• to RO_2• is dominated by the diffusion step (i.e. by the slowest step).

Radiation effects on polypropylene (PP) have been investigated by many authors, using various methods (1-10), including e.s.r. spectroscopy (4-10). However, no agreement between experimental data has been reported (10). Several radicals have been proposed for R• center. Some of them are given in Table 1.

Discrepancies in these data may arise from small differences in crystallinity, stereoregularity, irradiation and storing conditions. However, the nature of dominant radicals generated in polymer during irradiation may depend on integral and rate dose. The same situation holds for fibers of PP too; no agreement between experimental data is reported. Another feature of e.s.r. studies on PP concerns the effect of molecular motions on e.s.r. line intensity (11,12,13) and therefore on spin susceptibility. However, this dependence cannot be related to R• or RO_2• concentration, in a direct manner.

0097-6156/84/0260-0075$06.00/0

Table I. Radical R generated during gamma irradiation of polypropylene

Chemical Formula	Experimental Conditions	Observations	Reference
1. $-CH_2-\dot{C}(CH_3)-CH_2\sim$	room temp.		4,5
2. $-CH_2-CH(\dot{C}H_2)-CH_2\sim$	and heating studies	spectra depends on heating	6,7
3. $-CH(CH_3)-\dot{C}H-CH(CH_3)\sim$			
4. $\sim CH_2-\dot{C}H-CH_3$			8,9
5. $-CH_2(CH_3)-\dot{C}H-C(CH_3)=CH-CH(CH_3)\rightleftharpoons$			
6. $\rightleftharpoons -CH(CH_3)-CH=C(CH_3)-\dot{C}H-CH(CH_3)$			7

Experimental Methods

PP fibers have been irradiated using a ^{60}Co source. Three types of irradiation have been performed:

A. Irradiation at room temperature (r.t.), in air, denoted as A-I and in vacuum, denoted as A-II. Data concerning irradiation in air, at a dose rate of about 0.34 MRad/hour, up to about 4 MRads (A-I1) and 33 MRads (A-II2) respectively, are reported. Some samples have been irradiated in vacuum sealed glass tubes, at a dose rate of about 10 MRads/hour up to about 11 MRads. This irradiation is denoted as A-III.

B. Irradiation at r.t., in air, followed by a postirradiation at liquid nitrogen temperature (l.n.t.), in liquid nitrogen (l.n.). Experimental data obtained on fibers irradiated at a dose rate of about 0.5 MRads/hour, up to 11 MRads, and postirradiated at the same dose rate up to a total integral dose of about 12.5 MRads, are reported. In this irradiation, denoted as B-I, postirradiation has started as soon as irradiation has been finished.

C. Irradiation at l.n.t. in l.n.. The integral dose was about 5.0 MRads and the dose rate about 0.5 MRads/hour. This irradiation is denoted as C-I.

E.s.r. measurements have been performed using a JES-ME-3X spectrometer, operating in X band (about 9 GHz). Temperature dependence of e.s.r. spectra in the range 125 to 400 K has been investigated. Some data have been obtained on two fibers of PP, with higher diameters, irradiated according to A-III method, at r.t. Bulk PP has been also irradiated by gamma rays and measured by e.s.r. spectroscopy. Tentatively, some samples have been gamma irradiated using an accelerator.

The double integral of resonance spectrum is proportional to spin susceptibility and therefore is proportional to radical's concentration at a given temperature. The double integral of e.s.r. line is evaluated with an accuracy not higher than 10%. In order to avoid accuracy decrease, spin susceptibility is given only in relative units. The g value has been accurately estimated using freshly recrystallised D.P.P.H. as field marker. Immediately as irradiation stops, samples have been stored in l.n. Usually, samples are heated from l.n.t. to r.t. or high temperature (h.t.) (no more than 400 to 450 K). This branch of spin susceptibility dependence on sample's temperature is defined as positive "+". The reverse dependence of spin susceptibility (s.s.) on temperature, starting from high temperature or r.t. towards l.n.t., as soon as the positive branch is completely investigated, is reported. This dependence defines the negative "-" branch.

Some measurements, denoted by "0", have been started at r.t. and sample temperature has been decreased towards l.n.t. In this case, samples were free to oxygen attack, at r.t., prior to record e.s.r. spectra about 5 to 10 minutes. The glass transition range is reached after about 30 minutes (from the instant at which PP fibers were free to oxygen attack, above T_G).

Experimental data have been fitted by adequate equations presented in this paper using a SINCLAIR ZX 81 microcomputer. The double integral of some resonance spectra have been performed using a program written in Basic for ZX-81.

Experimental Data

Resonance spectrum of thin fibers, gamma irradiated according to A-I, B and even C methods is an assymmetrical singlet, represented in Figure 1a. No structure is observed and the axial symmetry may be correlated to fiber's axial symmetry. This resonance is close to the free electron g value (g = 2.043) and its resonance linewidth is about 45.0 Gs. We assign this line to peroxy radicals ($RO_2\bullet$) generated by the reaction of some radicals (R•) with the oxygen diffused in polymer before or during irradiation period. No any significant contribution from R• centers has been observed even in samples irradiated solely at liquid nitrogen temperature (in l.n.); C irradiation. Accelerator irradiation with gamma rays as well as with electrons accelerated up to few MeV, in air, at r.t. at a dose of about 10 MRads, leads to the same resonance spectrum.

E.s.r. data on fibers with higher diameters, irradiated according to A-III as well as irradiation of bulk PP according to methods A-I1 and A-II2 lead to a complex resonance spectrum, assigned to R• radicals. However, resonance spectrum of bulk PP exhibits a better resolved structure than the resonance spectrum of higher diameters fibers. In fibers, after 10 to 10^2 hours R• centers turn in RO_2• if irradiated fibers are stored at r.t., in air. The storing time necessary to remove the contribution of R• radicals from resonance spectra depends on fiber diameter. Accordingly, even after 10^2 hours, no significant effect of RO_2• centers on resonance spectrum is observed; after 10^3 hours all R• radicals are converted in peroxy RO_2• radicals, characterized by the resonance spectrum Figure 1a. Resonance spectra on fibers of high diameter as well as on bulk PP are represented in Figure 1b respectively Figure 1c. Small variations in resonance linewidths may be related to variations observed in the concentration of R• and RO_2• respectively (i.e. in changes of dipole-dipole or/and exchange interactions). For the resonance spectrum of R• center we suppose that radical (1) may be considered as responsible for spectra represented in Figure 1b and Figure 1c. However, our objectives are not directly related to the exact nature of radical R•.

For irradiations A-I1, A-II2, B-I and C on PP fibers, no significant dependence of RO_2• resonance line shape parameters (g value, resonance line width, resonance line shape) except resonance line intensity is observed. Under these circumstances, the temperature dependence of s.s. is anomalous and cannot be simple correlated to Curie, Curie-Weiss or even exciton spin susceptibility. In the case of A-I1 irradiation a weak increase of s.s. associated to RO_2• centers is observed around T_G (Figure 2). As the integral dose is increased, the number of RO_2• centers increases and the anomaly in s.s. around T_G is enhanced, as may be observed in Figure 3. After a B-1 irradiation, the anomaly observed in the temperature dependence of s.s. is dramatic, as may be observed in Figure 4. As C-1 irradiation is limited to few MRads, the anomaly in the s.s. dependence on temperature is spectacular, but the anomaly in s.s. around T_G is not further enhanced.

It is important to notice that all these anomalies have been observed only on positive branches of s.s. dependence on temperature, if samples are heated from l.n.t. For the same samples, the negative branch exhibits no anomaly or in some cases shows a "plateau" shifted towards lower temperature, as it is observed in Figures 2-4. Figure 5 points an important feature of this anomaly; starting from r.t. or h.t. negative branch and the next positive branch exhibits no anomaly. Another important feature is that only the first positive branch shows an anomaly around T_G. The next positive branch (observed after a negative one) shows no anomaly and this result is almost the same for the third, the fourth ... positive branch. Therefore, the first positive-negative cycle shows a hysteresis but not the next ones. The first negative-positive cycle shows no hysteresis (or a negligible one).

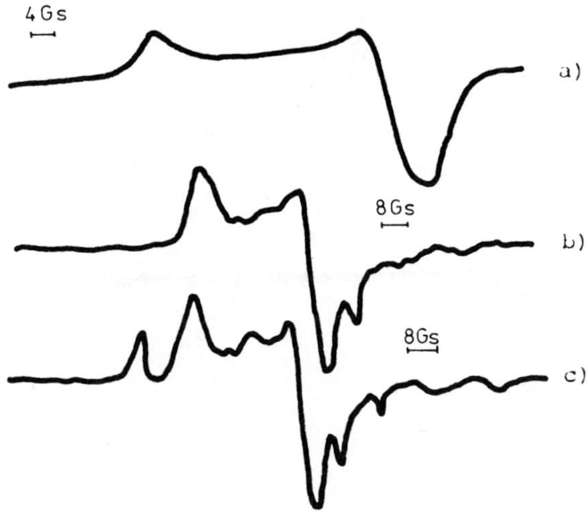

Figure 1. Electron spin resonance spectra of gamma irradiated PP. a) bulk, b) thin fiber, c) very thin fiber.

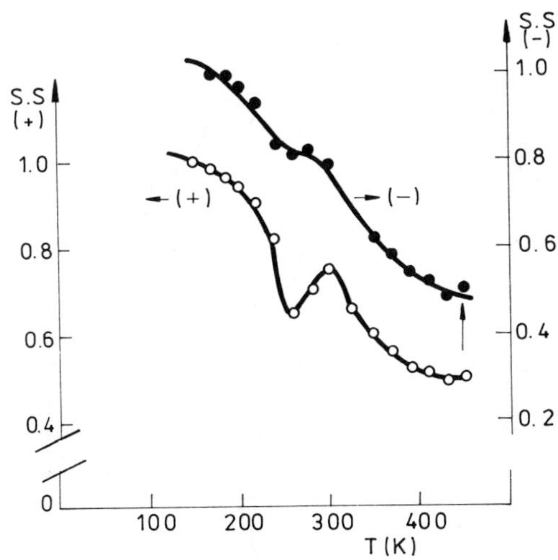

Figure 2. Temperature dependence of s.s. for A-II irradiation.

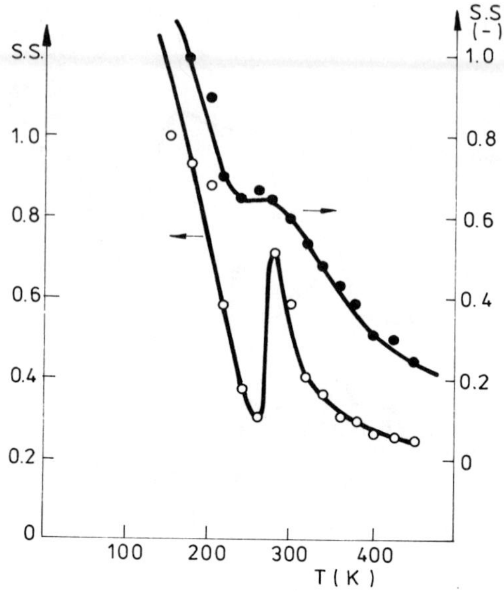

Figure 3. Temperature dependence of s.s. for A-II2 irradiation.

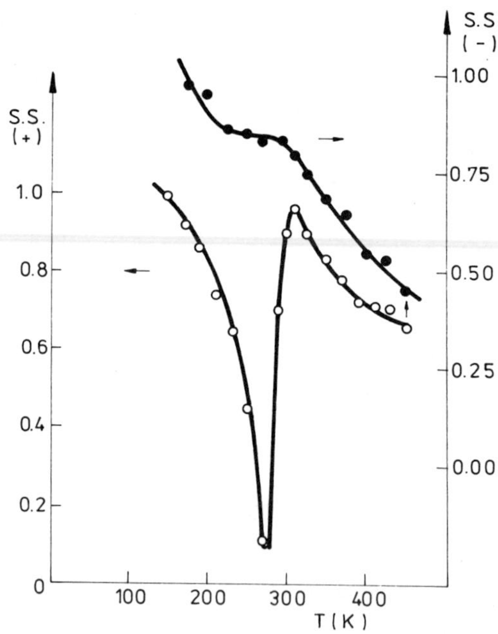

Figure 4. Temperature dependence of s.s. for B-1 irradiation.

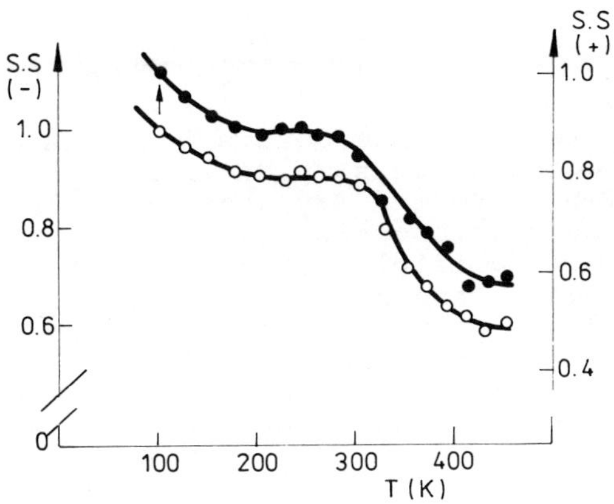

Figure 5. Temperature dependence of s.s. for AII-2 irradiation if heating/freezing cycle starts with sample heating ("0" measurements).

The anomaly reported by us is weak if fibers have high diameters or in bulk samples. However, a "plateau" around T_G, in gamma irradiated bulk PP is reported by several authors, and related to molecular motions (10-13).

Discussion

Experimental data on s.s. of $RO_2\bullet$ centers dependence on temperature, around T_G, exhibit an anomalous kinetic if the first branch is positive or a hystersis loop in the first cycle if measurements start on the positive branch. Next cycles show no significant hysteresis excepting a slight shifting of positive or/and negative branches. Samples stored in air, at r.t. for more than 30 minutes show no anomaly or hysteresis, even if measurements start on the positive branch.

All these features are consistent with an irreversible process occuring in samples, during the first positive branch. Two phenomena may account for this anomaly: 1. Strong exchange interactions among $R\bullet$ (or $RO_2\bullet$) centers which lead to a triplet state (S = 1) and give a low forbidden resonance line at g = 4.0, due to radical pairs. This radical-pair description may account for this anomaly supposing that during irradiation, many radical pairs $R\bullet - R\bullet$ or $RO_2\bullet - RO_2\bullet$ are generated. These pairs decay extremely rapidly to $R\bullet$ and $RO_2\bullet$ centers, respectively, as sample temperature increases over T_G, because the amplitude of molecular motions (at least in amorphous regions) is significantly enhanced if temperature is increased above T_G. According to this description the g = 2.00 resonance is due to $R\bullet$ or $RO_2\bullet$ centers. Many features of experimental data reported by us are consistent with this description. The increase of irradiation dose enhances the anomaly as $R\bullet$ or $RO_2\bullet$ concentration increases. The decrease of irradiation temperature enhances this anomaly as molecular motions of macromolecular chains are frozen and consequently the decay (rate) of radical-pairs to radical ($R\bullet$ or $RO_2\bullet$) is reduced. Although all these, we have failed to measure the $M_S = 2$ resonance, located at g = 4.00 as well as any satellite structure or shoulders superimposed on $M_S = 1$ resonance (of radicals), located at g = 2.00. It is important to notice that according to our experimental data, the number of $R\bullet$ or/and $RO_2\bullet$ centers is comparable (as order of magnitude) to the number of radical pairs, at least during B and C irradiations! This situation seems to be rather unusual. Samples stored in air, at r.t., for no more than 30 minutes show no anomaly, for any type of irradiation.

According to e.s.r. data on radical-pairs in gamma irradiated PP (11,12), samples storage in air, at r.t. for less than half of hour does not convert all radical-pairs (if present) in singlet radicals. Consequently, an anomalous dependence of spin susceptibility on sample temperature is expected. Unfortunately, experimental data are at variance with this conclusion. If radical-pairs are responsible for this anomaly, the anomaly is increased as fibers' diameter increases, as the concentration of R centers is enhanced and therefore the radical-pairs concentration is higher. E.s.r. data on gamma irradiated fibers of different diameters are in contradiction with this prediction. However, it is important to notice that this anomaly is observed as long as the resonance spectrum is due to peroxy-like radicals. We suppose that the transition from R• centers to RO_2• centers, due to chemical reactions between R• radicals and the oxygen diffused in polymer may be associated with some rearrangements of macromolecular chains. These motions reduce the number of radical-pairs significantly.

2. Reaction between oxygen diffused in polymer and R• centers, leading to peroxy (RO_2•) radicals. Under these circumstances, the changes in e.s.r. spectra are irreversible as during the first positive branch all R• centers are exhausted by the reaction with the oxygen diffused in polymer. In order to correlate this anomaly with the chemical reaction between R• and oxygen it is necessary to suppose that this reaction is frozen below T_G. Accordingly, we suppose that no sufficient oxygen is supplied to R• centers, in order to convert these centers in RO_2• centers, below T_G. Consequently, oxygen diffusion through PP fibers is anomalous below T_G (is frozen). Storing of irradiated samples in air, at r.t., even few minutes leads to a strong diffusion of oxygen from atmosphere in the fiber and converts R• centers in RO_2• ones. According to our description, the transition from R• to RO_2• occurs in two steps: a) Oxygen diffusion. b) Proper chemical reaction. The resonance line of R• centers is not observed in our cases because this line is broader than the resonance line of RO_2•, as may be easily observed from Figure 1a-c, and possible because the concentration of R• centers is lower than RO_2• concentration. However chains scissions due to oxygen attack may account too for the increases of RO_2• spin susceptibility. It is not impossible for some R• radicals to be coupled by exchange interactions, as radical-pairs although this situation does not affect significantly this description.

In conclusion, s.s. dependence on temperature, for gamma irradiated PP is anomalous around T_G, due to an anomaly in oxygen diffusion across the glassy transition region. However, as RO_2 concentration depends on time and temperature and because we have no reason to suppose a particular type of spin susceptibility dependence on temperature, is not possible to convert s.s. dependence on temperature in the dependence of RO_2• concentration on temperature. For usual Curie or Curie-Weiss spin susceptibility dependence on temperature, the anomaly reported by us is a true anomaly in the number of RO_2• centers.

However, it is difficult to establish if the anomaly around T_G is associated solely to the conversion of R• centers in RO_2• centers or if oxygen attack leads to significant scissions of macromolecular chains, increasing the number of RO_2• centers. In the first case the resonance line of R• centers, which is broader than the resonance line of RO_2• centers may be significantly broadened by various interaction (as dipole-dipole interaction). Under these circumstances, the double integral of resonance lines concerns practically only RO_2• centers, the contribution of R• centers is small and contained only in the base line.

A qualitative answer concerning the nature of spin susceptibility dependence on temperature may be obtained from an analysis of resonance lineshape; exchange interaction turns the resonance line in a lorentzian one or even in an "exchange lineshape", if exchange interaction are severe. For these line a Curie-Weiss s.s. dependence on temperature, with positive or negative asymptotic temperature, according to the nature of exchange interactions, ferro or antiferro respectively. If the resonance line does not exhibit significant departure for a pure Gaussian line, a Curie like dependence of s.s. on temperature is expected.

Mathematical Description

A very rough physical idea is to suppose that the dependence of RO_2• s.s. on temperature is given by a W.L.F. equation rather than by an Arrhenius one. The reason for such an assumption lies in the fact that "relaxation times" depends on temperature, around T_G, according to a W.L.F. equation, and an analogy between diffusion process and relaxation processes is possible. Accordingly,

$$RO_2\bullet(T) = RO_2\bullet(\exp(-\frac{E}{K_B(T-T_O)})) \quad (1)$$

where: T_O is a temperature close to T_G
K_B is Boltzmann constant
T is sample temperature
E represents the activation energy for the diffusion process
RO_2•(T) is s.s. of RO_2 centers at a given temperature, T
RO_2• is s.s. of RO_2 centers as T is extremely high ($T \rightarrow \infty$).

This rough description gives a correct qualitative description of the anomaly observed by us. The agreement between theoretical and experimental data is good for irradiation performed according to the method A (an acceptable agreement is obtained for A-I irradiation). For other irradiation, the agreement with equation (1) is poor. We suppose that these discrepancies are due to the low concentration of RO_2• centers produced during A irradiation.

In order to obtain a better description of experimental data, we have started from the reaction associated to the conversion of R• centers in peroxy radicals:

$$R\bullet + O_2 = RO_2\bullet \quad (2)$$

In order to keep the mathematics as simple as possible, we suppose that the maximum number of R• centers is equal to the maximum number of $RO_2\bullet$ centers. This means that we have neglected chains scissions induced by atmospheric oxygen. Accordingly, we must notice that this assumption is consistent with the hypothesis that only $RO_2\bullet$ centers give resonance signal at $g = 2.0$ (the concentration of R• centers is negligible, or their resonance is broadened over a large field, so that their contribution to the observed resonance is not significant). As our experiments have been performed in glass tubes open to atmospheric oxygen attack, we may suppose that the concentration of oxygen is constant. Under these circumstances, equation (1) turns to:

$$RO_2^*\bullet(T) = RO_2^*\bullet(1 - (\exp(-K_o t))) \quad (3)$$

where: K_o is the reaction rate for the oxidation process
t is the oxydation time

At this point it is necessary to introduce the temperature dependence. As K_o is more directly related to diffusion process then $RO_2^*\bullet$, equation (2) turns to:

$$RO_2^*\bullet(T) = RO_2^*\bullet(1 - (\exp(-K_o t(\exp - \frac{E}{K_E(T-T_o)})))) \quad (4)$$

In equation (2) and (3) RO_2^* represents the number of peroxy radicals, whereas $RO_2\bullet$ is associated to the s.s. of peroxy radicals. As e.s.r. data are related to $RO_2^*\bullet$, equation (4) becomes:

$$RO_2^*\bullet(T) = RO_2^*\bullet(1 - \exp(-K_o t(\exp - \frac{E}{K_B(T-T_o)})))(f(T)) \quad (5)$$

where f(T) gives s.s. dependence on temperature. Usual f(T) may have the following expression:

$$f(T) = C/(T-T_1) \quad (6)$$

where C is a constant

T_1 represents the asymptotic temperature. $T_1 = 0$ for a pure paramagnetic peroxy radical. For ferromagnetic coupling T_1 is positive whereas for an antiferromagnetic coupling T_1 is negative. However, T_1 is small in many situations, analogous to our one. Finally, the temperature dependence of the double integral of e.s.r. spectrum ($RO_2\bullet$) is:

$$RO_2\bullet(T) = CRO_2\bullet \frac{(1 - \exp(-K_o t(\exp - E/K_B(T-T_o))))}{T - T_1} \quad (7)$$

Equation (7) gives a satisfactorily description of the anomaly observed by us in gamma irradiated polypropylene, for various types of irradiation. Analogous behaviour for active centers has been experimentally reported in (16). The theoretical description for this case is suggested in (17).

Conclusions

An anomaly in s.s. of peroxy radicals around T_G, in gamma irradiated PP fibers is related.

The environmental conditions under which this anomaly is observed are carefully investigated (by experiments) and extensively discussed.

It is proved that this anomaly is related to the glassy transition phenomenon.

A phenomenological description of this anomaly is suggested. It is supposed that the anomaly concerns the conversion of R• centers in peroxy radicals. Accordingly, is proposed that this reaction occurs in two steps, diffusion and chemical reaction. Around T_G, the chemical reaction is controlled by the diffusion process. It is claimed that the diffusion depends on temperature according to a W.L.F. law.

A mathematical description of this anomaly is suggested. The agreement between theory and experiments is satisfactory.

Literature Cited

1. F. Bovey, "The effects of ionizing radiation on natural and synthetic polymers "Interscience Press, New York, 1958
2. Malcolm Dole (Editor) "The radiation chemistry of macromolecules" Academic Press, New York, 1973
3. N. Zolotova, T. Denisov, J. Polym. Sci., 1971, A-1, 9, 3311
4. H.P. Frank, Polypropylene, Gordon & Breach, 1968
5. P.B. Ayscough, K.J. Ivin, J.O. Donell, Proc. Chem. Soc., 1961, 71
6. H. Fisher, K. Hellwege, J. Polym. Sci., 1962, 56, 33
7. B.R. Loy, J. Polym. Sci., 1963, 1, 2251
8. L.J. Forrestal, W.G. Hodgson, J. Polym. Sci., 1972, A-1, 2, 1275
9. S. Ohnishi, Y. Ikeda, M. Kashiwagi, M. Nitta, Polymer, 1961, 2, 119
10. B. Ranby, J.F. Rabek, "E.S.R. Spectroscopy in Polymer Research" Springer-Verlag, Berlin, 1977
11. M. Iwasaki, T. Ichikawa, T. Ohmori, J. Chem. Phys., 1969, 50, 1991
12. P.O. Kinell, B. Ranby (Editors) "E.S.R. Applications to polymer Research"; Proceedings of the XXIInd Nobel Symposium, John Wiley & Sons, New York, 1973
13. S. Nara, H. Kashiwabara, J. Sohma, J. Polym. Sci., 1968, A-2, 6, 1435
14. R. Baican, M. Chipară, Proc. XXth Congress AMPERE, Tallinn, 1978, Springer Verlag Berlin, 1979, p. 214

15. R. Baican, M. Chipară, Rev. Roum. Phys., 1978, 23, 9, 1079
16. S. Pivovarov, A.I. Poljakov, Yu. A. Rjabikin, N.L. Philippov, M.I. Bitenbaev Rad. Effects, 1982, 59, 179
17. M.I. Chipară, S. Turbatu, L. Georgescu, L. Cojocaru, M. Velter-Stefănescu, VIIth Int. Congress of Radiation Research (I.C.R.R.), Amsterdamm, 1983

RECEIVED February 28, 1984

HIGH PERFORMANCE POLYMERS

6

A Partial Phase Diagram and Crystal Solvate for the Poly(*p*-Phenyleneterephthalamide)/Sulfuric Acid System

K. H. GARDNER, R. R. MATHESON, P. AVAKIAN, Y. T. CHIA, and T. D. GIERKE

Central Research and Development Department, E. I. du Pont de Nemours and Company, Experimental Station, Wilmington, DE 19898

It is now well established that certain aromatic polyamides form ordered complexes with their solvents (i.e., crystal solvates). A summary of the occurrence of these crystal solvates has been compiled in a recent review article by Iovleva and Papkov[1]. Among the crystal solvates that have been identified are poly(m-phenylene isophthalamide) with N-methylpyrrolidone[2] and with hexamethylphosphortriamide[2]; poly(p-benzamide) with sulfuric acid[3]; and poly(p-phenylene terephthalamide) (PPTA) with sulfuric acid[4] and with hexamethylphosphortriamide[5]. These solvates are all characterized by a discrete melting point and a crystalline diffraction pattern.

Although the presence of solvate phases has been established and qualitative phase diagrams have been published, to our knowledge, a detailed model for a polymer solvate and its phase behavior has not been presented. At this time we would like to present a partial phase diagram for the poly(p-phenylene terephthalamide) (PPTA)/sulfuric acid system and a model for the crystal solvate formed. In addition the structure of a model complex will be described.

Experimental

Experiments were carried out with low molecular weight poly(p-phenylene terephthalamide) (PPTA) produced by the acid degradation of commercial Kevlar aramid fibers (6N HCl, 20 hr)[6]. The inherent viscosity of the degraded polymer was ca. 1.0 which, based on an empirical correlation between molecular weight and inherent viscosity, corresponds to a molecular weight of ca. 7000. Solutions of PPTA in 99.7% sulfuric acid were prepared with concentrations ranging from 2% → 24% (w/w) under N_2 in a dry box. Homogeneous solutions were obtained by intensive mixing at

0097–6156/84/0260–0091$06.00/0

50-70°C for short times. This low molecular weight polymer was chosen for the study because of its (relatively) easy dissolution behavior and low solution viscosities as compared to those of higher molecular weight polymer.

Differential Scanning Calorimetry (DSC) of the PPTA/sulfuric acid system was carried out in order to establish the melting behavior of the system. Samples (5-10 mg) were sealed in gold-coated aluminum pans under N_2 and cooled at 10°C/min to -150°C. Subsequently the samples were heated to 100°C at 10°C/min and the melting thermograms were recorded.

The diffraction patterns of various PPTA/sulfuric acid solutions as a function of temperature were obtained using a Rigaku Theta-Theta Diffractometer run in the horizontal mode (CuK_α radiation). Digital data were taken at intervals of 0.02° in 2θ. The 2θ scale was calibrated with diamond and the temperature was measured by a thermocouple embedded in the sample.

RESULTS

Model Solvate Structure

In order to determine the manner in which sulfuric acid interacts with PPTA we are investigating the structure of PPTA oligomers complexed with sulfuric acid. We have recently determined[7] the structure of a N,N'-(p-phenylene)dibenzamide (PPDB)/sulfuric acid complex using single crystal x-ray methods and similar studies with longer PPTA oligomers are underway now. We expect that the general principles for the interaction of sulfuric acid and PPTA will become apparent from these studies.

The PPDB/sulfuric acid complex crystallizes in a triclinic unit cell with dimensions $\underline{a}$ = 9.75A, $\underline{b}$ = 10.31A, $\underline{c}$ = 7.88A, $\underline{\alpha}$ = 108.7°, $\underline{\beta}$ = 111.4° and $\underline{\gamma}$ = 89.2°. The space group is $P\bar{1}$ and the cell contains one PPDB molecule and four sulfuric acid moieties. A view of the molecule together with the four nearest H_2SO_4 groups in the plane of the phenyl groups is presented in Figure 1. Two of the sulfuric acid molecules can be seen to have protonated the carbonyl oxygens of the amide groups, thus the structure is actually a sulfuric acid/bisulfate salt of PPDB.

The neutral compound PPDB exists in two phases, and structural studies have been reported for each.[8,9] A comparison with these two structures and with the structure of N,N'-diphenylterephthalamide (DPTA)[10] is presented in Table 1. The protonated oxygens cause an expected lengthening of the C-O bonds from 1.22A (avg) in the neutral compound to 1.30A (avg) in the sulfuric acid salt. In addition the C-N amide bond is shortened from 1.36A (avg) to 1.30A (avg), indicating a substantial increase of double bond character in this bond.

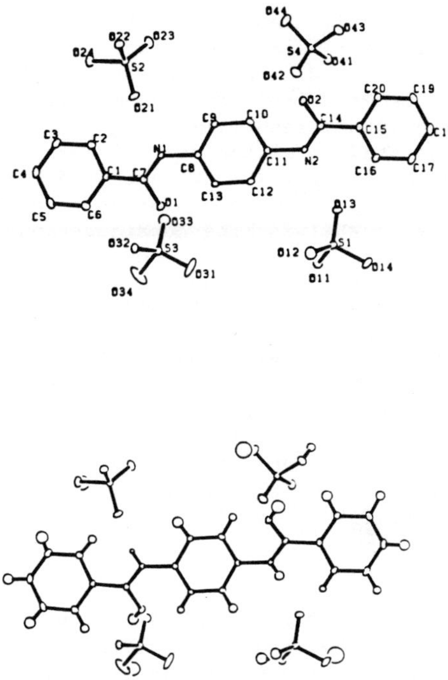

Figure 1. Molecular structure of N,N'-(p-phenylene) dibenzamide (PPDB) and the four nearest sulfuric acids; (a) atomic numbering, (b) with hydrogens.

Table 1

Structure	PPDB[8]	PPDB[9]	DBTA[10]	present
C-O distance(Å)	1.222	1.229	1.226	1.30, 1,293
C-N distance(A)	1.355	1.356	1.359	1.301, 1.303
amide...outer ring angle(°)	28.9	23.8	30.7	36.6, 37.9
amide...central ring angle(°)	35.7	55.4	30.4	-41.2, -44.4
outer...central ring angle (°)	64.4	79.2	60.9	5.7, 8.2

A pronounced flattening of the protonated PPDB structure is observed in comparison to the unprotonated compound. In the neutral compound the phenyl rings are rotated by 64° and 79° from each other, while in the present structure the phenyl rings are nearly coplanar. However, in all cases, the amide groups are severely rotated out of the plane of the phenyl groups. The torsional angles about the amide groups are ca. +30, +30 in the neutral compound and ca. +30, -30 in the acid salt. These conformers would be iso-energetic in the neutral state and, it would seem, of close if not identical energy in the protonated form. Both previous studies of the neutral compound indicate a significant influence of crystal packing on the conformation (see Table 1). It is possible that the conformer found in the acid salt was directed by crystal packing considerations which may not be a factor in the PPTA/sulfuric acid complex. In the crystal, the sulfuric acid and bisulfate groups are arranged in sheets with an extensive network of hydrogen bonding (Figure 2). These sheets alternate with sheets of PPDB consisting of stacks of the molecules with pi-pi* ring interactions (allowed by the planarity of the molecule). The sheets are inter-connected by strong amide to sulfate hydrogen bonds. Alternatively, the structure can be thought of as the close-packing of hydrogen bonded sheets containing PPDB and sulfuric acid/bisulfates (Figure 3). We believe a sheet structure of this or similar nature is the structural unit of the PPTA/sulfuric acid solvate phase.

Phase Diagram

The phase diagram for the (quasi-binary) PPTA-sulfuric acid system was compiled based on the melting point/composition information gathered from DSC-melting thermograms. In Figure 4, melting thermograms of a number of polymer-acid mixtures are presented. These thermograms were recorded for the temperature range -150° to 100°C at 10°C/min. The various thermograms were used to construct the melting point/composition diagram, which is shown in Figure 5. It can be seen that we are dealing with an apparent eutectic polymer-diluent system with a eutectic composition of 11-12% (w/w) of PPTA and a eutectic temperature of -8°C. The clearing temperature of the solutions as determined by optical microscopy have been indicated in Figure 5.

Figure 6 shows the diffraction pattern of a 22% PPTA/sulfuric acid solution as a function of temperature. The diffraction patterns can be seen to consist of two components - sharp diffraction peaks superimposed on a broad component. This pattern is consistent with a two phase system containing semicrystalline PPTA/sulfuric acid solvate and disordered components. As the temperature is raised the portion of the diffraction patterns attributable to the solvate phase decreases and finally disappears at temperatures consistent with the melting endotherm observed by DSC. This diffraction pattern agrees with that previously reported for the PPTA/sulfuric acid solvate.[1,4]

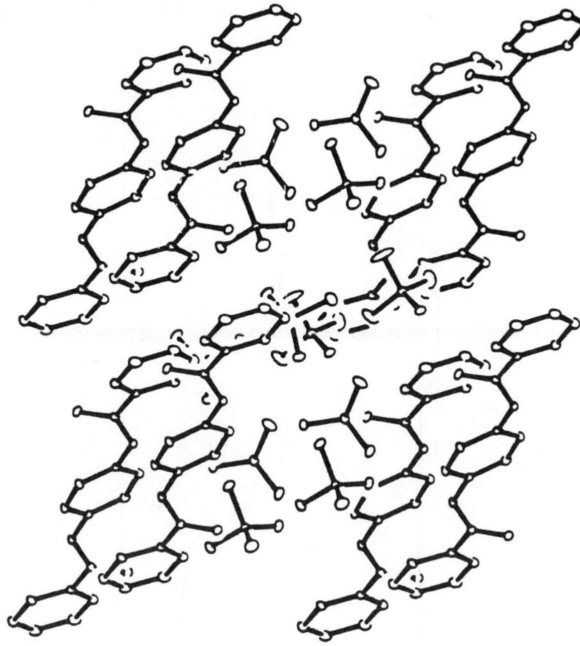

Figure 2. In the crystal, the sulfuric acids are arranged in sheets with an extensive network of hydrogen bonding. These sheets alternate with sheets of PPDB.

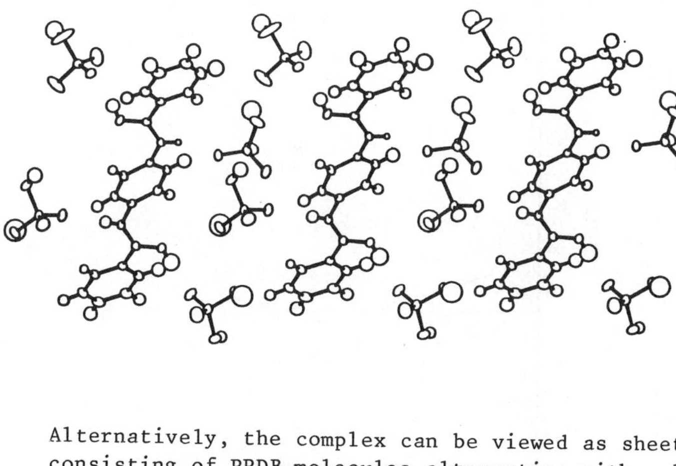

Figure 3. Alternatively, the complex can be viewed as sheets consisting of PPDB molecules alternating with sulfuric acids. Sheets of this nature are believed to exist in the PPTA solvate.

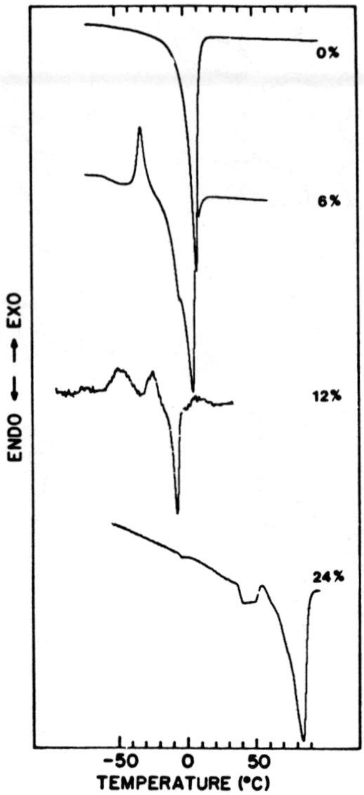

Figure 4. DSC-heating thermograms of PPTA/H_2SO_4 solutions at different concentrations.

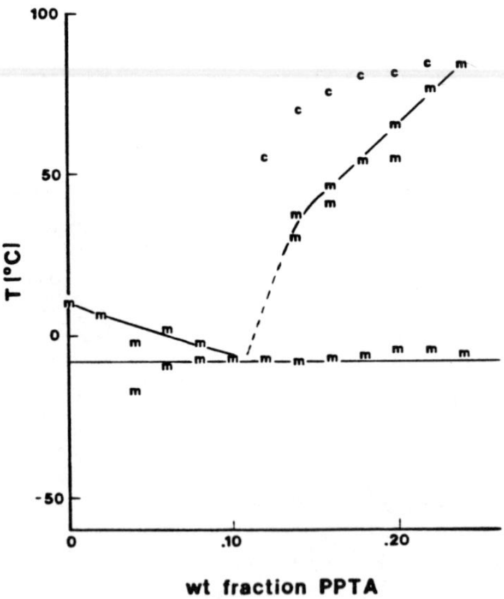

Figure 5. Phase diagram for PPTA/H_2SO_4. Experimental points were obtained by DSC (m) and optical microscopy (c).

Solvate Structure

The crystalline component of the diffraction pattern can be indexed by a metrically orthorhombic unit cell with dimensions $\underline{a}$ = 16.35A, $\underline{b}$ = 9.59A and $\underline{c}$ = 12.9A. The cell contains two chain fragments and ca. 9 sulfuric acid molecules per chemical repeat of the polymer. The quantity and organization of the sulfuric acid in the cell will be discussed below. Projections down the chain axis of the PPTA unit cell[11] and the proposed solvate structure are shown in Figure 7. Both structures have PPTA chains located at the corner and center of their respective unit cells. In the neutral structure, the chains form hydrogen bonded sheets parallel to the bc plane. The crystal solvate can be thought of as a "swollen" form of the neutral structure. The hydrogen bonded sheets parallel to the bc plane have been preserved but now sulfuric acid molecules participate in the sheet while other acid molecules lie in interstices between the hydrogen bonded sheets. We have modeled the "sheet" with two sulfuric acids per amide as was found in the PPDB/sulfuric acid complex. [The chain-chain separation distance in the polymer sheet is 9.59A as compared with 9.75A in the model structure.] However, with this chain-chain separation distance it is also possible to construct a plausable sheet structure with one sulfuric acid molecule per amide where a single sulfuric acid molecule bridges two amide groups (N-H...O-S-O...H...O-C). In both models the sulfuric acids between the sheets are not directly attached to the PPTA molecules.

The volume of the solvate unit cell is 2006Å^3. If we assume that the volume occupied by the PPTA chains in the solvate is the same as it would be in the pure PPD-T crystal, i.e., 526Å^3, and the volume of the sulfuric acid molecules in the hydrogen bonded sheet is the same as that in the PPDB/sulfuric acid complex (78Å^3) we can calculate the volume of the intersheet sulfuric acid molecules. These volume considerations and density measurements on undegraded samples (unpublished data) lead us to conclude that (i) there is a nonstoichiometric amount of sulfuric acid between the sheets and (ii) the acid is disordered with a molecular volume equivalent to that of liquid sulfuric acid (ca. 87Å^3). These observations lead us to believe that the space group of the solvate is actually monoclinic (or even triclinic) and the structural unit is a sheet.

Discussion

Our present understanding of the phase diagram in Figure 5 is at a semiquantitative level. If the eutectic-like pattern of the points labeled 'm' are interpretable as such, then the left hand branch reflects the freezing point depression of sulfuric acid (mp = 10°C). The slope of this branch might then be expected to obey the familiar[12] equation

$$\Delta T = k_f m \qquad (1)$$

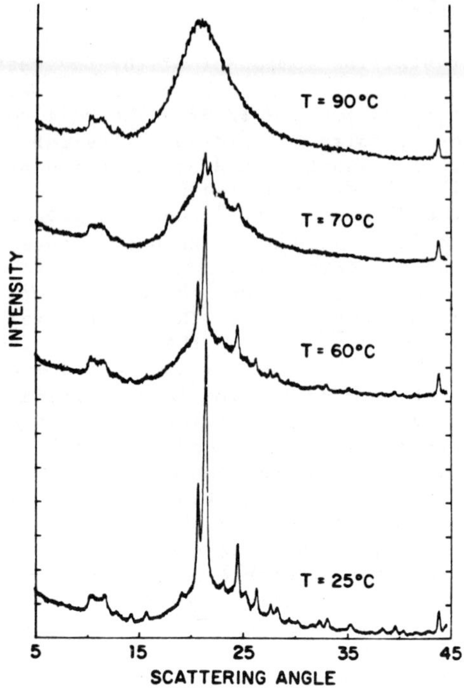

Figure 6. Diffraction pattern of a 22% (w/w) solution of PPTA/H_2SO_4 as a function of temperature.

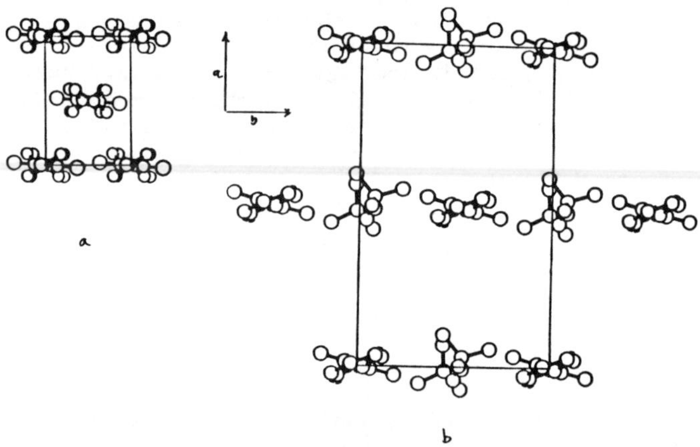

Figure 7. ab Projections of the PPTA and PPTA/H_2SO_4 solvate unit cells. (a) PPTA structure using Northolt coordinates, molecules form hydrogen bonded sheets parallel to the bc plane. (b) Proposed structure for the PPTA/H_2SO_4 solvate. Sulfuric acid molecules have been incorporated in the hydrogen bonded sheets and between the sheets (not shown).

in the limit of $m \to 0$. Here, ΔT is the difference in melting points between the pure acid and a solution of molarity m; and k_f is the cryoscopic constant. For sulfuric acid[13] $k_f = 68.1$ [$K \cdot 10^3$ g acid$\cdot$(moles of solute)$^{-1}$]. When the low concentration points in Figure 5 are fit with eq. 1 and this value of k_f, the desired number average molecular weight of the PPD-T solute is only ca. 500. This value is much smaller than the molecular weight derived from an empirical correlation between molecular weight and inherent viscosity; ca. 7000. We believe that this disparity reflects, in part, the electrolytic character of the dissolved PPTA[14].

The right hand branch of the locus of 'm' points in Fig. 5 is also interpretable as a freezing point depression. In this case, the most that we can say is that the data are consistent with plausible values for the heat of fusion and melting points of a stoichiometric sheet-like structure. If we extrapolate the limiting slope of the data [and suppose that this remains linear to the sheet melting point], and presume that the sheet stoichiometry is 2 acids/amide as suggested by the x-ray observations on PPDB (see Figure 3), then the sheets are computed to melt at 162°C with a heat of fusion of ca. 5.0 kcal/mole of amides. Assumed stoichiometries of 1 to 1 and 3 to 1 result in melting points of 205°C or 126°C and heats of fusion of ca. 4 or ca. 8 kcal/mole, respectively. All three sets are plausible. Their exact quantitative significance is questionable, since they are based on the implicit assumption that all of the PPTA present is incorporated into sheets. This assumption is suspect on the general grounds of inevitable imperfections in these quite viscous solutions, and the rather long extrapolation required; as well as on the specific grounds discussed in the next paragraph.

The points marked as 'c' in Figure 5 correspond to observed clearing points. We cannot claim that they are equilibrium clearing points; but, rather, indicate the temperature at which significant quantities of an isotropic solution phase form on the time scale of 10^1–10^3 seconds. Nevertheless, the biphasic character of the solutions lying near the locus of 'c'-points is indisputable. Consequently, it is of interest to ask about the predictions of theory[15,16] for nematic-isotropic biphasic stability. The appropriate equations have been published. The appropriate model is one of polydisperse chains of specified Kuhn length[17] in a nonathermal solvent. The results of some illustrative calculations are presented in Figure 8a and 8b, with the requisite parameters noted in the legends of that figure. It is important to note that the Flory-Huggins solvation parameter χ has been treated as a disposable parameter. The intention here is to merely demonstrate the qualitative correspondence between theory and observation; no claim is made for the actual value of χ. There is clearly predicted to be a biphasic regime in the concentration range corresponding to Figure 5. This is a direct consequence of the PPTA polydispersity, and is subject to only a minor degree to

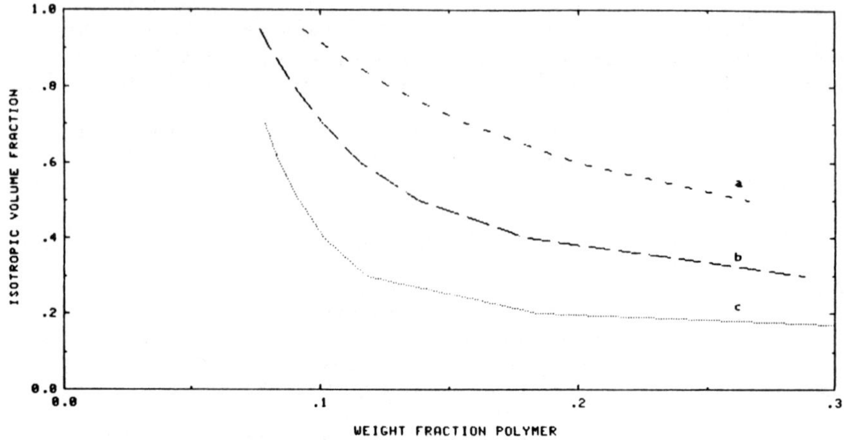

Figure 8a. The isotropic fraction of the total solution volume is plotted versus the overall polymer concentration (w/w). Parameters used in the calculation[17] are: most probable distribution of chain lengths with p = 0.9881; the longest axial ratio of the system, $\eta = 100$; the 1956 approximation[15], C = 0; and solvation parameters, $\chi = \chi'$ of (a) +0.005, (b) 0.0, (c) -0.005.

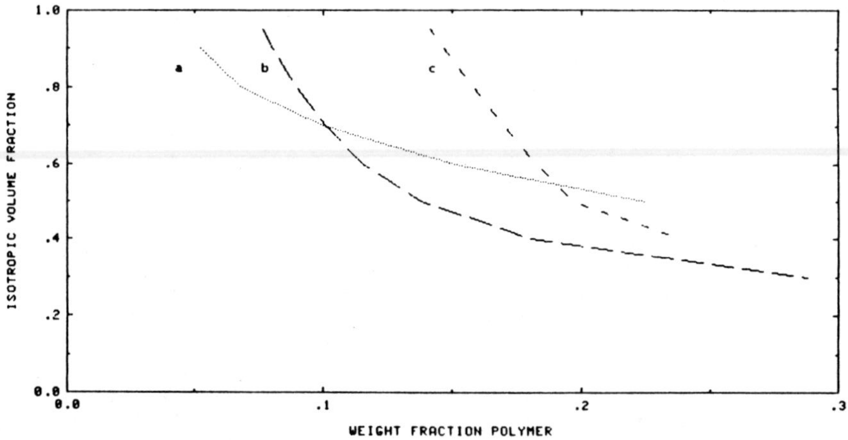

Figure 8b. The same plot as in Figure 8a but with $\chi = \chi' = 0$ and the longest axial ratio of the system, η = (a) 200, (b) 100, (c) 50.

the influence of the somewhat arbitrary parameterization of the calculations. It is not unreasonable that the isotropic phase will act differently than the predominent nematic phase at the freezing point. This vitiates the exact validity of the estimates of sheet melting parameters based on the data presented in Figure 5 and analyzed in the preceding paragraph.

In summary, all of the qualitative features of the phase diagram are explicable in terms of established physical principles. A quantitative description of the data is tentative, because the PPTA-sulfuric acid solution is very complicated. However, with plausible values for the various parameters a satisfactory semiquantitative description is attainable.

References

1. M. M. Iovleva and S. P. Papkov, Vysokoml. soyed. (1982) A24, 233 (translated in Polymer Sci. U.S.S.R. (1982) 24, 236).

2. Yu. A. Tolkachev, O. P. Fialkovskii and Ye. P. Krasnov, Vysokomol. soyed. (1976) B18, 563.

3. S. N. Pankov, M. M. Iovleva, S. I. Banduryan, N. A. Ivanova, I. N. Andreyeva, V. D. Kalmykova and A. V. Volokhina, Vysokomol. soyed. (1978) A20, 658. (translated in Polymer Sci. U.S.S.R. (1978) 20, 742).

4. M. M. Iovleva, S. I. Banduryan, N. I. Ivanova, V. A. Platonov, L. P. Milkova, Z. S. Khanin, A. V. Volokhina and S. P. Papkov, Vysokomol. soyed. (1979) B21, 351.

5. T. Takahashi, H. Iwamoto, K. Inoue and I. Tsujimoto, J. Polym. Sci., Polym. Phys. Ed. (1979) 17, 115.

6. M. Panar, et al., J. Polym. Sci., Polym. Phys. Ed. (in press).

7. J. Calabrese and K. H. Gardner - publication in preparation.

8. S. Harkema and R. J. Gaymans, Acta. Cryst. (1977) B33, 3609.

9. W. W. Adams, A. V. Fratini and D. R. Wiff, Acta Cryst. (1978), B34, 954.

10. S. Harkema, R. J. Gaymans, G. J. van Hummel and D. Zylberlicht, Acta Cryst. (1979) B35, 506.

11. M. G. Northolt, Eur. Polym. J. (1974) 10 , 798-804.

12. J. H. van't Hoff, Z. Physik. Chem. (1887) 1, 481.

13. International Critical Tables (1928) IV, 214.

14. D. G. Baird and J. K. Smith, J. Polym. Sci., Polym. Chem. Ed. (1978) 16, 61.

15. P. J. Flory, Proc. Roy. Soc. (London) (1956) A239, 73.

16. P. J. Flory and A. Abe, Macromolecules (1978) 11, 1119.

17. R. R. Matheson, Jr., and P. J. Flory, Macromolecules (1981) 14, 954.

RECEIVED May 4, 1984

7

Aromatic Azomethine Polymers and Fibers

PAUL W. MORGAN[1], TERRY C. PLETCHER, and STEPHANIE L. KWOLEK

Textile Fibers Department, E. I. du Pont de Nemours and Company, Experimental Station, Wilmington, DE 19898

An aromatic polyazomethine structure is shown in the first line of Table 1. The literature of polyazomethines is voluminous but in general these materials have been reported to have poor solubility and to be unmeltable, and therefore difficult to characterize (1-9). On the other hand, mono and bisazomethines with appropriate terminal groups were among the early substances found to exhibit a liquid crystalline state.

We set out to prepare tractable, high molecular weight, aromatic polyazomethines by depressing the melting points through ring substituents, flexible chain units, and/or copolymerization, as Schaefgen and co-workers (10) had done with aromatic polyesters. Since our work was published in patents in 1976 to 1978 (11), Millaud and co-workers (12) have examined several of these polymers, particularly the polymer from 2-methyl-1,4-phenylenediamine and terephthaldehyde, for viscosity-light scattering molecular weight relationships, persistence length, light absorption spectra, and the formation of lyotropic solutions in sulfuric acid.

Polymer Preparation

Table 1 shows four reactions for the preparation of polyazomethines. The diamine-dialdehyde reaction proceeds rapidly at room temperature or above without catalysts and the reactions of a diamine with a bisacetal or a bisazomethine proceed well at elevated temperature. We have used primarily the first and third reactions. Table 2 outlines several procedures using these reactions.

For the low temperature procedures, equivalents, or near equivalents, of the reactants are combined quickly in an anhydrous solvent. If water is to be removed, arrangement is made for distillation. After a short time the polymer usually precipitates.

[1] Current address: 822 Roslyn Ave., West Chester, PA 19380.

0097-6156/84/0260-0103$06.00/0

Some further gain in molecular weight may be attained in the precipitated state. Monoamines or monoaldehydes may be added to control molecular weight in the preparation and in subsequent processing steps.

The polymers from solution preparation can be further polymerized thermally in the solid state or as a melt. Of course, any of the preparative reactions can be carried out as a melt system with appropriate care in initiating the reaction. A well-controlled polymerization is obtained by melting together a diamine with a bisazomethine. An illustration of that process appears in Table 3 for the reaction of 2-methyl-1,4-phenylenediamine with the bisazomethine from aniline and terephthalaldehyde. The reaction was carried out in a small flask under nitrogen with stirring and distillation of byproduct aniline. Time was counted from the point when the ingredients were melted. The reaction was proceeding slowly after about forty minutes. The progress of the polymerization is controlled by aniline removal and would be accelerated by reducing the pressure.

Polymer Properties

Table 4 shows the inherent viscosities and melting temperatures of some mono-substituted polyazomethines. The unsubstituted polymers do not melt up to 400°C. One finds that the methyl group is more effective than the chloro group in lowering the melting point and that a substituent on the diamine ring has a somewhat greater melting point lowering effect than the same substituent on the aldehyde ring. The larger naphthalene and biphenylene units, surprisingly, yield polymers melting close to that from terephthalaldehyde. All of the polymers described in Table 4 and in Tables 5 and 6 yield liquid crystalline melts.

The diphenyl ether unit had a variable effect on melt temperature. The first polymer (Table 5) melted 32° higher than the 255° value for the phenylene polymer (Table 4) whereas the third polymer (Table 5) melted 30° lower than the corresponding phenylene polymer in Table 4. Flexible chain units, such as ethylene or dioxyethylene, lower the melting temperature and, in general, lengthening such segments increases the melting point lowering.

A wide variety of copolymers which give liquid crystalline melts were prepared. Three are shown in Table 6. In the third example a high proportion of hexamethylene units greatly depressed the melting point, yet the melted polymer was liquid crystalline.

Up to this point the melting temperature has been described as though it were an exact property. And it may be for a given polymer sample and testing method. However, we have found that the melting temperature may vary with inherent viscosity, the preparative method, the thermal history and the testing method. Differences in molecular weight, molecular weight distribution and in polymer crystallinity probably account for some of this variation.

Table 1. Reactions for Preparation of Polyazomethines

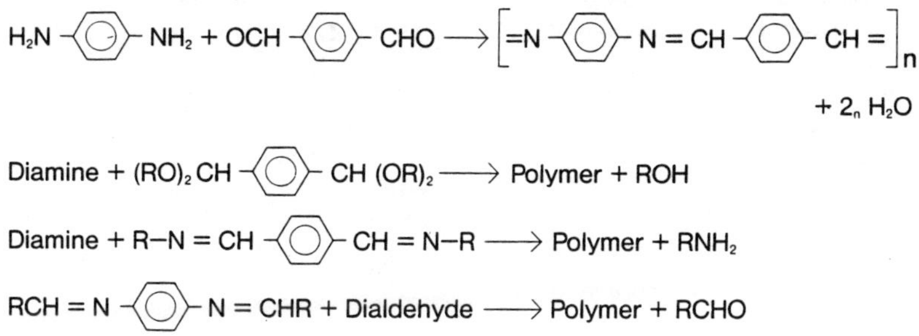

Table 2. Procedures for Polyazomethine Preparation

Diamine and Dialdehyde
Amide solvent plus LiCl
Ethanol, hot
Methylene chloride, azeotropic removal of water
Melt with distillation of water

Diamine and Bisazomethine
Melt with distillation of monoamine

Table 3. Melt Preparation of a Polyazomethine

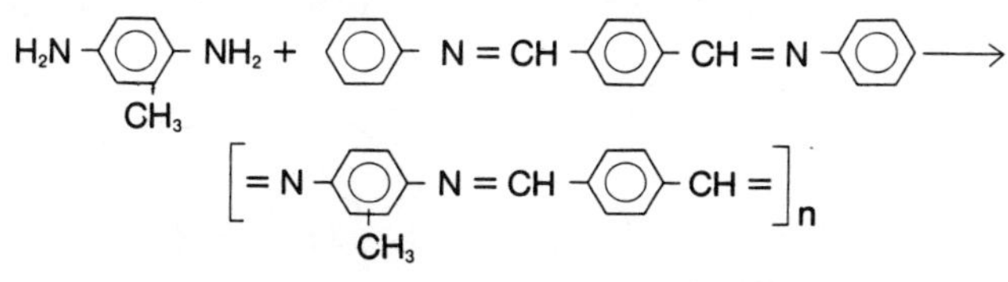

Reaction Time[a] Min	η_{inh} dL/g in conc. H_2SO_4
13	1.55
20	1.73
30	2.71
40	3.50
50	3.77
60	3.93
90	4.11

a) Equivalents of intermediate at 260°C and 760 torr with aniline removal.

Table 4. Aromatic Polyazomethines

Diamine Component	Dialdehyde Component	η_{inh} dL/g in H_2SO_4	Polymer Melt Temperature, °C
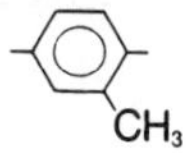		2.0	255
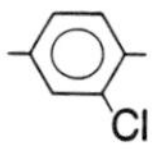	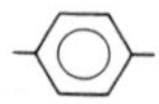	0.66	310
	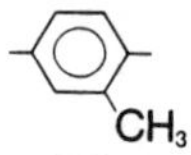	0.70	270
	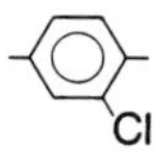	0.56	350
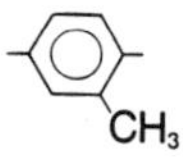	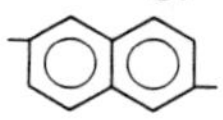	0.60	245
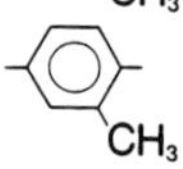	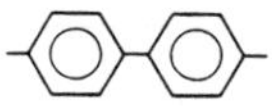	1.22	250

Table 5. Aromatic Polyazomethines with Angular and Flexible Units

Diamine Component	Dialdehyde Component	η_{inh} dL/g in H_2SO_4	Polymer Melt Temperature, °C
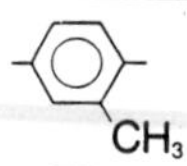	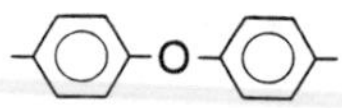	0.78	287
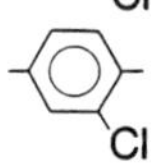	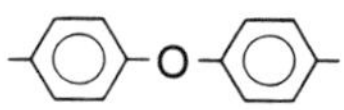	0.64	305
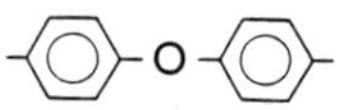	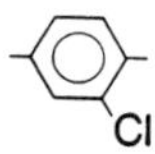	0.51	320
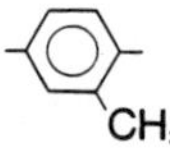	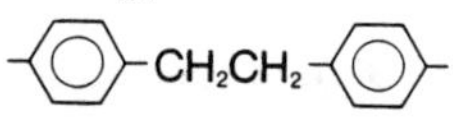	1.9	175
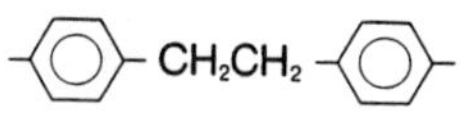	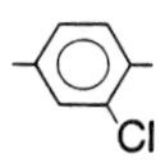	0.70	260
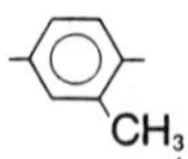	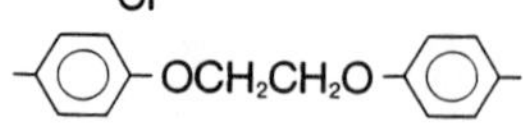	1.67	253

Table 6. Aromatic Azomethine Copolymers

Diamine Components	Dialdehyde Components	η inh dL/g in H_2SO_4	Polymer Melt Temperature, °C
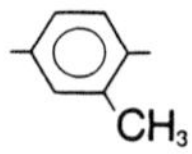	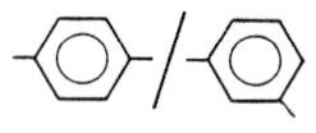 80/20	1.9	260
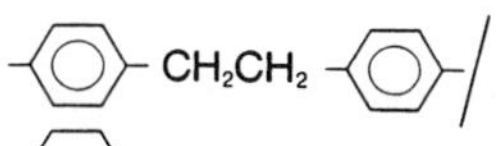 50/50	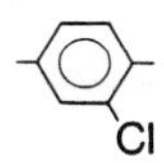	0.60	272
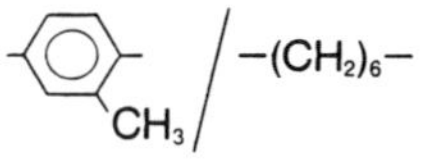 50/50	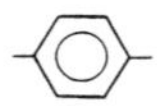	0.70	150

Figure 1 shows the type of variation that was observed for a series of copolymers from 2-methyl-1,4-phenylenediamine, 1,4-phenylenediamine and terephthalaldehyde. The methyl-substituted homopolymer is at the left with decreasing amounts of substitution toward the right. The samples were from different preparations and had differing inherent viscosities. The spread in melting temperatures is greatest in the middle range of compositions. There appears to be a rise in average melting temperature at low modification and then the usual depression in melting for copolymers with non-isomorphous units. With less than forty percent substitution of repeat units the polymers do not melt before decomposition.

The results for another copolymer series is presented in a different way in Figure 2. The partial plots show melting points based on DTA determinations and the temperature at which fibers could be pulled from the melt. The horizontal scale is the proportion of Y-component. In this series all of the polymers were prepared in a melt system by reaction of the diamines with a bisazomethine. There was some variation in inherent viscosity which appears not to have affected the regularity of the data.

As was stated earlier, the aromatic polyazomethines described form liquid crystalline melts. The presence of liquid crystallinity can be determined by observing the high light transmittance of the melt and its texture on the hot stage of a microscope with crossed polarizers. For a more quantitative examination the microscope can be equipped with a photometer and a recorder for plotting the transmitted light and the temperature of the sample on the hot stage (15). Because of the tendency of the polyazomethines to

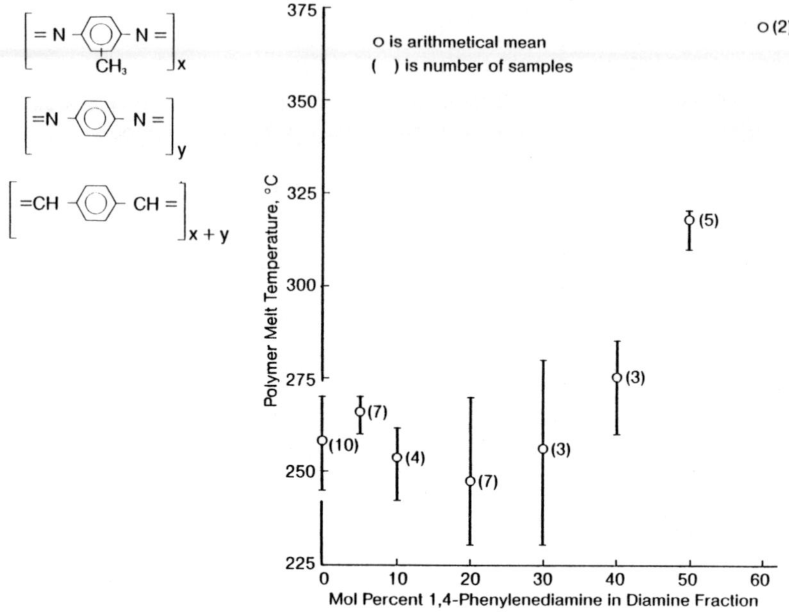

Figure 1 Relation of polymer melt temperature and composition for a series of copolyazomethines from 2-methyl-1,4-phenylenediamine, 1,4-phenylenediamine and terephthalaldehyde.

polymerize further at high temperatures, it was found desirable to pre-press thin films for this test and use a high heating rate. Figure 3 provides a typical plot with an anisotropic melt range, followed by a sharp decline in light transmittance. This latter region can be a nematic-isostropic phase transition, although at the high temperatures one must be careful to observe whether there is polymer decomposition. Most of the polymers have been examined by this method of thermal optical analysis.

The aromatic polyazomethines are colored yellow, orange, red or brown but not black as has been reported in some earlier studies. Bürgi and Dunitz (13) have examined the structure of benzylideneaniline by X-ray diffraction and determined the bond angles and ring arrangement in the solid state (see Figure 4). The angles at the trans azomethine unit are such that there is only a 2.8° deviation from parallel extension of the bonds to the phenyl rings. The phenyl ring attached to nitrogen is twisted out of the plane of the azomethine unit by 55.2° and the other ring has a 10.3° twist in the opposite direction. Substituents have an effect on these angles.

By analogy the polymers would be expected to have a moderate degree of color because of the limited range of conjugation brought about by the out-of-plane twist of the aromatic rings.

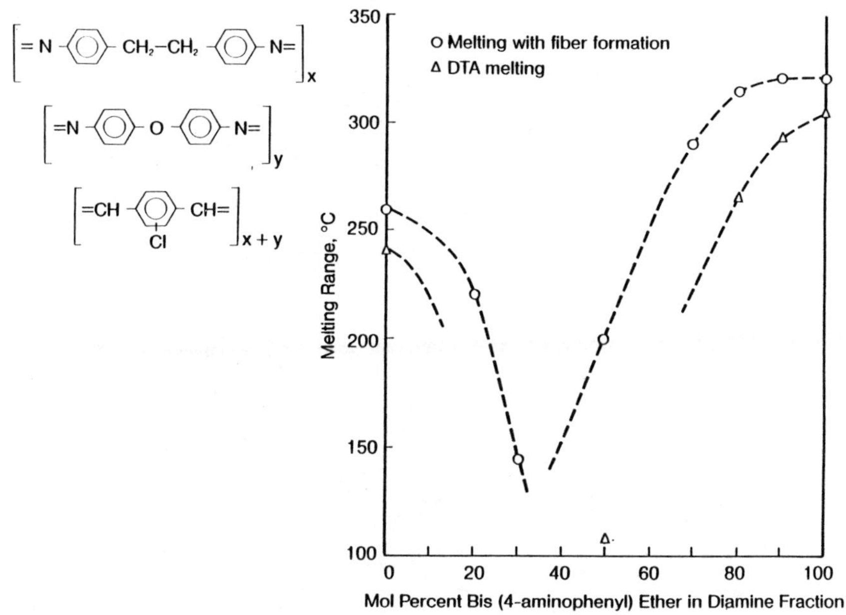

Figure 2 Relation of melting temperature and composition for a series of copolyazomethines from 1,2-bis(4-aminophenyl-ethane, bis(4-aminophenyl) ether and 2-chloroterephthal-aldehyde.

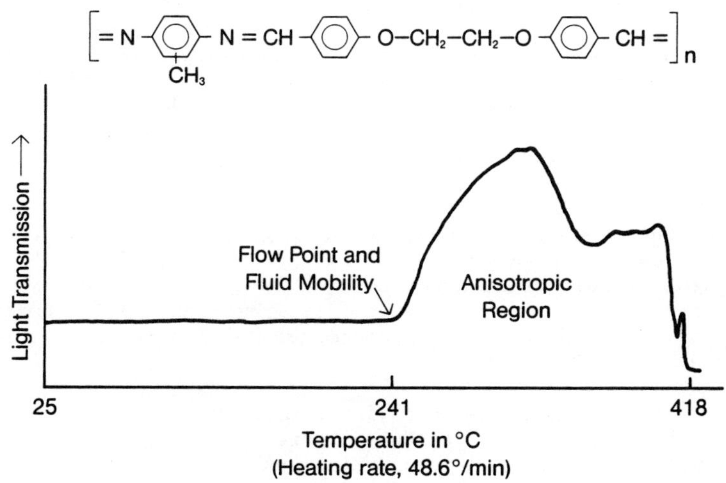

Figure 3 Thermal Optical Test on the polyazomethine from 2-methyl-1,4-phenylenediamine and 1,2-bis(4-formylphenoxy) ethane; particles pressed at 250°C; heating rate, 48.6°C/min.

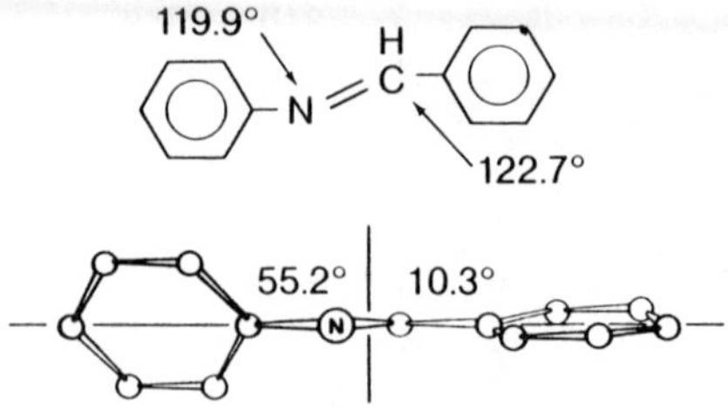

Figure 4 Diagrams showing the bond angles and ring twist angles for benzylideneaniline in the solid state (13).

Fiber Preparation and Properties

Now, let us turn to fiber preparation and properties.

For small-scale spinning experiments the granular polymers were molded into a small rod and then extruded through a narrow channel and out through a multi-hole spinneret. Table 7 gives a typical progression of inherent viscosities in the processing of a polyazomethine to heat-treated, high tenacity fibers. The starting polymer may have a lower or higher viscosity number. A gain in viscosity at each processing step is not essential or even desirable. However, a viscosity (molecular weight) gain during the heat treating step almost always accompanies a marked gain in tensile properties.

Because of the liquid crystalline nature of the melts in which the polymer chains are arranged in domains of parallel arrays, fibers are readily obtained with an appreciable degree of molecular orientation in the direction of the long fiber axis. This orientation of polymer chains leads to high tenacity and high modulus without a drawing step.

The orientation tendency is so great that fibers with high properties can be obtained by withdrawing them with forceps from a molten pool of polymer on a hot bar (Table 8). The tenacity of the heat-treated fibers in the example (9.8 grams/denier) is comparable to that of highly drawn 6-6 nylon and the modulus is about three times that of a glass fiber.

The preparative method for the polymer can have an influence on fiber properties. In Table 9 are described four polymer samples with different end-group relationships but with a fairly narrow range of inherent viscosities. Spinning was done under comparable conditions.

Table 7. Typical Progression of Inherent Viscosities During Processing of a Polyazomethine

Stage of Processing	η_{inh}, dL/g
Polymer as prepared	1.2
Molded rod	1.6
As-spun fibers	3.2
Heat treated fibers	7.5

Table 8. Fibers from a Melt-Pool of a Polyazomethine

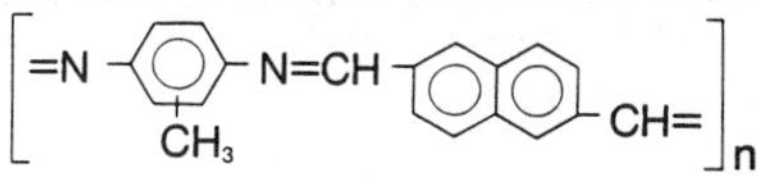

Fiber	Tenacity g/d	Elongation %	Modulus g/d	Denier/ Filament
As-spun (4 breaks)	3.2	1.1	300	67
Heat treated (3 breaks)[a]	9.8	1.3	764	15[b]

a) Heated on a bobbin 6 hr at 240°C under nitrogen.
b) Different fiber sample

Table 9. Monomer Balance and Properties of Fibers from MePPD-TA

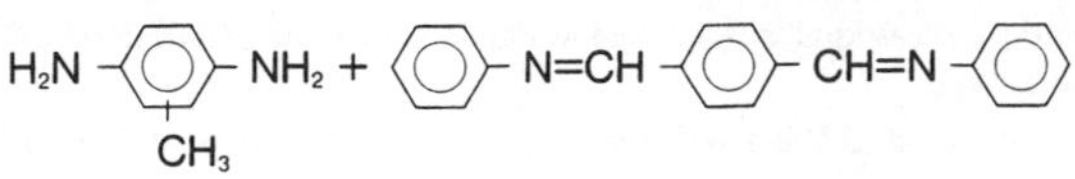

Reaction Conditions	Polymer η_{inh}, dL/g	T/E/Mi As Spun	T/E/Mi Heat Treated
Equivalence of Reactants	4.73	6.3/1.3/744	16.2/1.1/975
3 Mol % Excess Bisazomethine	3.44	3.0/1.1/833	9/0.9/637
3 Mol % Excess Diamine	4.72	7.5/1.4/735	19.8/2.5/908
8 Mol % Excess Diamine	3.57	5.7/1.2/648	30/1.8/868

The as-spun fiber tenacity parallels to some extent the initial inherent viscosity. Upon heat treatment, the polymer with chain ends capped with aniline (second preparation) responds relatively poorly. Increasing the proportion of free amine ends from a balanced system to 3% excess diamine and then to 8% excess diamine gives increasing levels of tenacity in the heat-treated fiber. At the same time there is a small opposing drop in tensile modulus. The amine groups on the polymer chain ends react readily in the solid state during heat treatment to increase the molecular weight and the strength of the fiber.

In Table 10 there are presented the properties of fibers from three different polyazomethines. The first example is an outstanding result with an average tenacity of 38 grams/denier and a modulus a bit above 1000 grams/denier. As note b indicates, another spin gave fibers with an average tenacity of 43.6 g/d (10 breaks). These tenacities are among the highest, perhaps the highest, which have been reported for fibers from condensation polymers. The third example shows that the introduction of flexible ethylene units in the chain will reduce the modulus of the resulting fibers. The X-ray orientation angles are typically in the 20 to 30° range for as-spun fibers and around 10° after heat treatment, indicating increased degree of orientation.

Table 10. Fibers from Several Polyazomethines

Polymer	Fiber Condition	Properties T	E	Mi	Den	O.A.[a]
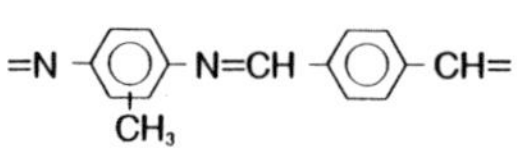	As spun	7.3	1.1	916	4.0	–
	Heat treated	38	4.4	1012	3.7	–[b]
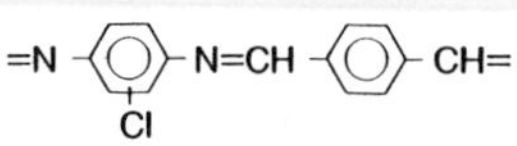	As spun	7.4	1.3	683	11	22
	Heat treated	15.3	1.9	844	11	10
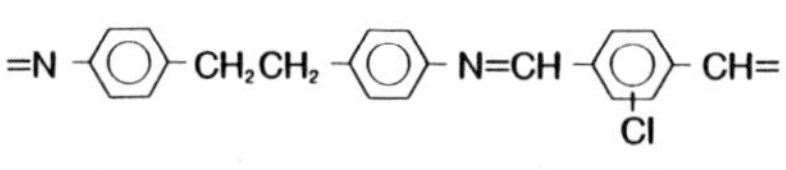	As spun	3.6	1.3	400	–	25
	Heat treated	21.4	4.3	447	–	9

[a]T: tenacity in g/denier; E: % elongation; Mi: initial modulus in g/denier; Den: denier/filament; O.A.: X-ray orientation angle.

[b]Another spin yielded heat treated fibers with average T: 43.6; E: 4.8; Mi: 944; Den: 3.2.

Summary

In conclusion then, we have synthesized a series of extended-chain, aromatic polyazomethines under non-degradative conditions. Fusible, tractable polymers were obtained by use of unsymmetrically placed substituents, copolymerization, and/or limited proportions of flexible chain units. Many of the polymers yield liquid crystalline melts which were spun into oriented, high tenacity, high modulus fibers. The fibers were further strengthened by heat treatment. The ease of preparation of the aromatic polyazomethines and the outstanding tenacity and modulus range of the fibers make these products excellent candidates for use as reinforcing fibers in resins and rubber.

Test Procedures

Dilute Solution Viscosity: Inherent viscosity, η_{inh} = $\ln(\eta_{rel})/c$, values were determined in 96-98% sulfuric acid at 30°C and a concentration, c, of 0.5 g of polymer per 100 ml of solution, unless otherwise noted.

Polymer Melt Temperature: Polymer melting was determined on dry, powdered polymer on a chrome-plated gradient temperature bar (14) and is noted as the temperature at which the polymer under moderate sliding pressure leaves a molten or waxy trail adhering to the bar.

X-Ray Orientation Angle: Wide angle X-ray diffraction patterns were obtained with a Warhus pinhole camera and Phillips generating unit No. 12045 having a copper fine focus diffraction tube and a nickel β filter. The distance from sample to film was 50 mm. The arc length in degrees between the half-maximum intensity points of a principal equatorial diffraction spot is reported as the orientation angle of the sample.

Fiber Tensile Properties: Tensile strength (grams per denier), percent elongation, and initial modulus (grams per denier) were determined on an "Instron" Tensile Tester (Instron Engineering Corp., Canton, MA). The fibers were conditioned for at least 16 hours at 21°C and 65% relative humidity. Single filaments were broken with 1-inch, initial jaw separation. Results are reported as the average of at least three breaks. Denier (the weight in grams of 9000 meters of filament) was determined for each filament by a resonance method described in A.S.T.M. D1577-66, Part 25, 1968.

Differential Thermal Analysis: DTA measurements were made with a Du Pont 990 Differential Thermal Analyzer in conjunction with a DSC cell. Samples were run at a heating rate of 20°C per minute in an atmosphere of flowing nitrogen.

Thermal Optical Test: A detailed description of this test is given in patents (11). The apparatus is similar to that described by Kirshenbaum et al. (15).

References

1. G. F. D'Alelio, Polyazomethines in Encyclopedia Polymer Sci. Tech., Wiley-Interscience, New York, 1969, Vol. 10, pp.659-670.

2. R. J. Cotter and M. Matzner, Poly(Schiff Bases) and Related Polymers in Ring-Forming Polymerizations, Vol. 13B, 1 of Organic Chemistry Monographs, Academic Press, New York, 1972, pp. 1-9, 30-53.

3. A. D. Delman, A. A. Stein and B. B. Simms, J. Macromol. Sci. Chem., A1 (1), 147-178 (1967).

4. J. H. Hodgkin and J. Heller, Macromolecules, 2, 556 (1969); J. Polymer Sci., C29, 37 (1970).

5. H. H. Hörhold, Z. Chem., 12 (2), 41 (1972).

6. W. P. Baker and R. A. Sahatjian, U.S. Defensive Publication T918,005 (Jan. 1, 1974).

7. W. G. Gall, U.S. Patent 3,493,541 (Feb. 3, 1970); assigned to the Du Pont Co.

8. R. Adams, J. E. Bullock and W. C. Wilson, J. Am. Chem. Soc., 45, 521 (1923).

9. C. S. Marvel and H. W. Hill, J. Am. Chem. Soc., 72, 4819 (1950); C. S. Marvel and N. Tarköy, Ibid., 79, 6000 (1957); 80, 832 (1958); C. S. Marvel and P. V. Bonsignore, Ibid., 81, 2668 (1959).

10. J. J. Kleinschuster, T. C. Pletcher, and J. R. Schaefgen, Belgian Patent 828,935 (1976); assigned to the Du Pont Co.

11. P. W. Morgan, U.S. Patents 4,048,148 (Sept., 13, 1977) and 4,122,070 (Oct. 24, 1978) and French Patent 2,310,426 (1976); assigned to the Du Pont Co.

12. B. Millaud, A. Thierry and A. Skoulios, Mol. Cryst. Liq. Cryst., Lett., 41, 263 (1978); B. Millaud and C. Strazielle, Polymer, 20, 563 (1979); B. Millaud, C. Strazielle and G. Weill, Polymer, 21, 639 (1980). See also C. Noel and J. Billard, Mol. Cryst. Liq. Cryst., Lett., 41, 269 (1978).

13. H. P. Bürgi and J. D. Dunitz, Chem. Commun., 1969, 472; Helv. Chim. Acta., 53, 1747 (1970).

14. W. Sorenson and T. W. Campbell, "Preparative Methods of Polymer Chemistry", 2nd Ed., Interscience Publishers, New York, 1968, pp. 57-59, Method A.

15. I. Kirshenbaum, R. B. Isaacson and W. C. Feist, J. Polym. Sci., Letters, 2, 897 (1964).

RECEIVED May 4, 1984

ELASTOMERS AND PROCESSABILITY

8

Silicone and Fluorosilicone Elastomers

Structure and Properties

K. E. POLMANTEER and J. R. FALENDER

Dow Corning Corporation, Midland, MI 48640

Silicones and fluorosilicones have traditionally been classified as high performance materials due to their suitability for use under extreme conditions and their processability has been particularly interesting as a result of the very flexible siloxane chain. The present paper discusses selected examples where more recent advances in science and technology have broadened the understanding and uses. Emphasis is given to relating the molecular chemistry and physics to the characteristics of the bulk materials.

Stability is discussed from the viewpoint of what can be done to broaden the performance limits. Aqueous dispersions are presented as an important method of enhancing processability even beyond that inherent in the siloxane chain. Foams involve an example where very interesting crosslinking and foaming chemistry is combined with rheological understanding to fulfill performance needs beyond that of the traditional compounds. Electrical features are finding use in many "high technology" applications and it is becoming more obvious that there is a need to be able to relate the properties to molecular structure. Finally, the optical properties of silicone composites are becoming more critical and the controlling physics is taking on new importance.

Polydiorganosiloxanes such as polydimethylsiloxanes, have established a very respectable position as high performance polymers. They have been successfully utilized in a variety of forms including low and intermediate molecular weight fluid polymers and matrix polymers for silicone elastomers. This paper focuses particularly on a few interesting high performance and processing aspects of

0097–6156/84/0260–0117$07.50/0

silicone and fluorosilicone elastomers. A description will be provided of some recent elastomer developments resulting from research at the Dow Corning Laboratories in Midland, Michigan, as well as other researchers' work pertaining to structure-property effects.

In 1979, a review paper (1) included over 1000 literature citations describing elastomer developments for the period 1967-1977, and it also listed earlier reviews. In 1981 a paper (2), "Current Perspectives On Silicone Rubber Technology" was published. These references (1-2) provide suitable background leading to the present paper.

Discussion

Influence of Structure of Silicon Containing Polymers on Stability. Certain applications require stability even beyond that found in polydimethylsiloxane. It has been recognized for many years that an approach to improve the stability of polydiorganosiloxanes would be to reduce the ionic character by replacement of some of the backbone oxygen atoms with oxidatively stable spacers such as arylene groups (e.g., p-phenylene) (3-13). These references describe much of the early synthesis development effort resulting in improved methods (12) (reported by Merker from Dow Corning's fellowship at Mellon Institute) giving 60-70% yields using the following in situ Grignard reaction for the preparation of p-(bisdimethylhydrogensilyl)benzene:

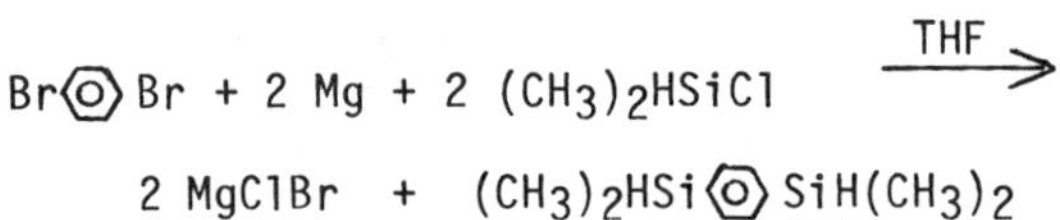

(I)

The hydrolysis of the dihydride (I) and condensation to form cyclic structures for subsequent polymerization similar to the polymerization methods used for the polydiorganosiloxanes resulted in poor yields as did alkaline thermal cracking methods. However, (12-13) hydrolysis of (I) to the corresponding dihydroxy compound followed by condensation using essentially non-equilibrating type catalysts such as tetramethylguanidine di-2-ethylhexoate easily provided high molecular weight homopolymers, $[(CH_3)_2Si(C_6H_4)Si(CH_3)_2\text{-}O]_n$, (TMPS, tetramethyl-parasilphenylene), or block copolymers containing blocks of $(CH_3)_2SiO$ (PDMS, polydimethylsiloxane). Both of these polymers are crystalline at ambient temperature. The homopolymer melts at about 148°C, while the block copolymers melt at temperatures dependent on the block size of the crystalline TMPS segments, with a high temperature approaching 140 °C. Thermal stability of the above homopolymer was shown (12) to be superior to a polydimethylsiloxane (catalysts carefully removed from each) by the reduced generation of volatiles and reduced levels of crosslinking resulting from oxidation. Weight loss for the catalyst free homopolymers was as follows:

Homopolymer Stability In Air

Temp. °C	Time Hours	% Weight Loss [PDMS]$_n$	[TMPS]$_n$
200	200	0.8	0.4
255	128	24.1	6.6
305	24	26.9	9.7

Although better stability was found in the absence of catalysts, Merker (13) noted that equilibrating catalysts such as potassium silanolate would rearrange a tough elastic block copolymer to an amorphous, clear gum while not reducing the molecular weight.

Merker, et. al. (13) reported that block copolymers of TMPS-DMS segments exhibited elastomeric properties and considerable toughness because of the crystalline TMPS blocks. The tensile strengths increased with molar concentration and block lengths of TMPS. Examples of some mechanical property data are as follows:

Effect of TMPS (B units) Content in Block Copolymers with DMS (A units)

TMPS Mole%	TMPS Wt.%	Min.Avg. Arrangement (A-B)	[η] dl/g.	Tensile MPa	Elong. %	Shore A Duro.
18	40	18-4	1.91	7.8	962	50
25	50	18-6	2.11	12.7	750	72
33	60	18-9	2.00	14.4	643	86
44	70	18-14	1.53	18.8	490	93

Kojima and Magill (14-15) have done extensive morphological characterization of TMPS-DMS block copolymers.

Another type of group studied as a relacement for some of the oxygen backbone atoms is the carborane cage structure [e.g., meta-$CB_{10}H_{10}C$-]. Papitti (16) et.al. reported the synthesis of meta SiB2 polymers:

$$[-(CH_3)_2Si-CB_{10}H_{10}C-Si(CH_3)_2OSi(CH_3)_2O]_n$$

The behavior of these polymers was studied and reported by Tobolsky and coworkers (17-18). They concluded that SiB2 had better high temperature properties than a polydimethylsiloxane based elastomer. Thermal stability decreased as the distance between carboranyl units was increased with dimethylsiloxane (DMS) units. In

1973 Roller and Gillham reported on the thermomechanical behavior in nitrogen (19) and in air (20) of a systematic series of linear poly(carborane-siloxane)'s containing icosahedral ($-CB_{10}H_{10}C-$) cages in place of some of the chain oxygen atoms:

$$A\text{-}O[R(CH_3)Si\text{-}Z\text{-}(Si(CH_3)RO)_x]_nA$$

where x=1,3,4,5, ; A=endgroups (reactive and inert); Z=meta-, para-carborane (for x=3); $R=CH_3$, $R=C_2H_4CF_3$ (for x=3), one in five R's=C_6H_5, with the remainder $-CH_3$ (for x=4); molecular weight= ∿10,000, ∿50,000 (for x=3). The results in nitrogen indicated the phenyl pendant groups improved stability. All of the meta-carborane polymer was more stable by about 50°C. The influence of the meta and para-carborane chain units on Tg, and Tm compared with polydimethylsiloxane (DMS) was as follows:

Key Chain Segment	°C Tm	°C Tg
meta-carborane	40	-68
para--carborane	110	-35
DMS	-40	-125

It was interesting that stability in air (20) indicated that for the all methylated polymers, PDMS was more oxidatively stable than the carboranes (e.g. about 350°C vs 280° to 300°C).

The work reported thus far has focussed on the substitution of either aryl groups or carborane cage structures for some of the backbone oxygen atoms. It will be shown next that fluorohydrocarbon alkyl spacers that possess greater thermal stability than their normal alkyl counterparts may also be substituted for some of the backbone oxygen atoms to provide polymers of not only improved stability, but good solvent resistant properties. This work was done at Dow Corning Corporation with Dr. Ogden R. Pierce as the Principal Investigator under the sponsorship of the Air Force Materials Laboratory, Wright-Patterson Air Force Base, Ohio. There are several publications (21-25) that describe the work in detail. The work led (25) to polymers of the following type,

$$(R'RSiCH_2CH_2ZCH_2CH_2SiR'RO)_n$$

where $R=R'=CH_3$, $CF_3CH_2CH_2$, and Z=perfluoroalkylene. The fluorohydrocarbon monomers were prepared by the following reaction sequence:

$$X(CF_2)_nX \xrightarrow[\text{init.}]{CH_2=CH_2} XCH_2CH_2(CF_2)_nCH_2CH_2X \xrightarrow{\text{base}} H_2C=CH(CF_2)_nCH=CH_2$$

x = Br or I, n = 1-10

Thermal initiation of ethylene addition to the diiodo perfluoroalkanes, n greater than 2, was the best high yield procedure. For n equal 1, the method of Henne and DeWitt (26) was used. The next step in the synthesis involved the hydrosilane addition to the above diene, for example:

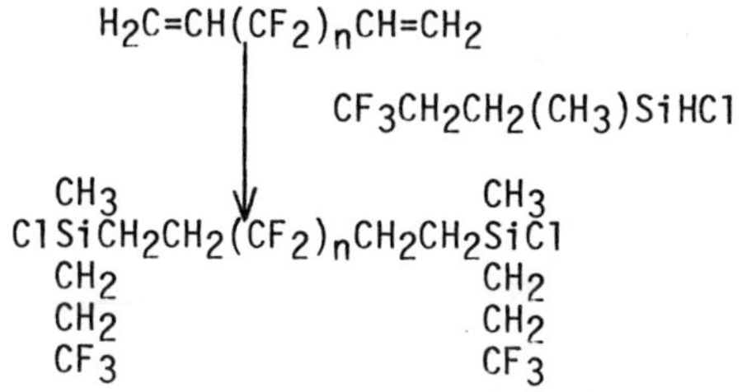

This particular hydrosilane addition catalyzed with free radicals from di-t-butyl peroxide gave high yields without complications. Homopolymers were prepared by first hydrolyzing the dichloride monomer using aqueous sodium bicarbonate in ether as solvent. Bulk condensation of the resulting monomer diol was catalyzed by tetramethyl-guanidine trifluoroacetic acid yielding high consistency elastomeric gum. This may be shown schematically as follows (25):

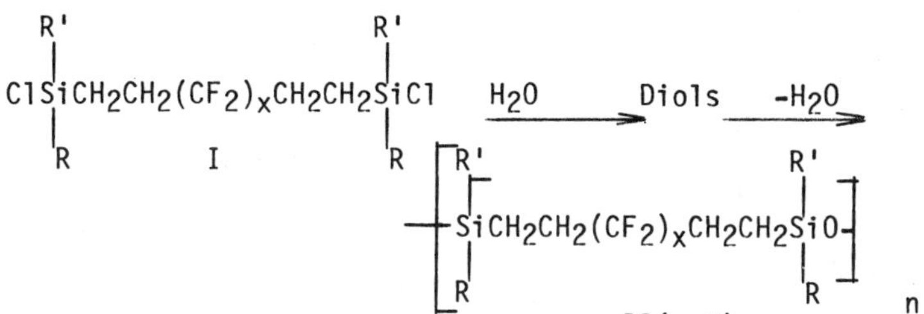

II(a-f)

II- a, x = 1
b, x = 2
c, x = 4
d, x = 6
e, x = 8
f, x = 10

R' = CH_3, R=$CF_3CH_2CH_2$

It is interesting that it was reported that no evidence of cyclization of the diols was observed during polymerization. The marked influence on the heterolytic stability by the fluorohydrocarbon alkyl spacers is amply pointed out by the following experiment. A sample of polymer IId for (x=6) formed no appreciable amount of volatile reversion products when heated at 270°C under high vacuum in the presence of potassium hydroxide. Under these conditions, both polydimethylsiloxane and polymethyl(3,3,3-trifluoropropyl)siloxane generate and lose volatile reversion products almost quantitatively.

Polymers II a-f were found by X-ray diffraction to be non-crystalline amorphous materials. Similar structured polymers were prepared for free radical vulcanization by the introduction of vinyl crosslinking sites. The polymers were formulated into high consistency elastomers reinforced with silica and were free radical vulcanized. The properties for only IIb and IId are shown in Table I with a commercial elastomer prepared from polymethyl(3,3,3-trifluoropropyl)siloxane (LS) shown for comparison. Also included is an elastomer prepared from the following copolymer (III),

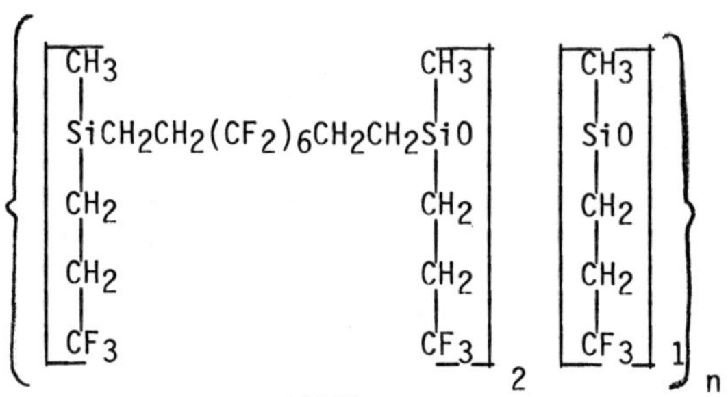

(III)

which was prepared to evaluate this as an approach to reducing Tg. The data in Table I indicate that the fluorohydrocarbon substitution for some of the oxygen atoms in polydisubstitutedsiloxanes does indeed improve the retention of physical properties both in air and under confinement at 250°C while maintaining solvent swell properties comparable with a typical fluorosilicone (LS) elastomer. However, low temperature flexibility as judged by Tg does suffer. The preparation of copolymers with methyltrifluoropropylsiloxane units such as in above copolymer III does reduce Tg by about 5°C.

A more effective method of lowering Tg can be achieved by replacing the perfluoroalkyl chain units with perfluoroether units (27). The influence of a variety of perfluoroether spacers and CF_2CF_2 on Tg is shown in Table II. Tg values of the perfluoroether containing polymer were lower than the CF_2CF_2 containing polymer (VIIa). The lower glass transition temperatures were exhibited by the polymers containing the more unsymmetrical fluoroether segments (e.g., VIIe in Table II), even though the difluoromethylene to oxygen ratio is comparatively high in these molecules. Thermogravimetric analysis provides a comparison of the oxidative and thermal stability of the raw polymers (Table III) with that of the CF_2CF_2 (VIIa) polymer. In general, the differences are quite small, indicating that it is possible to substitute a fluoro-ether segment for the fluorocarbon segment within the hybrid fluorocarbon-fluorosilicone polymer with relatively little loss of thermal and oxidative performance, while significantly extending low temperature capabilities.

Table I. High Consistency Elastomers

Polymer		IIB	IID	III[f]	LS[e]
M.W.[a]		13,000	100,000	70,000	800,000
Sp. G.		1.45	1.59		1.40
T_g[b] (°C)		-27	-24	-29	-71
Original, post cured at 200°C/ 8 hr	Durometer	62	58	53	45
	Tensile Strength, MPa	12.3	17.6	11.1	10.3
	Elongation, %	300	300	230	350
% Volume Swell[c]	Methylisobutyl Ketone	254	285	255	290
	Toluene	77	31	28	17
	Heptane	16	13	4	10
After 24 hr at 250°C in Air	Durometer	62	65	60	42
	Tensile Strength, MPa	7.0	16.1	10.0	3.2
	Elongation, %	200	260	220	260
	%Wt. Loss	2.7	2.2	1.6	1.1
After 24 hr[d] at 250°C in sealed glass container	Durometer	46	54	62	20
	Tensile Strength, MPa	5.8	9.3	3.4	0.3
	Elongation, %	300	300	140	100

[a]M.W. was difficult to obtain because of the insolubility of the polymer in common organic solvent. M.W. was determined either by vapor pressure or membrane osmometry.
[b]Determined by differential scanning calorimetry.
[c]Seventy-two hours immersion at room temperature
[d]Test for reversion resistance.
[e]Poly(3,3,3-trifluoropropyl)methylsiloxane
[f]Copolymer used was $[[(CF_3CH_2CH_2)(CH_3)SiCH_2CH_2(CF_2)_6CH_2CH_2Si(CH_3)(CH_2CH_2CF_3)O]_2[(CF_3CH_2CH_2)(CH_3)SiO]_1]_n$ (III).

Table II. Glass Transition Temperature (DSC Scan at 10°C/MIN)

$$\left[Si(CH_3)(CH_2CH_2CF_3)(CH_2)_2\ Z\ (CH_2)_2Si(CH_3)(CH_2CH_2CF_3)O \right]_n \qquad \text{(VII)}$$

POLYMER	Z	Tg (DSC), °C	CF_2 /O
VIIa	$-CF_2CF_2$	-26°	---
b	$-(CF_2)_3O(CF_2)_2$	-39°	5/1
c	$-(CF_2)_2O(CF_2)_2O(CF_2)_2-$	-40°	3/1
d	$-(CF_2)_2O(CF_2)_5O(CF_2)_2-$	-52°	4.5/1
e	mixture of $-CF(CF_3)O(CF_2)_5OCF(CF_3)-$ and $-(CF_2)_4OCF(CF_3)CF_2OCF(CF_3)-$	-47°	4.5/1

Table III. Thermal Gravimetric Analysis (10°C/MIN)

POLYMER*	Temperature at 10% Wt. loss, °C: AIR	Temperature at 10% Wt. loss, °C: NITROGEN	Temperature at 50% wt. loss, °C: AIR	Temperature at 50% wt. loss, °C: NITROGEN
IIIg	310°	--	355°	--
IVg	330°	412°	375°	497°
Vg	330°	450°	380°	502°
VIIa	324°	474°	442°	504°

$$* \left[OSi(CH_3)(CH_2CH_2CF_3)(CH_2)_2(CF_2)_\ell[O(CF_2)_m]_nO(CF_2)_2(CH_2)_2Si(CH_3)(CH_2CH_2CF_3) \right]_N$$

IIIg, $\ell = 3$, $n = 0$
IVg, $\ell = 2$, $m = 2$, $n = 1$
Vg, $\ell = 2$, $m = 5$, $n = 1$

Siloxane Elastomer Processing From Aqueous Dispersion

The intriguing scientific (28) and engineering features (29) (such as processing ease) associated with polymer-in-water emulsions have led to a significant amount of effort aimed at silicone elastomers. Ingredients incorporated during and after polymerization have included dimethylsiloxane oligomers, ionic surfactants and catalysts, organotin compounds, colloidal silica, and water. In one sense, the colloidal silica serves the same function as fumed silica does in more conventional silicone rubber technology. That is, the colloidal silica increases modulus, ultimate strength and swell resistance as seen in Figure 1 and Table IV.

The hypothesized chemistry is illustrated in Figures 2 and 3. Silicates graft to silanol groups on the polymer chain at the surface of emulsion particles. While still suspended in water, low molecular weight silicates (depolymerized from colloidal silica in the alkaline environment) could migrate into the particles and multiple graft sites would lead to crosslinking. Higher molecular weight silicates would remain at the oil-water interface and provide stabilization of the emulsion. Organotin compounds render the silicates hydrophobic and promote the transfer of these species to the oil-water interface. At the interface, the tin compound catalyzes reactions between silanols on the polymer and silicate. The result would be particles crosslinked before water removal. After water removal, the particles adhere to one another both through hydrogen bonding and siloxane bonds which would be expected to form under the alkaline condition of the composition.

The above chemistry is believed to lead to an inverted structure. Crosslinked polymer spheres are separated by a silica or silicate rich phase with siloxane bonds grafting the two phases together. This would lead to some of the unusual properties which have been observed including a high stress at low extension and high water transmission through a hydrophilic continuum. (28)

From a practical standpoint, the water based silicone provides material of lower viscosity than pure silicone elastomers and free of hazards associated with handling solvent dispersion. Because of processing advantages as well as the easy clean-up features (commonly associated with latex paints), the materials are being considered for commercialization for applications such as construction coatings, water containment and several other areas. (28)

Siloxane Elastomer Foams - Structure and Properties

The high performance properties of silicone foams depend not only on the siloxane polymer itself, but on some very interesting chemistry required following polymer synthesis (30-32). To achieve the desired mechanical properties and foam density, the kinetics of crosslinking and foaming reactions must be carefully balanced.

$$\equiv SiH + HOSi\equiv \xrightarrow{Pt} \equiv SiOSi\equiv + H_2$$

$$\equiv SiH + CH_2{=}CHSi\equiv \xrightarrow{Pt} \equiv SiCH_2CH_2Si\equiv$$

$$\equiv SiH + H_2O \xrightarrow{Pt} \equiv SiOH + H_2$$

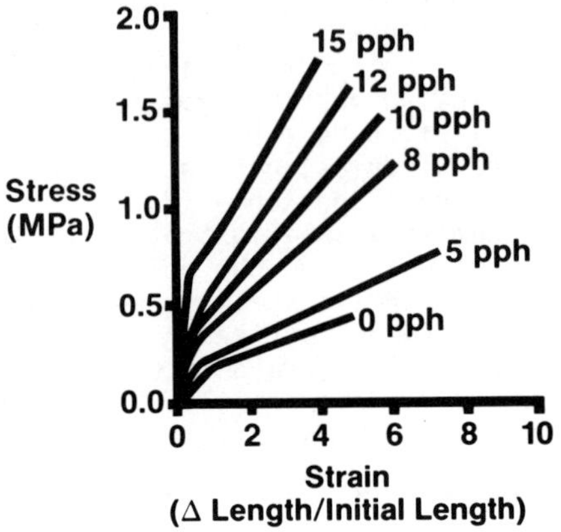

Fig. 1 Effect of silica on tensile properties of films cast from silicone latexes. The curve marked 0 pph was obtained from a film containing 2 pph of sodium silicate instead of collodal silica (pph = parts per hundred).

Table IV. Swelling Of Dried Films From Silicone Latexes Made With Sodium Silicate

Amount of sodium silicate phr	Gel fraction in cyclohexane, %	Swell in cyclohexane, %	Swell in water, %
0.10	46	12,000	180
0.50	73	3,200	210
1.00	77	2,700	290
2.80	83	1,800	260
3.50	84	1,700	280
4.50	83	1,800	349

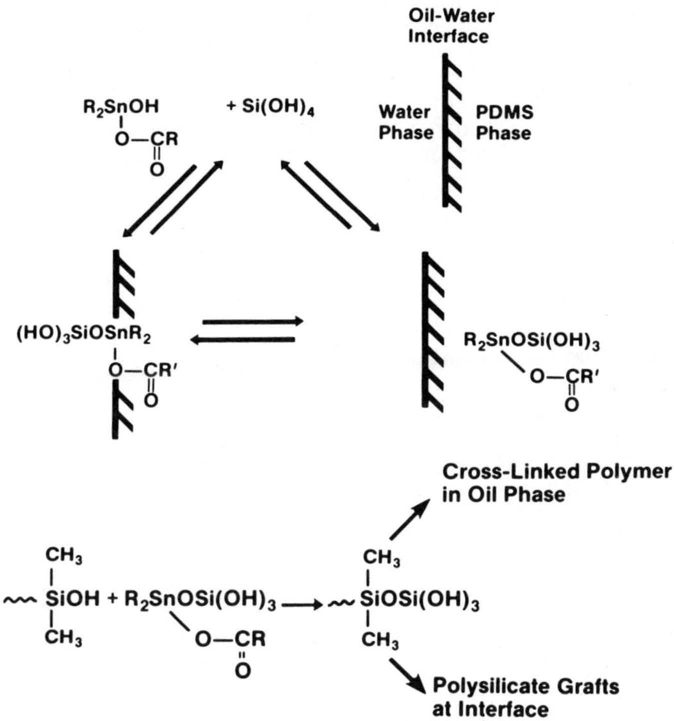

Fig. 2 Schematic representation of the process occuring while the system is still in dispersion.
Reproduced with permission from Ref. 28. Copyright 1981.

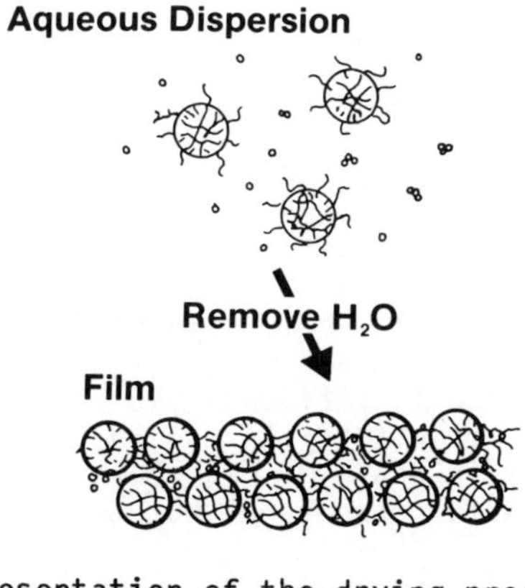

Fig. 3 Schematic representation of the drying process.
Reproduced with permission from Ref. 28. Copyright 1981.

In a number of commercially important examples, foaming enhances the performance of silicones or even makes them suitable where they would not otherwise function. For example, fire stop foam is used around wire and pipes penetrating through building walls and floors. The foaming reaction allows holes to be tightly sealed and the high thermal resistance greatly delays the spread of fire. In other applications, sound deadening, damping of mechanical vibrations, buoyancy, or mechanical flexibility are critical properties. The degree to which the cells are open or closed has a strong effect on properties such as sound deadening.

Recent work has shown that the nature of the foaming reaction can be controlled by the ratio of $\equiv$SiH to $\equiv$SiOH (designated as R_1) and by the ratio of polyfunctional $\equiv$SiH molecules to monofunctional $\equiv$SiH molecules (designated as R_2). For example, a high level of polyfunctional $\equiv$SiH molecules causes viscosity to increase more rapidly with time as the curing and foaming reactions take place (see Figure 4). It also causes the gel time to decrease and the amount of gassing before gelation to decrease (Figure 5). The amount of open cell structure goes down as the $\equiv$SiH to $\equiv$SiOH ratio increases. As the level of polyfunctional $\equiv$SiH increases, the amount of open cell structure first increases then decreases (See Figure 6). (32)

Influence Of Silicone Structure On Electrical Performance

Most silicone polymers act as very good insulators. Polarity of the siloxane backbone apparently is shielded by methyl groups. In addition, the hydrophobic nature of the surface helps to repel water (and dissolved ionic contaminates) which assures the retention of insulation properties under difficult conditions. One application where these properties have found recent importance has been for insulator sheds. Even after repeated exposure to salt fog and high voltage stress, the materials retain their resistance to conductive and arc failure (33).

The effect of humidity depends strongly on polymer structure. While the solubility of water is below 1 part per million in carefully purified, low silanol dimethylsiloxane polymers, at high humidity it is about 250 parts per million in commercial grade polymer (34). Here the water content goes up almost in direct proportion to the relative humidity (see Figure 7) (35), and this leads to an increase in dielectric loss (Figure 8) (35) and dissipation factor (Figure 9) (36). Table V reports the effect of silanol and metal ion content on volume resistivity at 50% relative humidity. Metal ions in the range of around 60 parts per million decreased volume resistivity (e.g. gum B), while silanol content exhibited the opposite effect (e.g., gums A and C). The side groups on the polymer itself can also strongly influence the electrical response of the material. Volume resistivity is considerably lower on polymers with phenyl side groups in place of methyl while dielectric strength is exceptionally high for the phenyl polymer (See Table VI) (37).

Optically Clear Silicone Elastomers From New Silica Technology

This paper has thus far focused principally on the structure-property behavior as it relates to the polymeric matrix with atten-

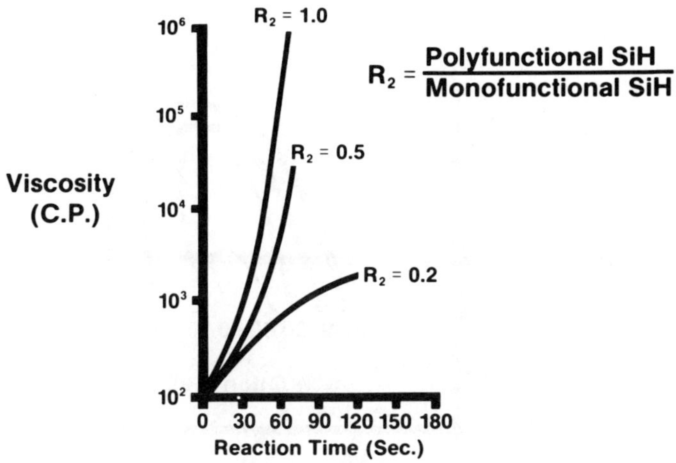

Fig. 4 Effect of R_2 on the cure rate at room temperature. Reproduced with permission from Ref. 32. Copyright 1982 J. Cellular Plastics.

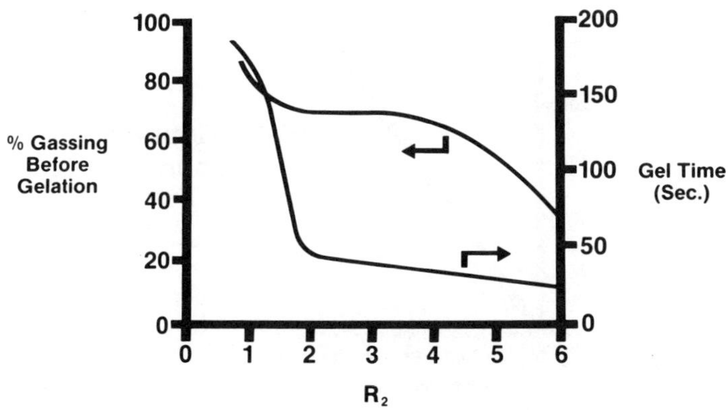

Fig. 5 Effect of R_2 on the percent gassing before gelation and gel time at room temperature.
Reproduced with permission from Ref. 32. Copyright 1982 J. Cellular Plastics.

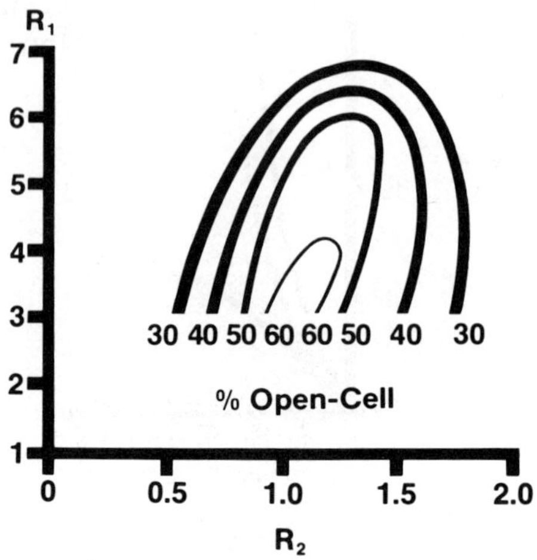

Fig. 6 Percent open-cell structure as a function of R_1 and R_2. Reproduced with permission from Ref. 32. Copyright 1982 J. Cellular Plastics.

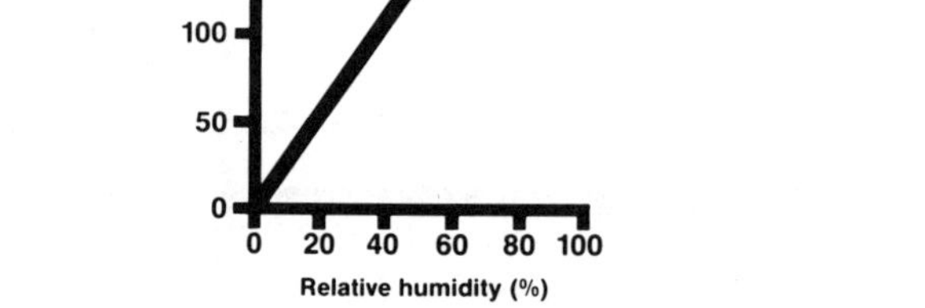

Fig. 7 Water content as a function of a relative humidity. Reproduced with permission from Ref. 35. Copyright 1977 IEEE.

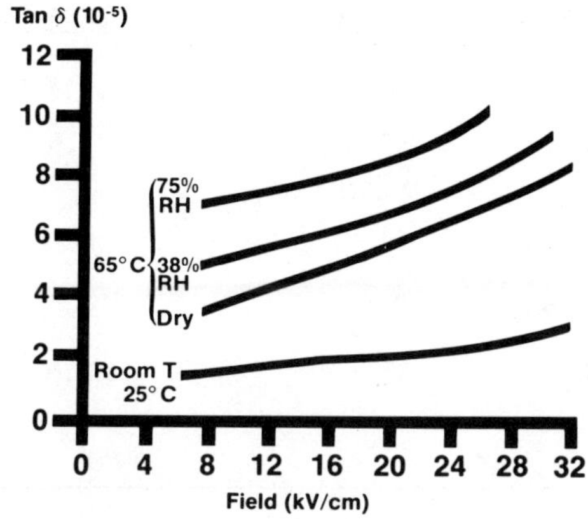

Fig. 8 The dielectric loss at high fields, at various temperatures and humidities.

Reproduced with permission from Ref. 35. Copyright 1977 IEEE.

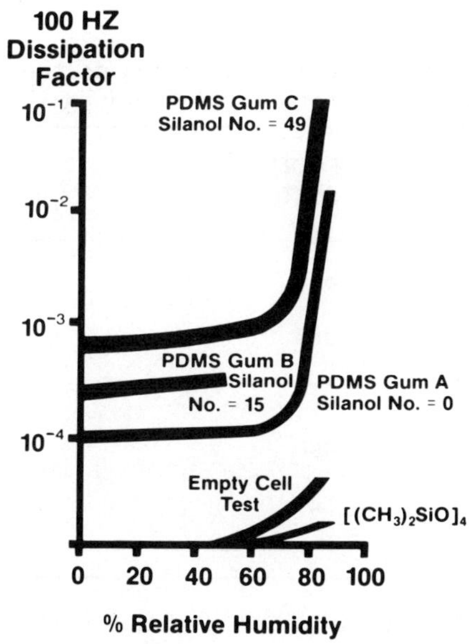

Fig. 9 The effect of humidity of the dissipation factor of polydimethylsiloxane. Silanol number is a measure of the silanol incorporated in polymer chains.

Reproduced with permission from Ref. 36. Copyright 1974 National Academy of Sciences

Table V. The Effect of Residual Silanol on the Volume Resistivity ofPolydimethylsiloxane

Reference	Metal Ion,ppm	Silanol* Number	50% RH Volume Resistivity, ohm-cm
Empty Cel	--	--	1×10^{17}
$[(CH_3)_2SiO]_4$	<0.5	<4.0 ppm Silanol	5.5×10^{16}
PDMS Gum A	<0.5	49	4.5×10^{16}
PDMS Gum C	<0.5	0	1.0×10^{15}
PDMS Gum B	60	15	2.0×10^{14}

*Silanol number is a measure of the silanol incorporated in polymer chains.

Table VI. Effect of Chemical Composition on the Dielectric Strength of Silicone Fluids of the Form $(CH_3)_3SiO(RCH_3SiO)_xSi(CH_3)_3$

SUBSTITUENT R	DIELECTRIC(A) STRENGTH VOLTS/MIL	VOLUME RESISTIVITY (OHM/CM)
CH_3	345	4.6×10^{14}
H	365	2.9×10^{14}
$CF_3(CH_2)_2-$	370	7.2×10^{12}
$CH_2 = CH_2-$	490	1.5×10^{15}
C_6H_5-	545	5.0×10^{13}

(A) Point to Plane Electrode 25 mil gap, 500 volts/sec. rate of rise.

tion being given to the processing of siloxane elastomers from aqueous dispersion. Now the discussion will be directed to the influence of reinforcement on properties, an essential requirement for most siloxane elastomers of commercial importance. A new silica has been developed that provides excellent reinforcement.

This was one of those instances where serendipity played a role. One of the authors (K.P.) was directing the development of a hydrophobic, high reinforcing silica which was achieved (38), but it was also discovered that it provided elastomers of unusually good optical properties (39). The chemistry for the synthesis (38) of the new silica filler is as follows:

Synthesis of New Silica

MeOH + NH_3 + H_2O

← Organosiloxane Hydrophobe material

← $Si(OCH_3)_4$

↓

Formation of Hydrophilic Silica Gel

↓

Formation of, MeOH Wet Hydrophobic Silica

The methanol wet, hydrophobic silica filler is ready for formulating with polymer. Mentioned above is the fact that the new silica possesses unusual characteristics relative to its ability to provide optically clear silicone compositions. A little background is in order to put this in perspective.

Optically clear filler reinforced silicone rubber was first developed during the mid-nineteen-fifties by Polmanteer and coworkers, working under support of the Air Force Materials Laboratory (40), for use as high temperature resistant interlayers for high speed aircraft windows. Since that time, other applications for optically clear reinforced silicone rubber have been identified. Some of the applications include contact lenses, medical tubing and protective gas masks. The silicone rubber for these applications is optically clear because the silicone polymer and the reinforcing silica used to prepare the rubber are isorefractive at a specified temperature, usually about 25°C. The polymers which have been used are copolymers with sufficient phenyl substitution to raise the refractive index, n_D^{25} from 1.4023 (for PDMS) to approximately 1.43 which is the refractive index of trimethylsilyl treated fumed silica (e.g., Cabosil® S-17). Such copolymers are prepared by copolymerizing $(Me_2SiO)_x$ and $(PhMeSiO)_x$ or $(Ph_2SiO)_x$ using equilibrium conditions.

It has been discovered that the new silica can be used to prepare a variety of optically clear silicone rubbers including such silicone polymers as polydimethylsiloxane η^{25}_{D} =1.4023) and poly(trifluoro-propylmethylsiloxane) η^{25}_{D} =1.38) as seen in Table VII. The term "optically clear" is intended to mean that a 2.54 millimeters thickness of the composition possesses a luminous transmittance value of at least 85% and a haze value of no greater than 4% at 23±2°C according to the procedure set out in ASTM D1003-61. The optical clarity demonstrated in both the earlier and new work may be explained in terms of the following equation for the intensity of scattered light by monodisperse particles suspended in a medium:

$$(1) \qquad I = \frac{8\pi^4 A^6 N}{X^2 \lambda^4} \cdot \frac{(m^2-1)^2}{(m^2+2)^2} (1+\cos^2\Theta) I_o$$

where
A = radius of scatterer
N = number of particles
x = distance for observation I
λ = wavelength of the light
I_o = intensity of the incident light
m = η^{25}_{D} (scattering particles) $\div \eta^{25}_{D}$ (Medium)

Referring to the above equation (1), an elastomer will be clear when either $(m^2-1)^2$ is zero (i.e., the indicies of refraction of polymer and filler are nearly equal) or when A^6/λ^4 becomes small (i.e., the aggregate or agglomerate size) become small relative to the wave length of visible light, 4000-7000 Angstroms). Qualitative results noted in Table VII, were obtained on elastomer bases pressed between pieces of glass to form laminates upon which haze and transmission measurements were made. Table VIII presents the results of some of these measurements. The haze and transmission values confirm the qualitative observations of Table VII. The new silica filled elastomer bases using polymers which were essentially PDMS (e.g., polymers A and D) gave optical properties comparable to bases prepared from polymer B and distinctly better than treated fumed silica in PDMS. It should be noted that optical properties are adversely affected by any foreign material inadvertently introduced into the elastomer base (e.g., note last two sets of data in Table VIII). It is virtually impossible to eliminate contamination of foreign material without going to clean room conditions. For the samples in Table VIII, all surfaces which came in contact with the polymer or elastomer bases were thoroughly cleaned. However, magnification did reveal numerous "lint" fibers in the bases in spite of these precautions. The technique of "sandwiching" the polymer between glass for optical measurements is less than satisfactory because the optical properties will be affected by the glass as well as the elastomer base. Percent transmission is reduced because of reflections by the different refractive index layers in the laminate. However, when curing the base by hot molding, surface distortions are transferred to the rubber if the mold surface is not perfectly smooth.

Table VII. Optical Clarity of Polymer/Filler Composition

Silica		Polymer		
Type and Treating Unit	25	Type*	25	Optical Clarity
Cabosil S-17 or MS-75, $(CH_3)_3SiO_{1/2}$	1.43	A	1.40	Milky haze
Cabosil S-17 or MS-75, $(CH_3)_3SiO_{1/2}$	1.43	B	1.43	Optically Clear
New Silica $(CH_3)_3SiO_{1/2}$ or $(CH_3)_2SiO$	1.428	A	1.40	Optically Clear
New Silica $(CH_3)_3SiO_{1/2}$ or $(CH_3)_2SiO$	1.428	B	1.43	Optically Clear
Cabosil S-17 or MS-75, $(CF_3CH_2CH_2)(CH_3)_2$-$SiO_{1/2}$ or $(CF_3CH_2CH_2)(CH_3)SiO$	1.418	C	1.38	Milky haze greater than first example
New Silica $(CF_3CH_2CH_2)(CH_3)_2$-$SiO_{1/2}$ or $(CF_3CH_2CH_2)(CH_3)SiO$	1.418	C	1.38	Optically Clear

*A: $((CH_3)_2SiO)_n$

B: $((CH_3)_2SiO)_m((C_6H_5)(CH_3)SiO)_n$ (random copolymer)

C: $((CF_3CH_2CH_2)(CH_3)SiO)_n$

Table VIII. Transmission and Haze Measurements (unvulcanized glass laminates- 100 mil thick interlayer)

Silica	Treating Unit	Filler Loading (phr)	Polymer*	% Trans.	% Haze (100 mil)
MS-75	$(CH_3)_3SiO_{1/2}$	50	D	81.9	14.9
S-17	$(CH_3)_3SiO_{1/2}$	50	D	80.1	10.2
S-17	$(CH_3)_3SiO_{1/2}$	50	D	85.4	8.6
New SiO_2	$(CH_3)_3SiO_{1/2}$	60	D	87.8	2.7
New SiO_2	$(CH_3)_2SiO$	60	D	88.2	2.0
New SiO_2	$(CH_3)_2SiO$	60	A	87.1	2.5
New SiO_2	$F_3CCH_2CH_2(CH_3)_2SiO_{1/2}$	40	C	88.0	3.3
S-17	$(CH_3)_3SiO_{1/2}$	55	B	86.0	4.1
*S-17	$(CH_3)_3SiO_{1/2}$	55	B	87.0	2.2

*(Prepared under a cleaner environment)

*A: $((CH_3)_2SiO)_n$

B: $((CH_3)_2SiO)_m((C_6H_5)(CH_3)SiO)_n$ (random copolymer)

C: $((CF_3CH_2CH_2)(CH_3)SiO)_n$

D: Similar to A plus a small % of $CH_3(CH_2CH)SiO$ incorporated

In an effort to obtain samples of cured rubber for testing optical properties, two standard stainless-steel molding plates were sand blasted. The surface of the plates was still not perfect. An elastomer base consisting of 60 phr of the new silica filler in polymer A was catalyzed with one part of Lupersol® 101 and cured by hot pressing using plates faced with smooth Teflon® film. A control using polymer B filled with treated fumed silica was catalyzed with one part Lupersol®101 and hot pressed in an identical manner. The optical propererties of these two samples were almost the same.

Composition	% Transmission	% Haze (100 mil thick)
New Silica in polymer A	93.6	3.8
Treated S-17 in polymer B	93.4	3.8

One other experimental observation demonstrates the difference between old and new technologies. When a composition such as treated fumed silica in polymer B is heated, haze develops. This haze is due to the fact that the refractive index of the polymer changes with temperature quite drastically while that for the solid filler does not. To demonstrate the change in refractive index with temperature, the refractive index, ^{T} , of a catalyst free devolatilized polymer A sample gave the following data:

Temperature, °C	η_D^T
25	1.4030 ± 0.0002
47	1.3955 ± 0.0002
65	1.3885 ± 0.0002
69	1.3872 ± 0.0002

Hence, extrapolating to 100°C, the refractive index would be about 1.375 which is a sizable change in refractive index units. Although the refractive indicies may be matched at room temperature, at elevated temperature there develops a mismatch and consequently there is an increase in haze. The new silica filler compositions show some temperature dependent haze, however not to as great an extent as compositions which depend upon matched refractive indices to achieve optical clarity. Table IX shows these differences in temperature dependent haze. For applications which require optical transparency over a range of temperatures, the new silica filled compositions offer a distinct advantage.

Summary

The thermal, oxidative, and depolymerization resistance of silicones has been shown to be significantly improved by the replacement of some (not all) of the chain oxygen atoms with oxidatively resistant organic spacers. This effectively reduced the overall or net ionic character of the alternating silicon-oxygen-silicon backbone of polydioganosiloxanes. Research was first directed toward the use of

Table IX. Temperature Dependent Haze

Qualitative Observations

Polymer*	Filler	Estimated % Haze at Room Temp.	Estimated % Haze at 100-105°C	Estimated % Haze at 150°C
D	S-17*	10	15	20
B	S-17*	2	10	15
D	New silica	2	2	7
B	New silica	2	2	7

*Treated S-17 Fumed silica

Quantitative Observations

Polymer	Filler	% Haze at room Temperature	% Haze at Approx. 100°C	%Trans-mission at room Temperature	%Trans-mission at Approx 100°C
B	Treated S-17	4.9	7.0	86.8	84.4
B	New Silica	3.1	3.1	88.4	88.2

B: $((CH_3)_2SiO)_m((C_6H_5)(CH_3)SiO)_n$ (random copolymer)

D: Similar to A plus a small % of $CH_3(CH_2CH)SiO$ incorporated.

phenylene spacers with results that did well in satisfying the objectives. Carborane cage spacers improved thermal stability in the absence of oxygen, but compared with catalyst free PDMS was actually about 50°C less oxidatively stable (20). This is believed to be due to a catalytic effect of the carborane cages on oxidation. Meta-carborane compared to para-carborane spacers gave about 33°C lower Tg values. It was also shown that fluorohydrocarbon alkyl spacers provide excellent resistance against depolymerization, even in the presence of catalyst. They also exhibit both very good solvent resistance as well as attractive thermal and oxidative properties at some sacrifice of Tg properties. . However, substituting fluorocarbon alkylether spacers instead gives improved Tg values compared with the fluorine substituted alkyl spacers with nearly as good thermal properties. The favorable alteration of properties by modifying the backbone chemical makeup of silicones is a good example of how research on polymer composition can have a sizeable impact on performance properties.

Recent research effort (28-29) made it possible to take advantage of the ease of water base processing techniques. An interesting facet of these silicone elastomer aqueous systems is the chemistry involved. The colloidal silica used as the means of reinforcement also takes a part in the crosslinking of these materials.

Silicone elastomer foam research and development (30-32) has provided a very interesting case where macrostructure rather than polymer microstructure plays a key role in properties and performance. The chemistry of silicone foams must be carefully balanced between gas generation (leading to foam formation) and crosslinking reactions so that foam integrity can be achieved as well as the desired foam density. Important is the fact that closed cells (holes) are formed. This type of foam structure affords excellent thermal and sound insulation. These foams provide excellent fire stop capability for sealing wiring and pipes passing through walls and floors in buildings.

The electrical performance of silicone elastomers and polymers is very much dependent on the influence of the polymer makeup as well as response to the electrical application environment. For example research has shown that at 50% relative humidity, dissipation factor and volume resistance is higher for polymers with higher levels of silanol groups (i.e., gums A, and C in Figure 9 and Table V), but an increased level of metal ions decreased volume resistivity (i.e., gum B in Table V). Dissipation factor increases with relative humidity and is higher for polymers having larger levels of silanol present. Dielectric strength is significantly increased by unsaturation, especially aromatic unsaturation compared with methyl group substitution in the polydiorganosiloxanes.

Changes in polymer makeup and structure have been shown to be important in altering the properties of polymers and elastomers. Reinforcing aids for silicone copolymers are often necessary to satisfy practical engineering needs. Recent research showed that it is possible to prepare reinforcing fillers that can also provide optical clarity to silicone elastomers. Analysis of the physics of light scattering has led to the conclusion that optical clarity can be enhanced through proper manipulation of the controlling parameters, indices of refraction and size of the scattering bodies. Newer filler chemistry has provided methods to make polymer filler composites which lose little, if any, optical clarity over many degrees of temperature change.

Acknowledgments

The authors would like to extend their appreciation to Dr. A. L. Logothetis for the invitation to prepare and present this paper at the Rubber Division Session of the Macromolecular Secretariat Symposium 1983, to Dow Corning Corporation for permission to write this paper, to the Fluids, Lubricants, and Elastomers Branch, Nonmetallic Materials Division, Air Force Materials Laboratory for supporting the fluorohydrocarbon alkyl spacer work in part under contract No. F33615-69-C-1301 and for permission to include this work in this paper, and to their many colleagues who carried out the described work (some of which has not been previously published outside the patent literature).

References

(1) E. L. Warrick, O. R. Pierce, K. E. Polmanteer, J. C. Saam, Rubber Chem. Technol. 52, No. 3, 437(1979).
(2) K. E. Polmanteer, Rubber Chem. Technol. 54, No. 5, 1051(1981).
(3) A. J. Barry (To Dow Corning Corp.) U. S. 2,557,931 (1951); Brit. 682,835 (1952).
(4) A. J. Barry, D. E. Hook and L. DePree (to Dow Corning Corp.) U.S. 2,511,820, (1950); Brit. 635,645 (1950).
(5) L. DePree, A. J. Barry and D. E. Hook (to Dow Corning Corp.) U.S. 2,580,159 (1951).
(6) M. Sveda (to E. I. DuPont) U. S. 2,561,429 (1951).
(7) M. Sveda (to E. I. DuPont) U.S. 2,562,000 (1951).
(8) G. C. Gainer (to Westinghouse) U. S. 2,709,692 (1955).
(9) L. W. Breed, W. J. Haggerty, Jr. and F. Baiocchi, J. Organic Chem., 25, 1633(1960)
(10) F. P. Price, J. Polymer Sci., 37, 71(1959).
(11) F. McKay, "Studies on Novel Linear and Cyclic Siloxanes", Doctoral Diss. Series, Pub. No. 19, 311(1956).
(12) R. L. Merker and M. V. Scott, J. Polymer Sci., A2, 15(1964).
(13) R. L. Merker, M. J. Scott and G. G. Haberland, J. Polymer Sct. Part A, 2,31(1964).
(14) M. Kojima and J. H. Magill, J. Polmyer Sci., Polym. Phys. Ed. 12, 317 (1974).
(15) M. Kojima and J. H. Magill, J. Macromol. Sci., Phys, B10(3), 419(1974).
(16) S. Papetti, B. B. Schaeffer, A. P. Gray and T. L. Heying, J. Polymer Sci. Part A-1, 4, 1623(1966).
(17) L. H. Sperling, S. L. Cooper, A. V. Tobolsky, J. Appl. Poly.Sci. Vol 10,1725(1966)
(18) E. J. Zaganiaris, L. H. Sperling, A. V. Tobolsky, J. Macromol. Sci.-Chem., A1 (6), 1111, Oct (1967).
(19) M. B. Roller and J. K. Gillham, J. Appl. Poly. Sci, 17, 2141(1973).
(20) M. B. Roller and J. K. Gillham, J. Appl. Poly. Sci., 17, 2623(1973).
(21) Y. K. Kim, O. R. Pierce, W. X. Bajzer and A. G. Smith, Polymer Preprints 12, (1), 482(1971).

(22) O. R. Pierce and Y. K. Kim, J. Elastoplastics, 3, 82, April (1971); Rubber Chem. Technol. 44,(3), 1350(1971).

(23) Y. K. Kim, O. R. Pierce, W. X. Bajzer and A. G. Smith, J. Fluorine Chem. 1, 203(1971/72).

(24) Y. K. Kim, O. R. Pierce, and D. B. Bourrie, J. Polymer Sci., A-1, 10, (3), 947 March (1972).

(25) O. R. Pierce and Y. K. Kim, Applied Polymer Symposium, No. 22, 103(1973).

(26) A. L. Henne and E. G. DeWitt, J. Am. Chem. Soc. 70, 1548(1948).

(27) M. O. Riley, Y. K. Kim and O. R. Pierce, J. Fluorine Chem. 10, 85(1977).

(28) J. C. Saam, D. Graiver, M. Baile, Rubber Chem. Technol., 54, 976(1981).

(29) J. C. Saam (to Dow Corning Corporation), U.S. Patent 4,244,849 (January 13, 1981).

(30) S. B. Smith (to Dow Corning Corporation), U.S. Patent, 3,923,705 (1975).

(31) C. L. Lee, R. G. Niemi and K. M. Kelly, J. Cellular Plastics, 13, 62(1977).

(32) C. L. Lee and G. M. Ronk, J. Cellular Plastics, 18, 178(1982).

(33) E. Reynaert, T. Orbeck and J. Seifferly, IEEE International Symposium on Electrical Insulation, p242,(1982).

(34) G. E. Vogel and F. O. Stark, J. Chemical and Engineering Data, 9,599(1964).

(35) R. M. Hakim, R. C. Oliver and H. St. Onge, IEEE Trans. Electr. Insul., EI-12, No. 5, Oct 1977.

(36) J. W. Swanson and K. E. Polmanteer, National Academy of Sciences, Conference on Electrical Insulation and Dielectric Phenomena, Washington, D.C., p373 (1974).

(37) R. M. Hakim, IEEE Power Engineering Society, 1974 Winter Meeting.

(38) M. A. Lutz (to Dow Corning Corp.), U.S. 4,344,800. (Aug. 17, 1982).

(39) K. E. Polmanteer and H. L. Chapman (to Dow Corning Corp.) International Publication Number: WO 81/03496 (10 Dec. 1981); U.S. Patent 4,418,165 (Nov. 29, 1983).

(40) K. E. Polmanteer, J. W. Cretzmeyer, J. W. Erwin, F. J. Campbell, T. L. Laur, J. Fenner, Contract Nos. AF 33(600)-23081,-27185,-34984 (April 15, 1953 to March 12, 1958).

RECEIVED February 6, 1984

9

Phosphazene Elastomers

Synthesis, Properties, and Applications

ROBERT E. SINGLER and GARY L. HAGNAUER

Polymer Research Division, Army Materials and Mechanics Research Center, Watertown, MA 02172

RICHARD W. SICKA

Firestone Tire and Rubber Company, Central Research Laboratories, Akron, OH 44317

The synthesis of a unique class of polymers with a phosphorus-nitrogen backbone is described, focusing on poly(dichlorophosphazene) and poly(organophosphazene) elastomers. Melt and solution polymerization techniques will be illustrated while briefly indicating the role of catalysts which give significantly improved rates of conversion and reproducibility in polymer properties. The elucidation of chain structure, molecular weight and polymer yield by various dilute solution techniques will be considered.

Poly(dichlorophosphazene) is the common, though hydrolytically sensitive, precursor of a large number of poly(organophosphazene) polymers. Examples of technologically significant elastomers are shown which are obtained through the use of selected side group substituents attached to the phosphorus-nitrogen backbone. This article concludes with a brief mention of alternate synthetic methods which may lead to useful phosphazene polymers.

The study of open-chain polyphosphazenes has drawn increasing attention during the past several years (1-4). The polyphosphazenes are highly flexible chains of alternating phosphorus-nitrogen (P-N) atoms with two substituents attached to phosphorus. Many different polyphosphazenes with a wide range of bulk properties and solubilities have been prepared by varying the type of substituent attached to the P-N backbone. Interest stems from the greater control achieved in the polymerization processes and appreciation of the technological potential of these polymers. This paper updates developments of the past several years (4) with an emphasis on the polymerization process and technological developments of the elastomers. Finally, a brief mention will be made of related polyphosphazenes which currently are attracting interest.

NOTE: This chapter is Part II in a series.

0097-6156/84/0260-0143$06.00/0

The Polymerization Process

The most facile route (Figure 1) for the preparation of the phosphazene polymers has been the ring opening polymerization of hexachlorocyclotriphosphazene (I) to give open-chain poly(dichlorophosphazene) (II), followed by direct replacement of the chlorine atoms with organonucleophiles to give hydrolytically stable polymers (IV, V, VI). It is now well recognized (3) that care must be taken in the purification of I and in monitoring of the polymerization process in order to avoid the formation of extensive branching and crosslinking (III) in favor of soluble polymer (II). The polymer must be soluble in order to replace the chlorine atoms with organo-substituents and thereby obtain a hydrolytically stable polymer. The conventional polymerization route used in the laboratory has been the high-temperature (250°C) melt polymerization of I. Unfortunately, for this type of polymerization, the time required for a suitable conversion to soluble polymer is markedly dependent on trimer purity, and minor impurities often lead to insoluble polymer. To avoid these problems a suitable catalyst system was needed. Various investigators (3,4) have shown water, Lewis acids and organometallic compounds, used both in the bulk and in solution, can serve as catalysts for the preparation of II. Some examples are illustrated in Figure 2.

Among the most promising catalyst systems are the boron halides and the boron halide-triphenylphosphate complexes studied by Fieldhouse, Graves and Fenske (5-7). Yields of II up to 83% could be obtained using the boron trichloride-triphenylphosphate complex with I in the melt at 220°C for 40 hr (6). Results from these studies (5-7) and independent studies by Horn and Kolkmann (8-10) suggest that the function of the Lewis acid catalysts is rather complex. In addition to a simple ring-opening initiation step, ligand exchange processes to form mixed phosphorus and boron compounds may be involved. Evidence includes phosphorus-31 NMR data and the observation that cyclic phosphazene esters will react with boron trichloride to give mixed substituted polyphosphazenes of low molecular weight (9).

In contrast to the use of cyclic phosphazene trimers, linear phosphonitrilic chloride polymers were reported from the reaction of open-chain phosphonitrilic chloride oligomers with ammonia or ammonium chloride (11). The polymers (II) obtained appear to have a lower MW than those described earlier, and they may be more useful for applications such as fire resistant coatings and foams which are discussed in a subsequent section of this paper.

Until recently, molecular weight and other dilute solution studies focused on the hydrolytically stable organo-substituted polymers. However, direct characterization of II can now be conducted on a routine basis using high performance liquid chromatography (12-15). The catalysts described above generally give lower molecular weights than those obtained in the uncatalyzed bulk process (Figure 2). Comparative measurements on both the chloropolymer (II) and derived products (IV, V, VI) have shown that the high molecular weight of the chloropolymer is retained during the substitution process (12), although different

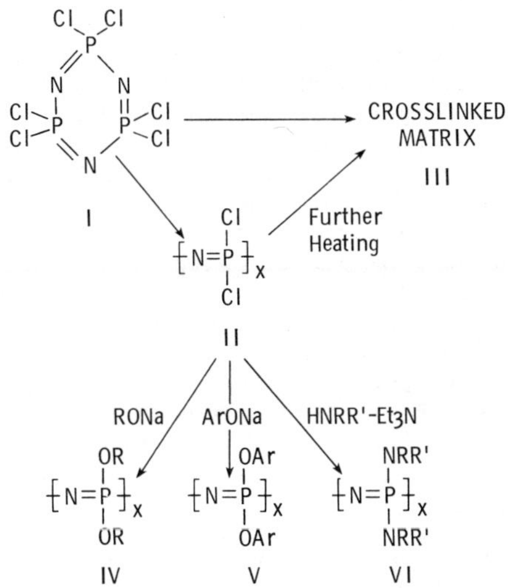

Figure 1. Synthesis of poly(dichlorophosphazene) and poly(organophosphazenes).

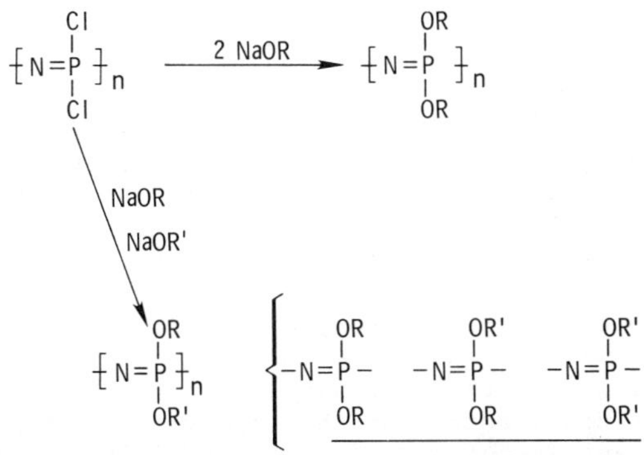

Figure 2. Contrast uncatalyzed and catalyzed $[NPCl_2]_3$ polymerization reactions. Examples are given of catalysts which are effective in producing high MW poly(dichlorophosphazene) (<u>3-5</u>, <u>7</u>).

nucleophiles can make minor modifications in the molecular weight (MW) and molecular weight distribution (MWD).

The polymerization kinetics and mechanism of the BCl_3 catalyzed polymerization of I are currently under investigation (16). Both solution and melt polmerization reactions are being studied, using size exclusion chromatography and laser light scattering techniques to monitor and characterize the products. Polymer yields close to 100% with little or no gel formation are obtained with BCl_3 at temperatures as low as 150°C. The MW and MWD of poly(dichlorophosphazene) samples obtained from the catalyzed polymerization reaction are different from those obtained for the high temperature uncatalyzed process. For example, uncatalyzed polymerization reactions run at 250°C typically yield very high MW polymer with broad MWD's; eg, $M_w = 2.4(10^6)$ and $M_n = 5.6(10^5)$ g/mol(15). By way of contrast, a BCl_3 catalyzed polymerization run at 170°C yielded a polymer with a lower MW and narrower MWD; eg, $M_w = 8.3(10^4)$ and $M_n = 2.2(10^4)$ g/mole. Preliminary studies also indicate that the BCl_3 catalyzed solution process is quite different from the catalyzed melt process (16).

Poly(organophosphazenes)

The Substitution Process. Poly(dichlorophophazene) serves as an important intermediate for the synthesis of an increasing number of poly(organophosphazenes). Property variations can be easily achieved by changing the nature and size of the substituent attached to the P-N backbone. The poly(alkoxy- and aryloxyphosphazenes) (IV,V) are elastomers or fiber forming semicrystalline thermoplastics, whereas the poly(aminophosphazenes) (VI) generally are amorphous, glassy materials. Initial work involved the use of simple nucleophiles during synthesis in order to demonstrate that hydrolytically stable polymers could be prepared (17,18). Subsequent workers have shown that using two or more nucleophiles during the substitution process could have a profound effect on the properties. This synthetic modification is outlined in Figure 3. If OR is OCH_2CF_3, a semicrystalline thermoplastic $[NP(OCH_2CF_3)_2]_n$ is obtained, whereas if OR = OCH_2CF_3 and OR' = $OCH_2C_3F_7$, the result is an amorphous "hybrid copolymer" (19).

If the nucleophiles used in the substitution process are comparable in reactivity, then a random distribution of substituents would be expected for the copolymers. When the nucleophiles have sufficiently different reactivity, such as with certain meta and para substituted phenoxides, then the preference for unsymmetrical, -NP(OR)(OR')-, rather than symmetrical, -NP(OR)(OR)-, substitution may occur under certain reaction conditions. Differences between nominally similar copolymers have been observed with differential scanning calorimetry (20), but, to date, spectroscopic measurements such as phosphorus-31 NMR have been unable to distinguish between possible copolymer structures.

Phosphazene Fluoroelastomers. The work of Rose (19) along with subsequent efforts at several laboratories has led to the commer-

$[NPCl_2]_3$ ⟶ $[NPCl_2]_n$

<u>BULK POLYMERIZATION</u>

250^oC, in vacuo, 40-100 hr

Soluble polymer $\leq$ 50% yield

$1 \leq [\eta] \leq 3$ dl/g

$M_n \sim 10^5$, $M_w \sim 10^6$, $M_w/M_n \sim 10$

<u>BULK POLYMERIZATION - CATALYSTS</u>

H_2O, BCl_3, $AlCl_3$, $Et_3Al_2Cl_3$

$CrCl_3 \cdot 6H_2O$

$(C_6H_5O)_3PO\text{-}BCl_3$

Lower Polymerization Temperatures
Higher Rates of Conversion
Improved Molecular Weight Control

US Patents: 3937790, 4123503, 4137330, 4226840

<u>CATALYZED SOLUTION POLYMERIZATION</u>

$H_3PO_4\text{-}P_2O_5$, $(CF_3SO_3Hg)_2O$ in $Cl_3C_6H_4$

Sulfur in Decalin/Tetralin

Atmospheric Pressure, Lower Temperatures, Inert Atmosphere

BCl_3, $(C_6H_5O)_3PO\text{-}BCl_3$ in Cyclohexane

Inert Solvents, Improved Process Control

US Patents: 4005171, 4139598, 4225567, 4327064

Figure 3. Contrasting synthesis of homopolymers and copolymers with possible copolymer structures.

cialization of the phosphazene fluoroelastomers by Firestone Tire and Rubber Company under the trademark PNF (1,4). The elastomer is a "terpolymer" (Figure 4) with a mixture of alkoxide substituents attached to the P-N backbone. Approximately 65% of the substituents are trifluoroalkoxide groups, and the remaining 35% are derived from a mixture of teleomer fluoroalkoxides. A small amount (ca. 1%) of a reactive pendant group is also added during the substitution step to provide more reactive curing sites for crosslinking and rubber processing. Extremely high molecular weights (ca. 10^6) and broad molecular weight distributions are distinctive features of these elastomers. The glass transition temperature is -68^oC (1).

The phosphazene fluoroelastomers can be compounded and processed to give an excellent balance of properties (Table I) which make them suitable for a variety of applications which require the combination of fuel and oil resistance with a wide service temperature range (4). Initial development work focused on demanding applications in aerospace, military, petrochemical and gas pipeline areas. The phosphazene fluoroelastomers are currently being used or evaluated in both military and commercial applications, and their continuing demand in these areas is now established. Some examples are shown in Figure 5 (21).

More recently there has been a surge of interest in PNF for biomedical applications. The work at the Gulf South Research Institute (GSRI) is quite promising for a practical dental application of the PNF materials (22,23). Examples of PNF soft denture liners are shown in Figure 6. PNF was solution blended with methyl methacrylate (MMA) monomer along with $BaSO_4$ and CdSSe pigment. Ethylene glycol dimethacrylate (EGDMA) crosslinker completed the MMA interpenetrating polymer network. The GSRI-PNF soft putty was cured at 100^oC using lauroyl peroxide as a crosslinking agent with MgO acid scavanger. This mixture was cured directly onto a hard poly(methylmethacrylate) PMMA baseplate. The mechanical properties of this composite, including peel bond strength, were reported to be quite suitable for the application. An SEM photomicrograph (Figure 7) of the peel test specimen indicates there is an excellent bond of the elastomer to the PMMA baseplate beads. Since the PNF and MMA are reacting simultaneously, an attempt was made to identify domains of either polymer at high magnification using SEM with an EDAX attachment. No concentrations of phosphorus or fluorine were found at magnifications up to 100.000X. Thus the extent of homogeniety in this polymer blend is unknown, but it may be high.

Toxicity screening indicated that these PNF soft denture compounds have very low acute toxicity compared to some currently used denture liner materials. Clinical trials are planned for the near future (23).

Another challenge in the biomedical materials area is the search for synthetic materials with improved blood compatibility for artificial heart devices and other organs. An early study by Wade (24) using a series of poly(organophosphazenes) showed these polymers in the unfilled state are as biocompatible as silicon materials. More recent blood compatibility studies using radiation crosslinked PNF showed excellent hemo compatibility

PNF^{R} GUM RUBBER

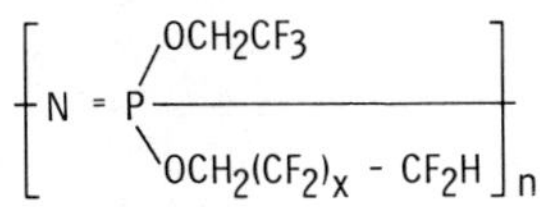

x = 1, 3, 5, 7, . . .

SPECIFIC GRAVITY = 1.75

SOLUBILITY - KETONES, ESTERS, THF, ALCOHOLS

T_g = -68°C

Figure 4. Chemical structure and properties of phosphazene fluoroelastomer, PNF gum rubber.

Table I

PHOSPHAZENE FLUOROELASTOMER

COMPOUNDING FORMULATIONS		VULCANIZATE PROPERTIES	
POLYMER	100	LOW TEMPERATURE FLEXIBILITY	
SILICA, CARBON BLACK, CLAY	30-60	BRITTLE POINT*	-65°C
		TR - 10†	-55°C
		PETROLEUM RESISTANT	
SILICONE GUM, FLUOROSILICONE	2-10	GOOD THERMAL STABILITY (LONG TERM)	175°C
SILANE COUPLING AGENT	2	SHORE A HARDNESS	35-90
MgO	2-10	COMPRESSION SET (%)‡ 70 hr AT 149°C	10-50
$(8\text{-}HQ)_2Zn$ (STABILIZER)**	2		
PEROXIDE OR SULFUR	1-3	TENSILE STRENGTH, MPa	6.9-17
		psi	1000-2500
PRESS CURE 20-30 min AT	170°C	ELONGATION, %	75-250
POSTCURE (OPTIONAL) 4 hr AT	175°C	100% MODULUS, MPa	2-10
		psi	290-1450

*ASTM D-746 †ASTM D-1329 ‡ASTM D-395 **ZINC 8-QUINOLINOLATE

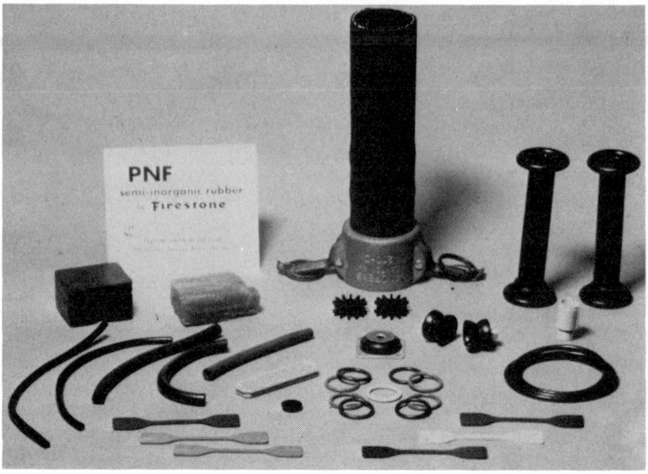

Figure 5. Phosphazene fluoroelastomer (PNF) compounds were used to prepare molded rubber parts for diverse applications. Shown here include O-rings fuel hoses, shock absorption and vibration damping mounts. Photograph courtesy of the Firestone Tire and Rubber Co.

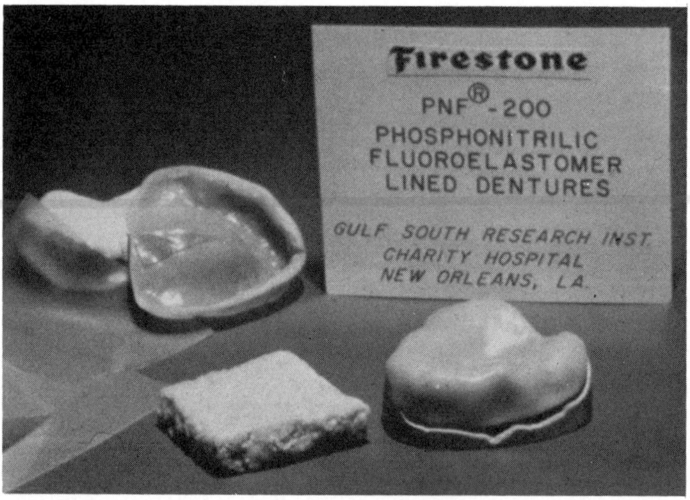

Figure 6. Phosphazene fluoroelastomer (PNF) gum stock and soft denture liners which were prepared at Gulf South Research Institute. Photograph courtesy of Firestone Tire and Rubber Co.

compared to currently used materials (25,26). Elastomer weight gains due to in vitro lipid sorption were measured for PNF candidates and several commercially available materials. The PNF materials were shown to be quite resistant to lipid absorption and can potentially serve as an effective barrier coating for soft implants (27).

Fire Resistant Materials. Much of the current impetus for polyphosphazene development stems from the need for new polymers with improved thermal stability and flame resistance combined with low smoke evolution and toxicity. Additives containing elements such as phosphorus, nitrogen and halogens are known to enhance fire resistance of organic polymers, but this often is accomplished at the expense of increased smoke evolution and formation of toxic combustion products in a fire situation. Compounded polyphosphazenes offer excellent potential for fire resistance in part because phosphorus, nitrogen and halogen are part of the polymer, which tends to reduce the evolution of smoke and toxic gases upon combustion. For example, the poly(fluoroalkoxyphosphazenes) can break down to give relatively stable products (28), and do not necessarily evolve large amounts of hydrogen fluoride during the degradation process (29).

It has also been demonstrated that polyphosphazenes can be prepared which offer fire resistance without incorporating halogen in the side chain (30-33). The halogen free poly(aryloxyphosphazenes) elastomers (APN) show excellent potential for applications such as flexible foams, coatings, and wire coverings (Figure 8). The feasibility of using the APN closed cell foams as fire retardant thermal insulation has been demonstrated by a Department of the Navy National Bureau of Standards Test Program (34). Further development and evaluation of the aryloxyphosphazene elastomers for applications such as shown in Figure 8 is likely to continue.

Alternate Synthesis Approaches Phosphazene

Thus far, the survey of phosphazene elastomers has been based on the formation and modification of poly(dichlorophosphazene). Although a large variety of polymers can be prepared by this approach, there are limitations in the preparation of polyphosphazenes with phosphorus-carbon bonds. The reaction of poly(dichlorophosphazene) with organometallic agents, such as RMgX or RLi, results mainly in decomposition and not the desired polymers $[NPR_2]_n$. There are three possible approaches to the preparation of polyphosphazenes with phosphorus-carbon bonds: polymerization of substituted trimers, poly(difluorophosphazene), and thermolysis of small linear molecules. These three approaches will be discussed in turn.

Polymerization of Organosubstituted Cyclophosphazenes. Whereas hexachlorocyclotriphosphazene (I) reacts to give open chain high polymer in good yield, the fully organosubstituted derivatives generally do not. There are intermediate cases such as mono- and diorganosubstituted cyclotriphosphazenes (trimers) which can form high MW polymers (38). One noteable example (39) is the

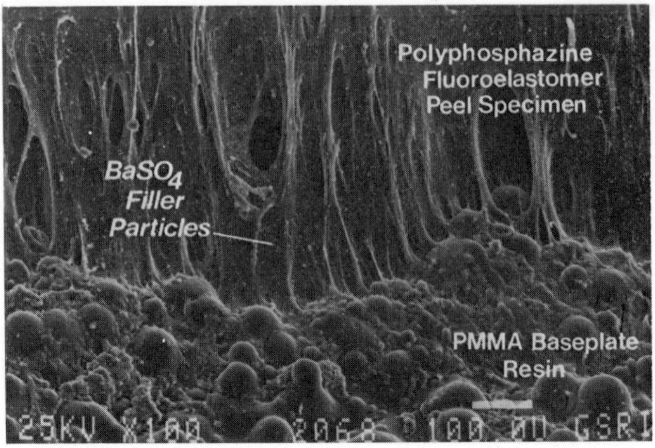

Figure 7. Scanning electron microscopy photo micrograph showing the boundary layer between the GSRI-PNF compound and a hard PMMA baseplate. Note the absence of a polymer domain structure in the PNF compound Photograph courtesy of Gulf South Research Institute (23).

Figure 8. Aryloxyphosphazene elastomers (APN) offer excellent potential for applications such as : A, closed cell foam thermal insulation with high fire retadancy and low smoke generation (32,34); B, pigmented APN coatings in aluminum substrates with low flammability, low flame spread, and low smoke (35); C, APN insulation and cable jacketing (36); D, open cell APN comfort cushioning (37). Photograph courtesy of Firestone Tire and Rubber Company (4).

polymerization of monomethylpentachlorotrimer (VII), which after chlorine replacement gives a hydrolytically stable elastomer (VIII) with a $T_g = -50^oC$. The estimated molecular weight was $1.5(10^6)$. The presence of the methyl group is sufficient to disrupt crystallinity which is found in the related homopolymer $[NP(OCH_2CF_3)]_n$.

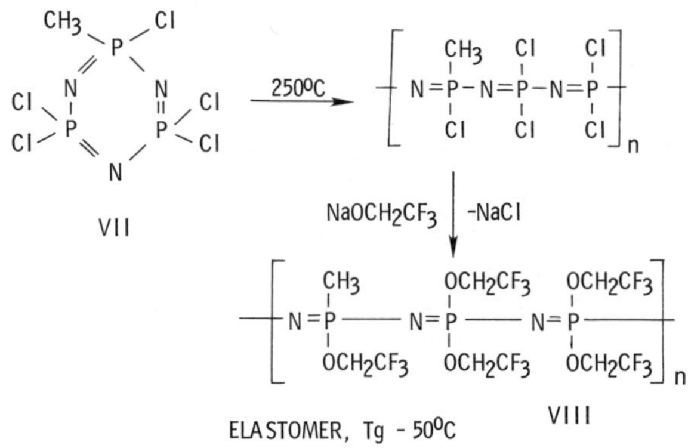

The reason that the organo-substituted trimers do not readily undergo a ring opening polymerization has been explained on the basis of thermodynamic considerations (38). However, in the presence of Lewis acids such as $AlCl_3$ or BCl_3, the formation of low to medium MW polyphosphazenes from organo-substituted trimers has been reported (9,40). Apparently, ligand exchange reactions occur to form the more readily polymerizable chlorotrimers.

Poly(difluorophosphazene). Hexafluorocyclotriphosphazene will polymerize in the bulk state at 350^oC to give poly(difluoro phosphazene). It is an elastomer with a glass transition temperature of -96^oC and a crystalline melting temperature of -68^oC (41). Poly(difluorophosphazene) can be prepared in the uncrosslinked state and will undergo substitution reactions with amines, alkoxides, and aryloxides (42). More important is that in contrast to poly(dichlorophosphazene), poly(difluorophosphazene) will react with organometallic agents to prepare poly(organophosphazenes) with phosphorus-carbon bonds such as poly(diphenylphosphazene) (43).

N-Silylphosphinimines. A third approach, which has been investigated by Neilson (44,45), involves the synthesis of suitably constructed N-silyphosphinimines. Upon heating, compound IX yields poly(dimethylphosphazene) with a moderate molecular weight. Other N-silylphosphinimines have been synthesized, and these may eventually yield new phosphazene elastomers with novel properties (45).

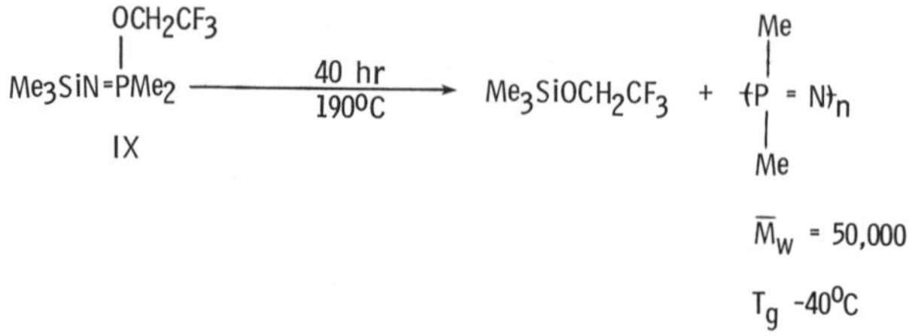

Related Phosphazenes

Although they are not elastomers, other phosphazenes also offer potential in areas requiring heat and fire resistance. Mixed fluoroalkoxy - aryloxy substituted cyclic phosphazenes are currently attracting interest for fire resistant fluid applications (46). In other studies, cyclophosphazene matrix resins have been prepared which have potential for high temperature adhesive applications (47) and as composite matrix materials (48). These studies serve to further demonstrate the enhanced thermal stability and fire resistance which can be achieved with these phosphorus - nitrogen systems.

Future Trends

This paper has reviewed the most recent developments of phosphazene elastomers and touched briefly in some phosphazenes of related interest. At the time of this writing, Firestone was the supplier of the PNF and APN elastomers. Ethyl Corporation recently licensed the phosphazene technology from Firestone (49), and thus the continued availability of these elastomers is assured for further commercial development.

The examples shown in this paper represent only a fraction of the known polymers based on the phosphazene (P-N) backbone. Sulfamic acid catalysts hold some promise for facilitating the polymerization process (50) and phase transfer catalysts can aid in the substitution step (51). Poly(organophosphazenes) having bioactive sites and organometallic substrates have been prepared and are under investigation. Amino acid esters (52), steroids (53), and biologically active amines such as benzocaine (54) and dopamine (55) have been attached to the phosphazene chain. Water soluble polymers that bear glucose side groups have also been synthesized (56). Poly(organophosphazenes) containing reactive side groups have been used to bind transition metal units (57,58) to form polymers which are potentially useful for hydrogenation catalysts and other applications.

The large number of different pendant groups with widely varied chemical functionality which can be attached to the P-N backbone demonstrate the unusual molecular design potential of this class of polymers. Undoubtedly, some of these will hold promise for future research and development.

Acknowledgments

The authors wish to acknowledge the helpful discussions and assistance provided by Dr. Fred Lohr, Firestone, and Dr. Lawrence Gettleman, Gulf South Research Institute, during the preparation of this manuscript.

Literature Cited

1. Tate, D. P. and Antowiak, F. A. *Kirk-Othmer Encycl. Chem. Technol. 3rd Ed.* 1980, *10*, 939.
2. Allcock, H. R. *Makromol. Chem. Suppl.* 1981, *4*, 3.
3. Hagnauer, G. L. *J. Macromol. Sci.-Chem.* 1981, *A16*, 385.
4. Singler, R. E.; Hagnauer, G. L.; Sicka, R. W. In "Elastomers and Rubber Elasticity"; Mark, J. E.; Lal, J., Eds.; ACS SYMPOSIUM SERIES No. 193, American Chemical Society: Washington, DC, 1982; p. 229.
5. Fieldhouse, J. W.; Graves, D. F., U.S. Patent 4226840, 1981. In "Phosphorus Chemistry; Quin, L. D.; Verkade, J. G., Eds.; No. 171, AMERICAN CHEMICAL SOCIETY: Washington, DC 1981; p. 315
6. Fieldhouse, J. W.; Graves, D. F. In "Phosphorus Chemistry:; Quinn, L. D.; Verkade, J. G., Eds.; ACS SYMPOSIUM SERIES, 1981, No. 171, American Chemical Society: Washington, DC, 1981; p. 315.
7. Fieldhouse, J. W.; Fenske, S. L., U.S. Patent 4327064, 1982.
8. Horn, H. G.; Kolkmann, F. *Makromol. Chem.* 1982, *183*, 1833.
9. Horn, H. G.; Kolkmann, F. *Makromol. Chem.* 1982, *183*, 1843.
10. Horn, H. G.; Kolkmann, F. *Makromol. Chem.* 1982, *183*, 2427.
11. Li, H. M., U.S. Patent 4374815, 1983.
12. Hagnauer, G. L. and Singler, R. E. *Coat. Plast. Chem. Pap.* 1979, *41*, 88.
13. Hagnauer, G. L. In "Size Exclusion Chromatography"; Provder, T., Ed.; ACS SYMPOSIUM SERIES No. 138, American Chemical Society, Washington, DC, 1980; p. 239.
14. Adams, H. E.; Valaitis, J. K.; Henderson, C. W.; Strauss, E. J. In "Size Exclusion Chromatography (GPC)"; Provder, T., Ed.; ACS SYMPOSIUM SERIES No. 138, American Chemical Society: Washghinton, DC, 1980; p. 255.
15. Hagnauer, G. L. and Koulouris, T. N. In "Liquid Chromatography of Polymers and Related Materials - III" Jack Cases, Ed.; Marcel Dekker Inc.: New York, 1981; p. 99.
16. Sennett, M. S.; Hagnauer, G. L.; Singler, R. E.; Davies, G. *Polym. Mat. Sci. Eng. Proc.* 1983, *49*, 297.
17. Allcock, H. R.; Kugel, R. L.; Valan, K. J. *Inorg. Chem.* 1966, *5*, 1709.
18. Allcock, H. R.; Kugel, R. L. *Inorg. Chem.* 1966, *5*, 1716.
19. Rose, S. H. *J. Polym. Sci. B*, 1968, *6*, 837.
20. Beres, J. J.; Schneider, N. S.; Desper, C. R.; Singler, R. E. *Macromolecules*, 1979, *12*, 566.
21. Lohr, D. F. and Beckman J. A. *Am. Chem. Soc. Rubber Division Meeting,* Cleveland, OH, Oct 1981, Paper No. 34; *Rubber Chem. Technol.* 1982, *55*, 271.
22. Gettleman, L.; Farris, C. L.; LeBoeuf, R. J.; Rawls, H. R.;

Dillingham, E. O. Biomaterials, Medical Devices, and Artificial Organs 1981, 9, 345.
23. Gettleman, L.; LeBoeuf, R.; Rawls, H. R. Meeting of the American Association for Dental Research, Cincinnati, OH., March 1983, Abstr. No. 725.
24. Wade, C. W. R.; Gourlay, S.; Rice, R.; Hegyeli, A.; Singler, R.; White, J. In "Organometallic Polymers"; Carraher, C. E.; Sheats, J.; Pittman, C. U., Eds.; Academic Press, New York, 1981.; p. 289.
25. Reichert, W. M.; Filisko, F. E.; Barenberg, S. A. J. Biomed. Materials Res. 1982, 16, 301.
26. Reichert, W. M.; Filisko, F. E.; Barenberg, S. A. In "Biomaterials: Interfacial Phenomena and Applications"; Cooper, S. L.; Peppas, N. A., Eds.; ADVANCES IN CHEMISTRY No. 199, American Chemical Society: Washington, DC, 1982; p. 177.
27. Owen, D. R.; Zone, R.; Armer, T.; Kilpatick, C.; In "Biomaterials: Interfacial Phenomena and Applications"; Cooper, S. L.; Peppas, N. A., Eds.; ADVANCES IN CHEMISTRY No. 199, American Chemical Society: Washington, DC, 1982; p. 395.
28. Valaitis, J. K.; Kyker, G. S. J. Appl. Polym. Sci. 1979, 23, 765.
29. Knight, G. J.; Wright, W. W. Polym. Deg. Stab. 1982, 4, 465.
30. Quinn, E. J. and Dieck, R. L. J. Cell. Plast. 1977, 13, 96.
31. Lawson, D. F. and Cheng, T. C. Fire Res. 1978, 1, 223.
32. Thompson, J. E. and Reynard, K. A. J. Appl. Polym. Sci. 1977, 21, 2575.
33. Lieu, P. J.; Magill, J. H.; Alarie, Y. C. J. Combustion Technol. 1981, 8, 242.
34. Widenor, W. M. "Model Fire Tests on Polyphosphazene Rubber and Polyvinylchloride (PVC)/Nitrile Rubber Foams" David W. Taylor, Naval Ship R&D Center, Annapolis, MD 1978, NASA TM-78523, N79-12029; Widenor, W. M. "Polyphosphazene Foam Thermal Insulation" 1981, DTNSRDC TM-28-81-324.
35. Chattopadhyay, A. K.; Hinrichs, R. L.; Rose, S. H. J. Coat. Technol. 1979, 51, 87.
36. Peterson, T. C. "Polyphosphazene Wire and Cable Insulation," Defense Technical Information Center, Cameron Station, Alexandria, VA, 1982; AD A091410.
37. Gerber, A. H. and Peterson, T. C. "Preparation and Testing of Polyaryloxyphosphazene Open Cell Foams," Horizons Research Inc., Cleveland, Ohio. July 1979, David Taylor Naval Ship R&D Center, Contract N0060-78-C-0253.
38. Allcock, H. R.; Polymer, 1980, 21, 673.
39. Allcock, H. R.; Ritchie, R. J.; Harris, P. J. Macromolecules 1980, 13, 1338.
40. Rieck, H. P.; Wegel, S., European Pat. Appl.; Chem. Abs. 1982, 97 216475.
41. Allcock, H. R., Kugel, R. L.; Stroh, E. G. Inorg. Chem. 1972, 11, 1120.
42. Allcock, H. R., Patterson, D. B.; Evans, T. L. Macromolecules 1979, 12, 172.

43. Allcock, H. R., Evans, R. L.; Patterson, D. B. Macromolecules, 1980, 13, 201.
44. Neilson, R. H. and Wisian-Neilson, P. J. Am. Chem. Soc. 1980 102, 2848.
45. Neilson, R. H. and Wisian-Neilson, P. J. Macromol. Sci.-Chem. 1981, A16, 425.
46. Singler, R. E.; Koulouris, T. K.; Deome, A. J.; Lee, H.; Dunn, D. A.; Kane, P. J.; Bieberich, M. J. "Synthesis and Evaluation of Phosphazene Fire Resistant Fluids," Defense Technical Information Center, Cameron Station, Alexandria, VA, 1982, AD-A117298.
47. Devadoss, E. J. Appl. Polym. Sci. 1983, 28, 921.
48. Kumar D.; Fohlen, G. M.; Parker, J. A. Macromolecules 1983, 16, 1250.
49. Rubber World 1983, 188, 8.
50. Sinclair, D. P., U.S. Patent 4,242,316, 1980.
51. Austin, P. E.; Riding, G. H.; Allcock, H. R. Macromolecules 1983, 16, 719.
52. Allcock, H. R.; Fuller, T. J.; Mack, D. P.; Matsumura, K.; Smeltz, K. M. Macromolecules 1977, 10, 824.
53. Allcock, H. R.; Fuller, T. J. Macromolecules 1980, 13, 1338.
54. Allcock, H. R.; Austin, P. E.; Neenan, T. X. Macromolecules 1982, 15, 689.
55. Allcock, H. R.; Austin, P. E. Macromolecules 1981, 14, 1616.
56. Allcock, H. R., Scopelianos, A. G. Macromolecules 1983, 16, 715.
57. Allcock, H. R.; Scopelianos, A. G.; Whittle, R. R.; Tollefson, N. M. J. Am. Chem. Soc. 1983, 105, 1316.
58. Allcock, H. R.; Lavin, K. D.; Tollefson, N. M.; Evans, T. L. Organometallics 1983, 2, 267.

RECEIVED February 6, 1984

10

Curing of Vinylidene Fluoride Based Fluoroelastomers

W. W. SCHMIEGEL and A. L. LOGOTHETIS

E. I. du Pont de Nemours and Company, Experimental Station, Wilmington, DE 19898

Vinylidene fluoride based fluoroelastomers have remarkable resistance to flames, chemicals, solvents, heat and oxidation. These fluorolastomers can be cured either by bis-nucleophiles such as bis-phenols or diamines or by peroxides. The bis-phenol curable stocks provide excellent processability and compression set resistance. In addition to the bis-phenol, an accelerator such as a quaternary phosphonium or ammonium salt and a metal oxide or hydroxide such as $Ca(OH)_2$ or MgO are required. The mechanism involves base attack on the polymer backbone to generate olifinic unsaturation through HF elimination, followed by nucleophilic attack on these double bonds. Compared to nucleophilic cures, peroxide cures provide advantages in acid and steam resistance but require a cure site on the polymer such as a bromine or iodine atom that is susceptible to radical attack. In addition to an organic peroxide, a radical trap, such as triallylisocyanurate or cyanurante, and an acid acceptor such as a metal oxide are needed. The mechanism of curing involves peroxide decomposition to generate polymeric radicals to the trifunctional radical trap to form a crosslinked network.

Most commercially available fluoroelastomers consist of copolymers of vinylidene fluoride (VF_2) with hexafluoropropylene (HFP), and, optionally, tetrafluoroethylene (TFE)(1,2): these highly fluorinated polymers of 62-70 wt % fluorine have remarkable resistance to flame, chemicals, solvents and oxidative attack. This stability has been attributed to the strength of the carbon-fluorine bond compared to that of the carbon-hydrogen bond, to steric hindrance due to the presence of fluorine, and to strong van der Waals forces between hydrogen and fluorine atoms present in the macromolecules.

0097-6156/84/0260-0159$07.00/0

Fluoroelastomers are commercially very important, particularly in the manufacture of sealing elements such as o-rings, shaft seals, valve stem seals, fuel hoses, and flue duct expansion joints for power generating plants. Fluoroelastomer products are used in the aircraft, automotive and chemical industries and in power plants and oil exploration. They provide the high performance required by these industries that is not available from the standard rubber compounds (Table 1).

The polymers are prepared in aqueous emulsions at 100-120°C and 5-7 MPa pressure with a free radical initiator. The polymers are prepared at high monomer conversions and display normal solution viscosity behavior. Vinylidene fluoride is much more reactive than hexafluoropropylene, and, under the polymerization conditions, no (HFP)n blocks are formed, whereas $(VF_2)n$ sequences are present. The main polymer structure, therefore, is represented by the following blocks:

$$-(CH_2CF_2)_a-(CF_2\overset{CF_3}{\overset{|}{C}}FCH_2CF_2)_b-(CH_2F_2)_c-(CF_2CF_2CH_2CF_2)_d-$$

Although tetrafluoroethylene sequences can be formed, the monomer feed is adjusted to minimize such sequences, which might lead to crystallinity. Solution viscosity data, ^{19}F NMR, spectra, and radioassay of polymers made with labelled initiators and chain transfer agents indicate that these polymers are predominantly linear.

Like all thermosetting elastomers, in order to make useful products, these elastomers have to be cured. They can be cured by nucleophiles such as diamines (3) or bisphenols (4,5) or by organic peroxides (6). The original cure system was based on diamines and was introduced in the late 1950°s along with the introduction of the VF_2 based elastomers. Its mechanism has been discussed in length by previous investigators (3). The importance of diamine curing has diminished with the discovery of the bis-phenol cures in the late 1960°s and their commercialization in 1970. Bis-phenol cure systems are presently predominant. Peroxide curing of VF_2-based fluoroelastomers was introduced in the late 1970°s and is slowly gaining commercial acceptance, particularly in applications that require exceptional acid and steam resistance. These two cure systems provide versatility in processing the elastomeric compounds and in designing products of high performance.

This paper discusses the chemical reactions involved in the curing of fluoroelastomers with bis-phenols and with peroxides. The mechanism of curing with peroxides is based on our work where bromine was introduced in the polymer by copolymerization. The experimental procedures and materials have been described in detail in previous publications (5,6c).

Results and Discussion

Curing of Vinylidene Fluoride Based Fluoroelastomers with Bis-Phenols. We have investigated the reactions of basic nucleophiles with vinylidene fluoride (VF_2) copolymers in solution and have determined some reactivity principles which can be used to interpret the vulcanization behavior of related fluoroelastomers.

It is the purpose of this section to consider only the reactions of normal VF_2 copolymers toward basic nucleophiles in solution and to develop some structure-reactivity generalizations which can be compared to experience with practical vulcanization.

In general, the results of dehydrofluorination and gelation of dimethylacetamide (DMAC) solutions of the VF copolymers were evaluated by ^{19}F NMR spectroscopy at 94.1 MHz.

VHF₂/HFP Reactivity. A standard system for VF_2/HFP copolymers intended for o-ring use and, therefore, optimized for maximum resistance to high temperature compression set consists of the ingredients shown in Figure 1, which also depicts the progress of a 177° cure as measured by a Monsanto Rheometer. All of the curatives are solids, and only the bis-phenol has a melting point (165°) which is below the cure temperature. The heterogeneous nature of the curing system confers extreme scorch safety upon such compounds: Mooney scorch curves at 121° show no rise within 30 minutes. The ODR response is characterized by a distinct induction period whose length can be regulated easily by maniuplation of the levels of inorganic bases or benzyltriphenylphosphonium chloride (BTPPC) if a fairly constant cure state is desired, or by the level of the bisphenol, if sufficient freedom exists with respect to cure state. High Bis-phenol-AF (Bp-AF) levels markedly increase the length of the induction period and lead to high cure states. The formation of crosslinks, once the process begins, is extremely rapid. In the compound shown, over 75% of the crosslinks are formed within one minute. Maximum cure state is reached rapidly and without reversion. After the press cure, to obtain optimum resistance to high temperature compression set, the vulcanizate is subjected to an air oven cure between 200 and 260°C for 8 to 24 hours.

When BTPPC is omitted from the standard recipe, no cure occurs within one hour at 177° as judged from ODR response. Figure 2 shows the dependence of the ODR state of Bp-AF concentration in the presence of the standard concentrations of $Ca(OH)_2$, MgO, BTPPC, and carbon black. The slightly curving line has a continuously decreasing slope from 0.5 to 4.0 phr Bp-AF but generally indicates a linear relationship between Bp-AF level and ODR cure state. Swollen compression modulus studies also show that crosslink density varies linearly with bis-phenol AF concentration (11a).

We have shown previously (5) that dehydrofluorination of poly(VF_2/HFP) by a hydroxylic base or an amine results in highly specific structural changes. Figure 3 shows the ^{19}F NMR spectrum of a 3.5/1.0 VF_2/HFP polymer in DMAC solution.

TABLE 1

IMPORTANT FLUOROELASTOMERS BASED ON VINYLIDENE FLUORIDE

FLUOROELASTOMER	WT % HYDROGEN	PROPERTIES	USES
VF_2/HFP	~2.0	COMPRESSION SET RESISTANCE	O-RINGS/SHAFT SEALS
VF_2/HFP/TFE	~1.5	OIL RESISTANCE	SHAFT SEALS/EXPANSION JOINTS
VF_2/HFP/TFE/X	~1.1	STEAM AND ACID RESISTANCE	CYLINDER LINER SEALS
VF_2/PMVE/TFE/X	~2.0	LOW TEMPERATURE	SEALING APPLICATIONS

VF_2 = VINYLIDENE FLUORIDE
HFP = HEXAFLUOROPROPYLENE
TFE = TETRAFLUOROETHYLENE
X = CURE SITE MONOMER

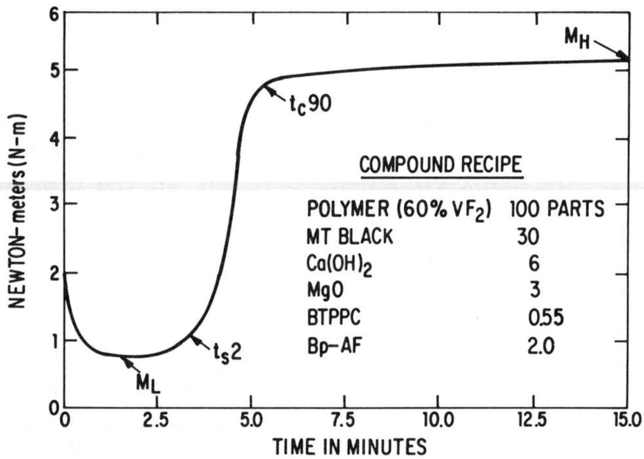

Figure 1. Cure response by oscillating disc rheometry at 177 °C of a compound optimized for use in o-rings.

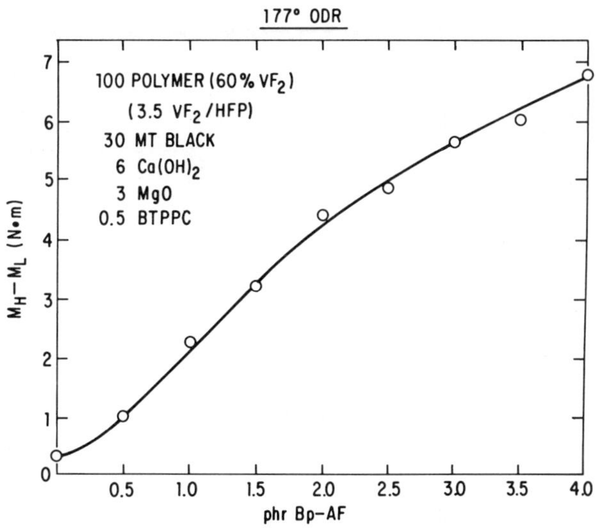

Figure 2. Dependence of cure state (M_H-M_L) on bis-phenol AF concentration at fixed benzyltriphenylphosphonium chloride (BTPPC) concentration.

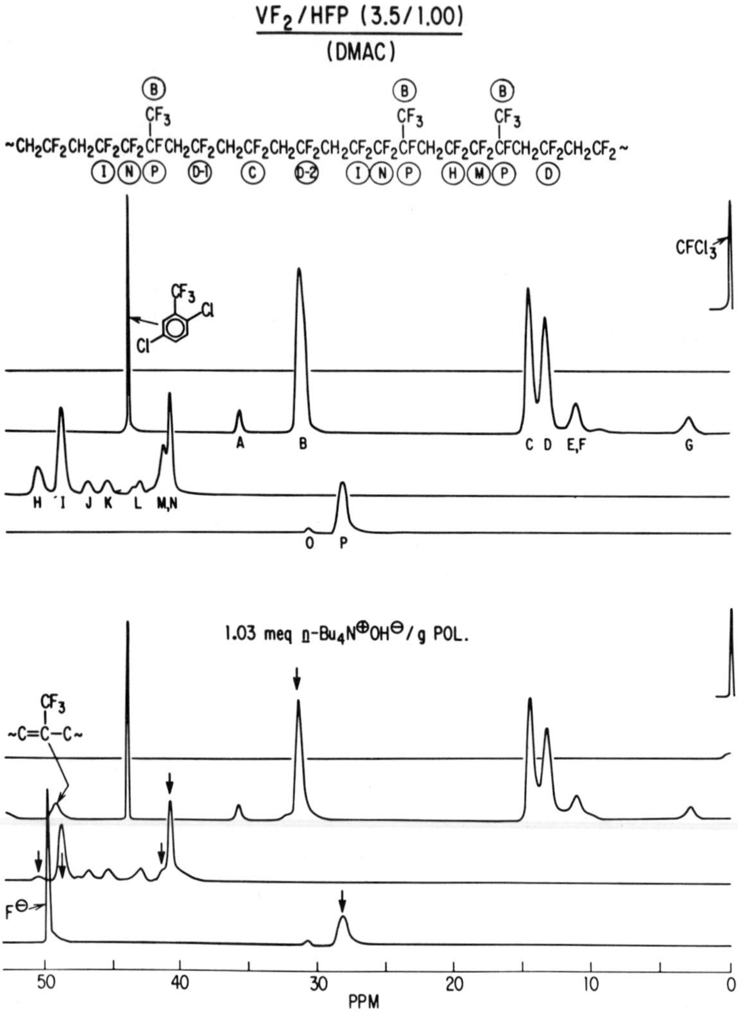

Figure 3. 94.1 MHz ^{19}F NMR spectra of VF_2/HFP polymer before (top) and after (bottom) treatment with hydroxylic base in DMAC solution at 20 °C (2,5-dichlorobenzotrifluoride internal standard). Arrows indicate changes in peak intensities.

Structural assignments (13) of the important resonances of a 3.5 VF_2/HFP polymer are shown in the upper part of Figure 3. The small resonances A, G, and O are due to HFP inversions, whereas F, J, K and L are due to VF_2 inversions. The lower spectrum shows the highly selective intensity reductions of resonances B, H, I, M, N, and P upon addition of *n*-Bu_4NOH. These changes have occurred due to the virtual exclusion of changes in C and D, which are due to contiguous VF_2 units. Furthermore, the reductions of integrals H + I and M + N are equal and the reduction of H is equal to that of I. These observations can only be accommodated by the highly selective dehydrofluorination of isolated VF_2 units, i.e., HFP-VF_2-HFP structures. Fluoride ion, which is conveniently determined by comparison with the internal standard 2,5-dichlorobenzotrifluoride, is formed in a 2:1 ratio to hydroxide added, while 2 equivalents of hydroxide are consumed per equivalent of isolated VF_2 site, and an allylically deshielded CF_2 groups appears at 56.9 ppm. These results can be accommodated by the reaction scheme shown in Figure 4. Figure 4 shows the proposed reaction path of the HFP-VF_2-HFP site. The concentration of this site in a 3.5 VF_2/HFP polymer is about 0.6 mol/kg. Attack by hydroxide is followed by fluoride ion initiated rearrangement of the initial double bond. The resulting allylic hydrogen is abstracted by fluoride, a relatively strong base in dipolar aprotic solvents, and is followed by elimination of a second fluoride. This results in the formation of bifluoride and a formally conjugated diene. Nucleophilic attack by hydroxide on the diene forms an enone (*via* a dienol) and subsequent attack of fluoride ion on the highly acidic hydrogen of the tertiary carbon gives a dienone.

In order to determine the structural changes that occur when poly(VF_2/HFP) is treated with bis-phenols in the presence of bases, solutions of the polymer in the dipolar, aprotic solvent DMAC were treated with a cyclic amidine base and Bp-AF at room temperature. Whereas most bases strong enough to dehydrofluorine the VF_2 copolymer are also highly nucleophilic toward the formed unsaturation, a suitable base would be strong enough to dehydrofluorinate the polymer and ionize the phenol but be sterically hindered from being an efficient competitor of phenoxide for the fluoro-olefin. The cyclic amidine DBU meets these requirements approximately.

Whereas bis-phenols dissolved in the polymer solution rapidly gel the polymer upon addition of a solution of the base or when a solution of the bis-phenol and the base are added to the polymer sollution, addition of only the base or substitution of the bis-phenol by the corresponding monophenol fails to gel the polymer. Addition of base to the polymer, followed after several minutes by addition of the bis-phenol, also fails to gel the polymer. The requirement for the proper sequence of addition in this hydroxide-free system, where the enone cannot be formed by ketonization of an intermediate dienol, requires that a fluoro-olefin is the active substrate and that the free phenol cannot add to it.

The NMR spectra of the base-treated soluble polymer and the gel produced by base plus bis-phenol show very similar changes compared to the spectrum of the original polymer solution. Both show an allylically deshielded CF_3 resonance at 56.7 ppm and a much smaller allylically shifted CF_3 resonance at 61.6 ppm, which may be due to an isomeric structure. The pattern of selective dehydrofluorination described earlier is also evident in these samples. The spectrum of the gelled sample shows a much broader fluoride resonance as well as the resonance of the geminal CF_3 groups of Bp-AF at 64.0 ppm. The amount of base in the gelled ample had to be reduced to half of that of the soluble sample in order to delay gelation sufficiently long to be able to transfer the sample to an NMR tube.

To demonstrate that phenols actually become attached to the polymer chain and do not merely promote another process which forms crosslinks when a VF_2/HFP polymer is treated with bis-phenol in the presence of base, the monofunctional analog of Bp-AF and DBU were used to treat the polymer (Figure 5). After twice precipitating the polymer with an excellent solvent for free phenol or any unreacted phenolate, the washed polymer clearly showed the presence of the geminal trifluoromethyl groups. Under the conditions shown in Figure 5, about 40% of the phenolate is incorporated based on the internal standard p-fluoroanisole.

In view of the demonstration of the attachment of the mono-hydroxy analog of bis-phenol AF to the VF_2/HFP polymer upon addition of its potassium salt in the presence of DBU and because of nucleophilic attack by hydroxide ion on the unsaturated polymer, it is concluded that in practical vulcanization a bis-phenol-derived phenolate probably also attacks the intermediate diene and leads to the dienic phenyl ether crosslinks shown in Figure 6. Attack on the diene by phenoxide, unlike attack by hydroxide, is not expected to proceed beyond vinylic nucleophilic substitution and, therefore, the expected product is the phenyl ether that corresponds to the dienol proposed as an intermediate in attack by hydroxide.

The phosphonium ion, originally present as a chloride, is believed to undergo several cycles of conversion from fluoride or bifluoride to intermediate hydroxide to phenoxide to fluoride before exhaustion of the bisphenol. Ultimately, the phosphinium ion is converted to triphenylphosphine oxide.

VF_2/HFP/TFE Reactivity. It is well known to those familiar with the characteristics of fluoroelastomers that polymers of unusually high fluorine content, e.g., VF_2/HFP/TFE terpolymers which contain about 45 wt% VF2 and about equal amounts of HFP and TFE, exhibit some compromise in cure rate and high-temperature compression set resistance in order to achieve their increased fluid resistance. In order to determine whether there are discernible chemical reasons for these differences, the behavior of VF_2/HFP/TFE polymers toward bases in solution was investigated.

The question of terpolymer reactivity is quite complex because four environments can exist about isolated VF_2 units

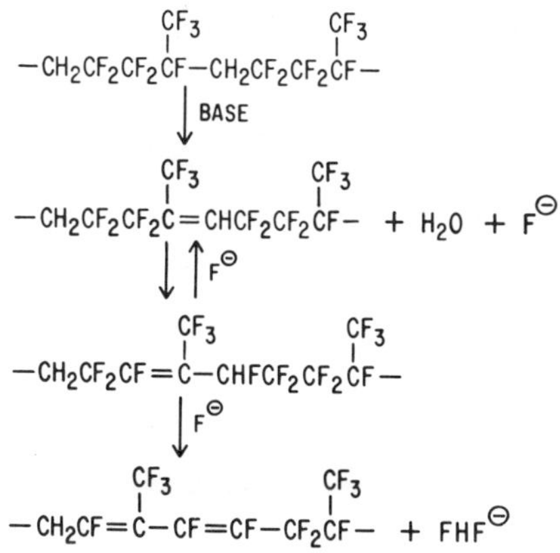

Figure 4. Proposed reaction sequence of selectively base-sensitive site in VF_2/HFP polymers.

Reproduced with permission from Ref. 5. Copyright 1978 Kaut. Gummi Kunst.

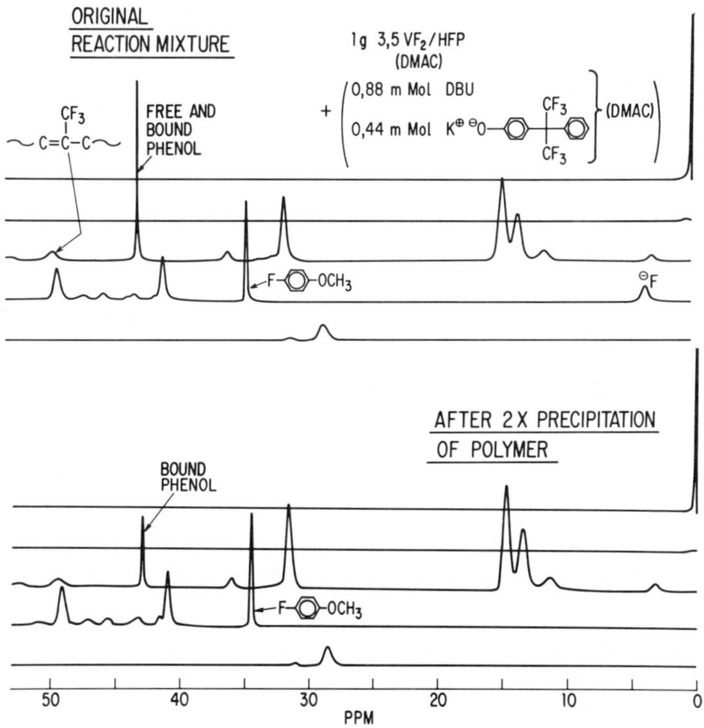

Figure 5. 94.1 MHz ^{19}F NMR spectrum of a VF_2/HFP polymer in DMAC after treatment with DBU and the monofunctional analog of Bp-AF: reaction mixture (top) and purified polymer (bottom).

Reproduced with permission from Ref. 5. Copyright 1979 Angew. Makromol. Chem.

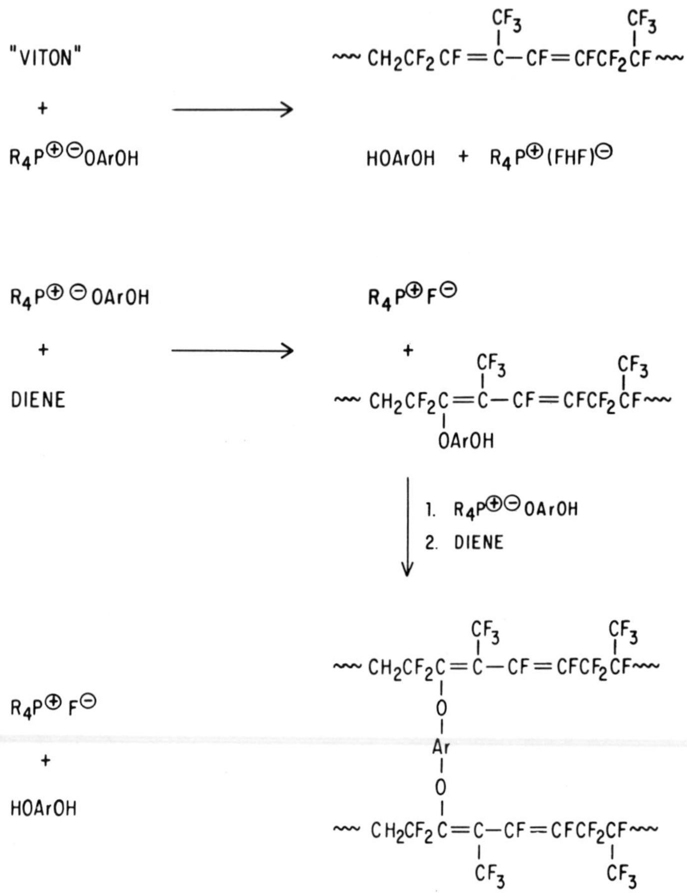

Figure 6. Proposed scheme for crosslinking VF_2/HFP polymers with bis-phenols in basic media.

(Figure 7). From competitive dehydrofluorination of the VF_2/HFP and the VF_2/TFE polymers, it is known that Sequence 1 and Sequence 3 undergo dehydrofluorination with similar ease. Structurally, the environment of the CH_2 group in Sequence 1 is highly similar to that of Sequence 2, whereas that of Sequence 3 is highly similar to that of Sequence 4. This would suggest that the CH_2 group in all four environments would participate with similar ease in reactions with bases, as is indeed seen. Figure 8 shows the spectra of a VF_2/HFP/TFE polymer before and after reacton with n-Bu_4NOH. The appearance of allylically shifted CF_3 resonances and intensity reductions of peaks B, H, *H + *H', I + H^r + H", *M, M, N, *M', M' and P are consistent with virtually exclusive reaction of, and competitive reaction between, Sequences 1-4. The above resonance signals are insufficiently resolved to obtain accurate integrals; however, it appears that the reactivity difference between Sequence 1 and 3 is no larger than that of any other pair of sequences.

<u>VF_2/PMVE/TFE Reactivity</u>. The motivation for copolymerizing VF_2 with PMVE and, optionally, TFE is based on the considerable improvement in low temperature properties which can be attained at equivalent fluid resistance compared to VF_2/HFP and VF_2/HFP/TFE polymers.

When VF_2/PMVE/TFE terpolymers are compounded with either diamine precursors and metal oxides or with bis-phenols, accelerators, and metal oxides and hydroxides, the resultant vulcanizates possess fissures and considerable porosity when removed from a hot mold. Moderate success in avoiding fissures can be attained with the diamine system when the sample is cooled while still compressed in the mold and when higher than normal levels of curatives are used with high viscosity polymers. This highly impractical situation suggests that considerable quantities of a polymer-insoluble condensable vapor is produced during the cure cycle and that this vapor is produced before substantial crosslinking has occurred. To gain an understanding of the chemistry which leads to such undesirable consequences, the behavior of VF_2/PMVE/TFE polymer solutions toward bases was examined.

Treatment of a 10.1 VF_2/PMVE polymer in DMAC solution with n-Bu_4NOH results in specific changes in the NMR spectrum (Figure 9). Comparison of the spectrum of the untreated polymer with those of VF_2/HFP polymers indicates a remarkable parallelism and allows the use of a formally equivalent peak assignment scheme. Because of the additional deshielding effect of oxygen, the trifluoromethyl resonance of PMVE units, relative to HFP units, is shifted downfield from about 76 to 53 ppm relative to the internal standard $CFCl_3$. The lone fluorine of PMVE, which corresonds to the tertiary fluorine of HFP, is deshielded even more substantially by the geminal oxygen, and shifts from about 185 ppm in HFP to 126 ppm in PMVE. Curiously, inversion of PMVE units results in a 19 ppm upfield shift for the lone fluorine, whereas inversion of HFP results in a 2.5 ppm downfield shift. As in VF_2/HFP polymers, treatment with base results in specific intensity reductions of peaks B,

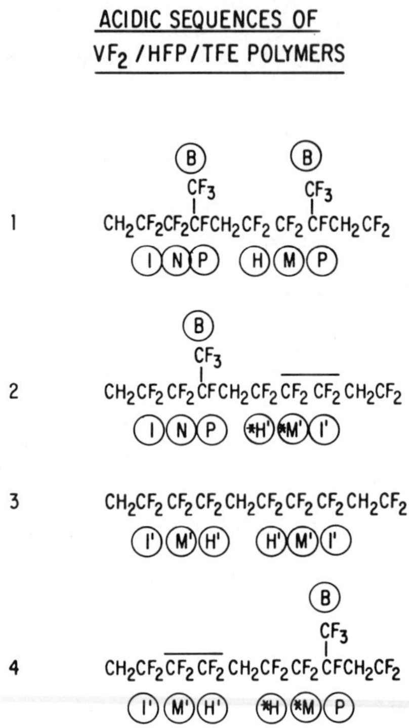

Figure 7. Selectively base-sensitive sites of VF_2/HFP/TFE polymers.

Reproduced with permission from Ref. 5. Copyright 1979 Angew. Makromol. Chem.

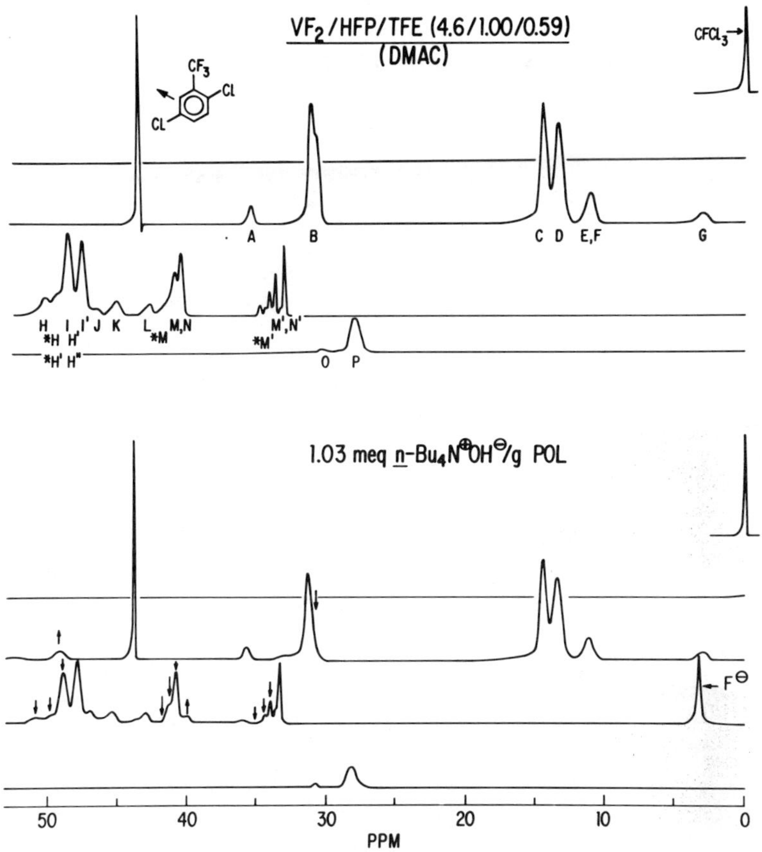

Figure 8. 94.1 MHz ^{19}F NMR spectrum of a VF_2/HFP/TFE polymer in DMAC before (top) and after (bottom) treatment with a hydroxylic base.

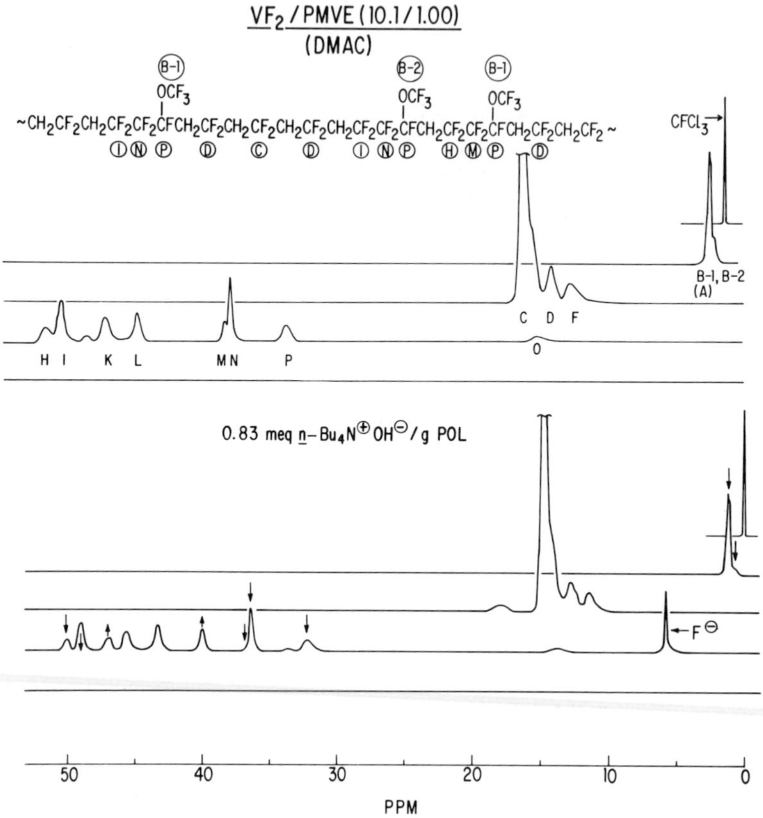

Figure 9. 94.1 MHz ^{19}F NMR spectrum of VF_2/PMVE polymer in DMAC before (top) and after (bottom) treatment with hydroxylic base.

H, M, N, and P. B is partially resolved into B-1 and B-2 and it is evident that B-2 corresponds to the base-sensitive structure. A critical difference, however, between spectra of VF_2/PMVE and VF_2/HFP polymers which have been treated with base is that no allylically deshielded OCF_3 resonance appears when the B-2 peak, which corresponds to original OCF_3 groups, disappears. This results requires that the entire OCF_3 groups be lost from the polymer and undergo decomposition. Given the choice of eliminating fluoride or trifluoromethoxide, the carbanion formed from isolated VF_2 groups apparently prefers to eliminate the perflouroalkoxide.

Treatment of the TFE-containing terpolymer with base also results in specific changes in the NMR spectrum. The spectrum of the unreacted terpolymer is highly similar to that of VF_2/HFP/TFE polymers and examination of the spectral changes accompanying base treatment indicates that reaction has occurred at all four possible sites which contain isolated VF_2 units.

In view of the results of base treatment of the VF_2/PMVE dipolymer, however, it is likely that whenever base attacks either Sequence 1 or 2, elimination from the carbanion occurs toward the adjacent oxygen-bearing carbon, and presumably results in the net loss of the elements of trifluoromethanol.

Confirmation of the elimination of trifluoromethanol from the VF_2/PMVE/TFE polymer when treated with base is provided by the efficient formation of dibenzylurea when a solution of the polymer is treated with benzylamine. After 40 hours at 20°C sufficient dibenzylurea was isolated to correspond to about 60% of the benzylamine added, based on the consumption of 5 equivalents of benzyl-amine/urea formed. Elimination and degradation of the elements of trifluoromethanol produces 3 equivalents of HF/simple olefin. It is therefore not surprising that the formation of nucleophilically quite unreactive unsaturation, excessive HF evolution and subsequent water formation in vulcanizates containing metal oxides and hydroxides, and higher than usual curative levels all combine to lead to fissure and porosity problems when VF_2/PMVE/TFE polymers are subjected to standard diamine or bis-phenol cures. Moreover, the costly PMVE, which was copolymerized to confer improved low-temperature properties compared to VF_2/HFP polymers, is considerably degraded.

This analysis of the vulcanization problems that arise when VF_2/PMVE/TFE polymers are subjected to conventional curing systems has led to development of specifically peroxide-curable polymers. Thus, the entire problem of undesirable response to basic curatives has been bypassed and excellent vulcanization behavior and vulcanizate properties, especially low temperature service fluid resistance, have been obtained.

<u>Curing of Vinylidene Fluoride Based Fluoroelastomers with Peroxides</u>. The co- or terpolymers that give elastomeric properties typically contain 20-70 wt % vinylidene fluoride, 20-60 wt % hexafluoropropylene, and 0-40 wt % tetrafluoroethylene. In order to make these elastomers peroxide curable, a cure site susceptible to free-radical attack has been introduced.

Bromine-containing monomers such as bromotrifluoroethylene, 1-bromo-2-,2-difluoroethylene, 3,3,4,4-tetrafluoro-1-butene provide such cure sites. It is desirable that the bromo-monomers copolymerize to high conversion with minimum chain transfer and inhibition. Recently Daikin Kogyo introduced iodo-containing fluoroelastomers which are peroxide curables. For good curing performance, an aliphatic peroxide, a suitable radical trap, and an inorganic acid acceptor are needed. Among the various peroxides tried, 2,5-dimethyl-2,5-di-t-butylperoxyhexane and 2,5-di-t-butylperoxyhex-3-yne were found to work best overall.

The preferred coagent or radical trap is triallylisocyanurate. The isomeric triallylcyanurate and some other common radical traps are simmilarly effective in formation of crosslinks, but the resulting networks are less stable.

Metal oxides like $Ca(OH)_2$, CaO, Mgo, ZnO, and PbO are useful in absorbing any traces of hydrogen fluoride generated during the curing process. The rate and degree of crosslink development was followed by measurements with a Monsanto Oscillating Disk Rheometer (ODR). A typical curve is similar to that shown in Figure 1. Both cure rate and state are directly proportional to the amount of cure site monomer in the polymer (Figure 10). The compression set resistance after press cure improves as the cure site monomer concentration increases and then levels off.

When the cure site concentration is kept constant and the peroxide and radical-trap concentrations vary, the cure rate is proportional to both of these variables. The state of cure, however, depends on the triallylisocyanurate concentration and only up to a point on the peroxide. Increasing the peroxide concentrations above the 3 phr level causes no change in the state of cure. In order to establish the effect of the rate of the peroxide decomposition on the curing characteristics of these polymers, the samples were cured at various temperatures. Table 2 shows the half-life of two peroxides at various cure temperatures as well as the ODR data obtained at these same temperatures. The results indicate that the rate of cure, as expressed by the $t_S 2$ and $t_C 90$ values, is greatly affected by the cure temperature. However, the state of the cure as expressed by $M_H - M_L$ values remains nearly constant, which indicates that a chain mechanism is involved and that the peroxide acts as the initiator.

Premature cure at processing temperature ("scorch" at 121°C) is conveniently delayed or prevented by small amounts of various phenolic and amine free-radical inhibitors and antioxidants. When the inhibitor concentration is properly chosen, the retarding effect disappears on the curing temperature (177°C).

Fluoroelastomer vulcanizate properties are improved by oven post curing. This is true for diamine, bis-phenol, and peroxide cures. For example in Table 3 both bis-phenol and peroxide cured black stocks of a VF_2/TFE/HFP terpolymer show a 50% increase in M100 and TB, a 50% decrease in elongation at break, and a substantial improvement in compression set resistance.

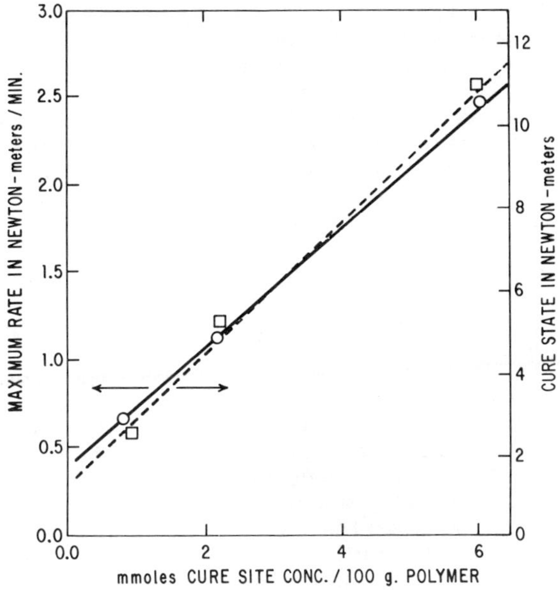

Figure 10. The cure rate (open circles) and cure state (squares) as obtained from ODR measurements at 177 °C plotted as functions of cure site concentration.

TABLE 2

EFFECTS OF PEROXIDE TYPE AND CURE TEMPERATURE

CURE TEMP °C	PEROXIDE HALF-LIFE[1] MIN	ODR VALUES t_S2 MIN	t_C90 MIN	CURE STATE (M_H-ML) N-m
2,5-DIMETHYL-2,5-Di-t-BUTYLPEROXYHEXANE				
160	4.8	4.0	24	8.0
177	0.80	1.6	8.3	8.6
190	0.24	1.4	4.6	8.2
204	0.07	0.8	2.8	7.9
2,5-DIMETHYL-2,5-Di-t-BUTYLPEROXYHEX-3-YNE				
160	18.7	7.1	41[2]	-
177	3.4	3.4	14	7.7
190	1.0	2.1	7.5	7.7
204	0.30	1.2	4.2	7.5

[1] IN DILUTE BENZENE SOLUTION

[2] EXTRAPOLATED VALUE

Compression set resistance is further improved when the specimens are exposed to nitrogen while under compression.

One of the essential functions of the oven cure in the nucleophilic diamine and bis-phenol vulcanizates is the removal of water, a by-product of the cure whose presence prevents full development of the diamine cure and causes reversion of the bis-phenol cure. This is not an important factor in the peroxide cure.

From our results and by analogy to the literature (11,14,15), we conclude that during post cure there is indeed a thermally induced bond-breaking and bond-making process that results in a more stable network (Figure 11) shows that peroxide and bisphenol vulcanizates may develop comparatively fewer additional crosslinks and fewer ruptures than the diamine vulcanizates. Thus, their overall vulcanizate crosslink density before and after oven post curing remains similar while that of the diamine vulcanizates increases substantially. In view of the differences in the chemistry among these vulcanizate systems, it is surprising the physical properties after oven post curing are so similar (Table 3). The actual mechanism of bond-breaking and bond-making is not well understood and additional work is needed to get a better understanding of the various steps involved.

The most informative technique in following reactions that involve free radical chemistry is electron spin resonance spectroscopy (ESR). By this powerful technique, we showed that the simplest way of generating a fluoroalkyl radical involves the abstracton of a bromine atom from a fluoroalkyl bromide using 2-bromoperfluorobutane as a model compound. Upon ultraviolet irradiation of a dilute solution of 2-bromoperfluorobutane and di-*t*-butyl peroxide in cyclopropane in an ESR cavity at room temperature, an intense spectrum (Figure 11) resulted. Much weaker spectra were obtained if the peroxide was omitted or if the temperature was lowered. The direct abstraction of a bromine atom by the di-*t*-butoxy radical would seem to be, at best, a thermoneutral reaction and, therefore, unlikely to occur. Consequently, it was concluded that bromine abstraction in these ESR experiments is caused mainly by methyl radicals produced by β-scission of the *t*-butoxy radical. The lower radical concentrations at lower temperatures support the above conclusion since β-scission is a strongly thermally activated process and takes place less readily at lower temperatures. Equations (1)-(3) below depict the most likely chemistry involved:

$$(CH_3)_3COOC(CH_3)_3 \xrightarrow{h\nu} 2(CH_3)_3CO^{\bullet} \quad (1)$$

$$(CH_3)_3CO^{\bullet} \longrightarrow CH_3^{\bullet} + CH_3\overset{O}{\overset{\|}{C}}CH_3 \quad (2)$$

$$CH_3^{\bullet} + CF_3CF_2\overset{Br}{\overset{|}{C}}FCF_3 \longrightarrow CH_3Br + CF_3CF_2\overset{\bullet}{C}FCF_3 \quad (3)$$

With this information in hand, we investigated the curing process of these fluoroelastomers by examining the amount and

TABLE 3

POST CURE EFFECT ON PROPERTIES

PROPERTIES	BISPHENOL PRESS CURE	BISPHENOL POST CURE	PEROXIDE PRESS CURE	PEROXIDE POST CURE
M_{100}, MPa	5.0	7.9	5.0	7.9
T_B, MPa	10.0	13.8	9.7	15.9
E_B %	225	175	165	150
COMPRESSION SET (70 HRS/200°C)				
O-RINGS	63	25	50	27
PELLETS	85	20	52	20

PRESS CURE - 10 MINS at 177°C

POST CURE - 24 HRS at 232°C

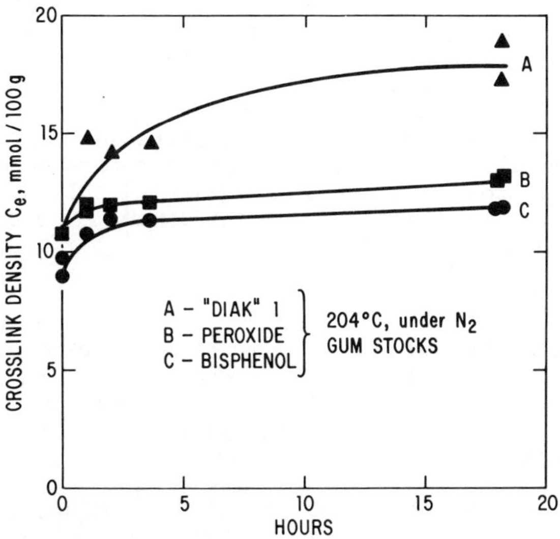

Figure 11. The crosslink density of various cure systems plotted as a function of post-cure time at 204 °C under nitrogen.

nature of the volatile by-products generated during a peroxide cure. Fully compounded stocks containing the fluoroelastomers, carbon black, metal oxide, peroxide, and radical trap were used. The total amount of volatile materials is proportional to the amount of peroxide present. A typical analysis is shown in Table 4. Among organic fragments, acetone is the largest component. Surprisingly, methane and, especially, methyl bromide are present in the gas mixture in relatively low concentrations. The water formed comes from the calcium hydroxide and from water dissolved in the polymer. When the radical trap, triallylisocyanurate, is omitted, the methyl bromide concentration increases by 100% but still remains relatively low. If the peroxide is omitted, water is the only volatile product. If the peroxide is retained and the metal oxide and the triallylisocyanurate are both omitted, the isobutene concentration increases dramatically. The isobutene is formed by dehydration of t-butanol. Figure 12 shows the most probable reactions taking place in a fully compounded stock.

The primary decomposition product is the t-butoxy radical which may abstract a hydrogen atom to give t-butanol (minor reaction) or decompose into acetone and a methyl radical (major reaction). The methyl radical, in turn, can abstract a bromine atom from the polymer to give methyl bromide (a minor reaction) or add to the triallylisocyanurate to give a more stable radical intermediate. These intermediate radicals abstract bromine from the polymer to generate polymeric radicals (see Figure 13). The driving force for the chain reaction during propagation is the transfer of a bromine atom from the electron-poor fluoropolymer to an electron-rich hydrocarbon radical on the coagent. Crosslinking takes place when the polymeric radicals add to the allylic bonds of the trifunctional coagent. The coagent, therefore, becomes the crosslinker and in order to give optimum properties it should be (a) an efficient radical trap, (b) multifunctional, and (c) thermally and oxidatively stable.

Conclusion

Peroxide-curability of vinylidene fluoride-based fluoroelastomers becomes possible by incorporating a bromine or iodine-containng monomer. An organic peroxide, coagent, and an acid acceptor are necessary to obtain good cures. These results support the following cure mechanism:

Initiation: Generation of alkyl radicals via peroxide decomposition.
Propagation: Generation of radicals on the polymer via a chain reaction in which a bromine or iodine atom is abstracted.
Crosslinking: Addition of the polymeric radicals to the trifunctional coagent.

Since the crosslinking molecule is the coagent, its structure determines how effective it is as a radical trap and how stable the network will be to thermal and oxidative attack.

TABLE 4 OFF-GAS ANALYSIS

RECIPE:	POLYMER	100 (5.6 mmoles BROMINE)
	MT BLACK	30
	$Ca(OH)_2$	4
	2,5-DIMETHYL-2,5-Di-t-BUTYLPEROXYHEXANE	4 (5.5 mmoles)
	TRIALLYL ISOCYANURATE	4
VOLATILES:	ACETONE, mmoles	9.45
	tBuOH "	0.78
	CH_3Br "	0.18
	CH_4 "	2.68
	C_2H_4 "	0.26
	$C_3H_6 + C_3H_8$ mmoles	0.41
	C_4H_8 mmoles	0.39
	H_2O	16.6

CONDITIONS - 40°C/MIN TO 190°C, HELD AT 190°C/10 MINS UNDER HELIUM ATMOSPHERE

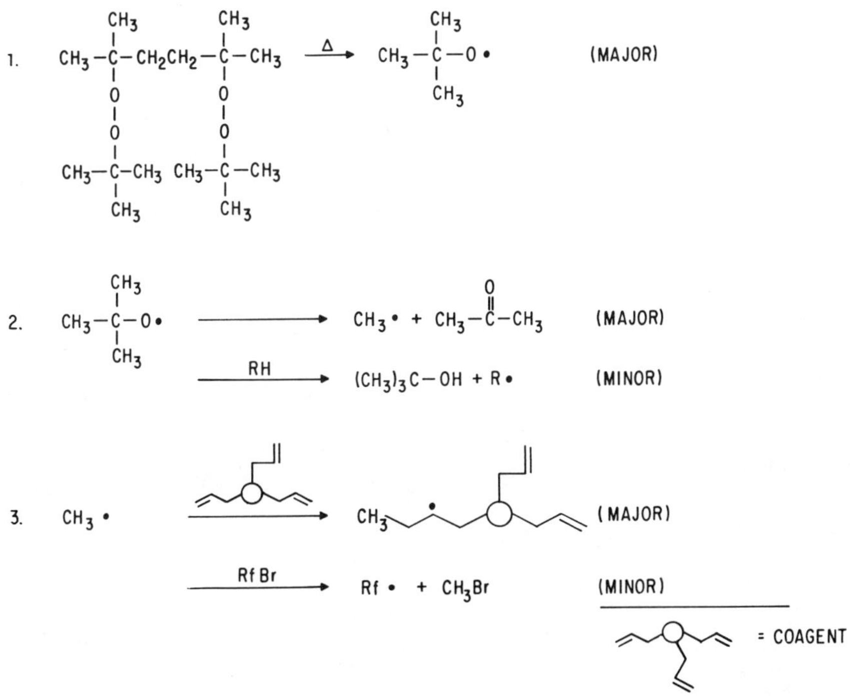

Figure 12. The most probable reactions resulting from peroxide decomposition.

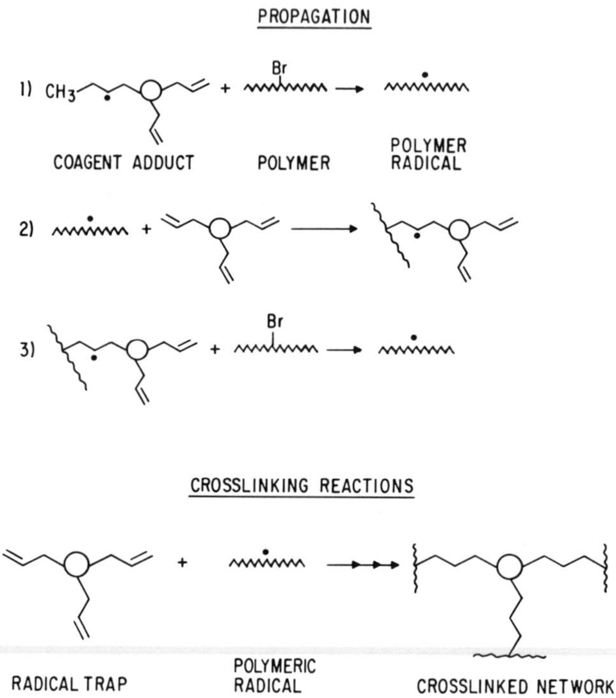

Figure 13. The main reactions of the proposed crosslinking mechanism.

Reproduced with permission from Ref. 6c. Copyright 1982 _Rubber Chem. Technol._

Curing with nucleophiles does not require any specific cure site. As it was shown in the case of bis-phenols, curing requires the additional presence of an accelerator and a metal oxide base. The results described in this paper support the following mechanisms:

Step 1 - Reaction between bis-phenol, phosphonium accelerator and metal oxide to give R_4P OArOH.
Step 2 - Attack by R_4P OArOH on the polymeric backbone to generate double bonds.
Step 3 - Nucleophilic attack by R_4P OArOH on the unsaturated sites to form crosslinks.

The selection of the bis-phenol is important because its structure determines the cure rate and the stability of the resultant network.

Literature Cited

1. D. C. Thompson and A. L. Barney, Kirk-Othmer Encyclopedia of Polymer Science and Technology, 1971, 14, 611.
2. R. G. Arnold, A. L. Barney and D. C. Thompson, Rubber Chem. Technol., 1973, 46, 169.
3. J. F. Smith, Rubber World, 1960, 142, 103; J. F. Smith and G. T. Perkins, J. Appl. Polym. Sci., 1961, 5, 460; K. L. Paciorek, L. C. Mitchell, and C. T. Lenk, J. Polym. Sci., 1960, 45, 405; K. L. Paciorek, W. G. Laginess, and C. T. Lenk, J. Polym. Sci., 1962, 60, 141; D. K. Thomas, J. Appl. Sci., 1964, 8, 1415; T. L. Smith, J. Polym. Sci., Part A-2, 1972, 10, 133.
4. A. L. Moran and D. B. Patterson, Rubber Age, 1971, 103, 37.
5. W. W. Schmiegel, Kaut. Gummi Kunst., 1978, 31, 137; Angew. Makromol. Chem., 1979, 76/77, 39.
6. (a) J. W. Finlay, A. Hallenbeck, and J. D. MacLachlan, J. Elastomers Plast., 1978, 10, 3.
   (b) D. Apotheker and P. J. Krusic, U.S. Patent 4 035 565, 1977; U.S. Patent 4 214 060, 1980.
   (c) D. Apotheker, J. B. Finlay, P S. Krusic and A. L. Logothetis, Rubber Chem. Technol., 1982, 55, 1004.
7. P. Tarrant and J. P. Tandon, J. Org. Chem., 1979, 34, 864; A. H.. Fainberg and W. T. Miller, Jr., J. Am. Chem. Soc., 1957, 79, 4170.
8. G. E. Decker, R. W. Wise, and D. Guerry, Rubber Chem. Technol., 1963, 36, 451.
9. P. J. Krusic and J. K. Kochi, J. Am. Chem. Soc., 1971, 93, 846.
10. K. S. Chen, P. J. Krusic, and J. K. Kochi, J. Phys. Chem., 1974, 78, 2030.
11. (a) A. W. Fogiel, J. Polym. Sci. Symp., 1975, 53, 333.
    (b) U. Flisi, G. Giunchi and S. Geri, Kaut. u. Gummi Kunst., 1976, 29, 118.
12. D. B. Pattison, U.S. Patent 3 876 654, 1975.
13. R. C. Ferguson, J. Am. Chem. Soc., 1960, 82, 2416.

14. S. H. Kalfayan, R. H. Silver, and S. S. Liu, *Rubber Chem. Technol.*, 1976, 49,, 1001.
15. S. H. Kalfayan, R. H. Silver and A. A. Mazzeo, *Rubber Chem. Technol.*, 1975, 48, 944.
16. P. Meakin and P. J. Krusic, *J. Am. Chem. Soc.*, 1973, 95, 8185; B. E. Smart, P. J. Krusic, P. Meakin, and R. C. Bingham, *J. Am. Chem. Soc.*, 1975, 96, 7385; K. S. Chen, P. J. Krusic, P. Meakin, and J. K. Kochi, *J. Phys. Chem.*, 1974, 78, 2036; P. J. Krusic and R. C. Bingham, *J. Am. Chem. Soc.*, 1976, 98, 230.

RECEIVED February 6, 1984

11

Rubber Processing Through Rheology

JEAN L. LEBLANC

Monsanto Europe S.A., 270-272, Avenue de Tervueren, B-1150 Bruxelles, Belgium

This paper gives some insights into the growing understanding of the particular flow phenomena associated with rubber processing. Attention is focused on two particular operations, i.e. internal mixing and extrusion. Using simple capillary rheometry experimental results, the importance of elongational flow and energy absorption processes is demonstrated in the mixing of natural rubber. The rôle played by the converging zone at the entrance of short extrusion dies is underlined and estimated using experimental data. In addition, some particular effects typical of heterogeneous rubbery materials are presented and their significance in rubber processing discussed.

The discussion in rheological terms of the problems associated with rubber processing is relatively recent and results from theoretical and experimental progress in understanding flow properties of pure polymers, and particularly thermoplastics. However the rheology of elastomers is further complicated by the necessary presence of fillers, plasticizers and other ingredients. These lead to peculiar flow properties associated with heterogeneous matter and therefore not yet well understood.

The aims of this presentation are to give some insights into the growing understanding of the particular flow phenomena involved in rubber processing. Rather than give a general survey of the numerous flow situations in rubber processing, attention will be focused on two particular operations, i.e. internal mixing and extrusion. Using experimental results, the main rheological effects associated with rubber processing will be demonstrated and their significance underlined. In addition, some particular effects typical of heterogeneous rubbery materials, such as thermoplastic elastomers, will be presented and their significance in rubber processing discussed.

Most of the results discussed in this paper have been obtained through a suitable use of the Monsanto Processability Tester, and therefore this paper will also consist in an implicit demonstration of the usefullness of this versatile instrument.

0097-6156/84/0260-0183$07.00/0

Rheology and mixing rubber

Preparing rubber compounds by blending of raw elastomers and mixing fillers, oils and other ingredients into rubbers is a fundamental operation in the rubber industry which, surprisingly, has been the object of little work until recently. The aims of mixing are to incorporate and uniformly disperse the various ingredients within the elastomer matrix, in such a way that easy processing, efficient curing and adequate end-use properties will be obtained. This operation is generally achieved in internal mixers which are thoroughly described elsewhere (1, 2).

Physics of mixing. The mixing of rubber compounds is a very complex operation about which it is difficult to have an overall and clear understanding (3-5). However, it is useful to consider that four basic physical operations take place during the mixing cycle (1). For the sake of discussion, those four operations can be best described with respect to the mixing power curve, as shown in Figure 1:

1. The incorporation of ingredients, either solid or liquid, in the elastomer takes place at the beginning of the mixing cycle. More specifically, this operation concerns the 'wetting' of solid particles which, according to Tokita and Pliskin (6) and Nakajima (8), can occur following two mechanisms:
(i) large deformation of the elastomer increases the contact area with filler particles and seals them inside,
(ii) the elastomer breaks down into small pieces, mixes with the filler agglomerates and once again sealing occurs.
If the former mechanism can easily be observed in an open mill, the latter is not necessarily apparent because these phenomena occur on a micro-scale. The incorporation involves a decrease in the specific volume of the mix (6).
2. The dispersion results in progressive spreading of filler particles. This operation involves a reduction in size of filler agglomerates, possibly to their ultimate particle size and, generally takes place between the two power peaks. An attempt in theoretically predicting the dispersive mixing process from fundamental considerations has recently been published (7).
3. The plasticization modifies the rheological properties of the mix mainly a viscosity decrease by mechano-chemical degradation of the polymer and modification of its viscoelastic properties.
4. The distribution or simple mixing involves the moving of particles throughout the compound, without changing their physical shape, in order to increase the randomness of the spatial distribution and therefore the entropy of the mix. This elementary physical process takes place during the whole mixing cycle.

To describe practical mixing with a set of four basic physical operations is clearly an oversimplification since, to a certain extent, all actions occur simultaneously. This description helps however in introducing the rheology of mixing.

Mixing rheology - lubrication theory. When considering only the nip region, it is possible to develop a rheological approach based on the lubrication theory (11). Analyses by Bernhardt (3), Bolen and

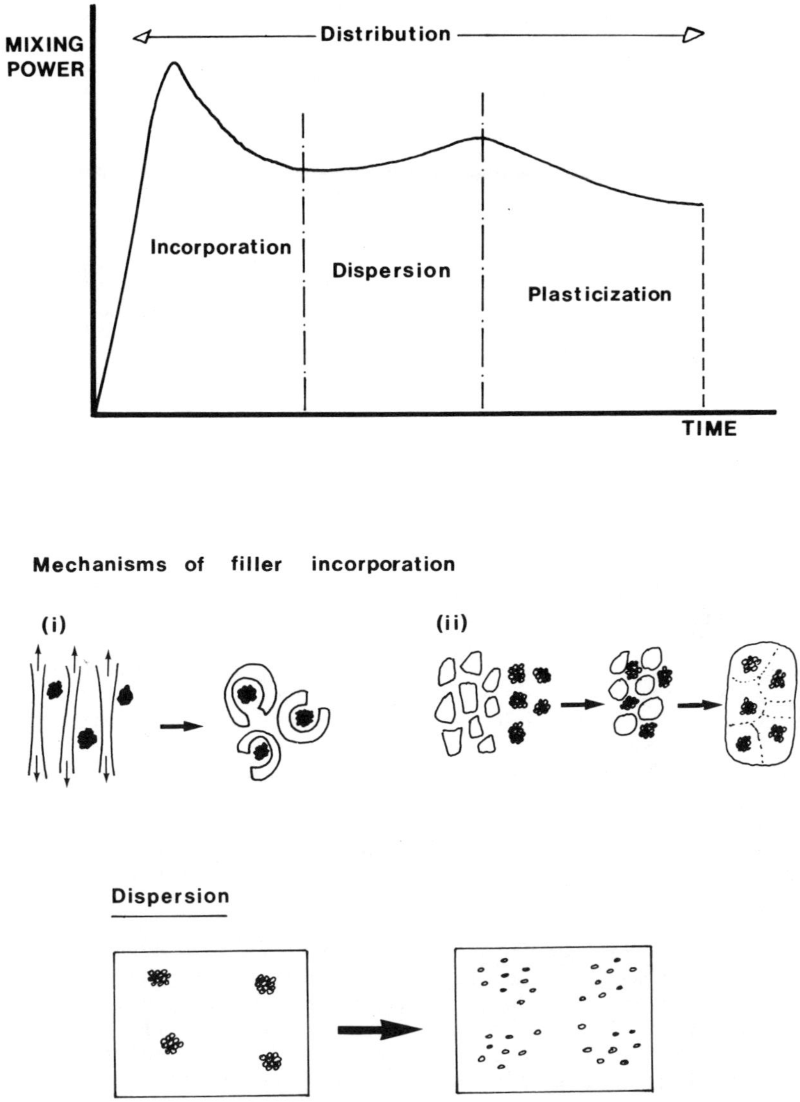

Figure 1 : Basic physical operations in mixing rubber compounds

Colwell (12) and others (1, 2, 4) are all based on this theory, with the usual assumptions of isothermal and steady flow, the condition of no wall-slippage, and negligible normal stresses. Details of these analyses can be found elsewhere and, with respect to the simplified geometry considered by Bolen and Colwell (12) (see Figure 3), only essential results will be given here.

In the simplified case of a Newtonian fluid, the material viscosity is not changed by shearing in the nip (i.e. $\eta_a = \eta_r$ in Figure 3) and the following relationships are established for the maximum shear rate ($\dot{\gamma}_{max.}$) and shear stress ($\sigma_{max.}$) at wall (of the chamber):

$$\dot{\gamma}_{max} = \frac{\pi D_a \Omega}{e} + \frac{e \Delta P}{2\eta_a W} \quad (1)$$

$$\sigma_{max} = \eta_a \dot{\gamma}_{max} = \frac{\eta_a \pi D_a \Omega}{e} \left\{ 1 + \frac{3\left(\frac{D_r h}{D_a} - e\right)}{\left(\frac{W h^3}{\pi e^2 D_r} + e\right)} \right\} \quad (2)$$

where D_a, D_r, h, e and w are defined in Figure 3, Ω is the rotor speed and ΔP the pressure drop in the nip.

Whilst oversimplified, these rheological relationships help to understand how the rotor speed and the nip width directly affect the shear rate in the unidirectional flow region. This corresponds well to the empirical experience that the higher the rotor speed, the lower the effective mixing time. In addition, equation (1) explains why rotors wearing decreases the shear rate (since e increases) and therefore affects the mixer efficiency.

Equation (2) shows that the maximum shear stress mainly depends upon the rotor speed, the nip and the fluid viscosity. With polymer materials, the viscosity decreases with increasing shear rate and temperature. Consequently, increasing the shear stress by increasing the rotor speed is counterbalanced by the associated viscosity decrease. Therefore, the viscosity function of the mix, i.e. $\eta = f(\dot{\gamma}, T)$ plays an essential rôle in controlling the actual shearing field.

Mixing rheology - elongational flow. The above considerations are obviously oversimplified and neglect the particular behaviour of viscoelastic materials. On one hand, we have already pointed out that it is difficult to introduce the more realistic viscosity function $\eta = f(\dot{\gamma}, T)$, and,on the other hand, it is quite clear that the mixing process is not isothermal. Furthermore, Tokita and White (13, 14) demonstrated that the lubrication theory can be applied to a nip of length L and depth H providing that

$$\frac{\Delta (N_1)_w}{(\sigma_{12})_w} \gg \frac{L}{H} \quad (3)$$

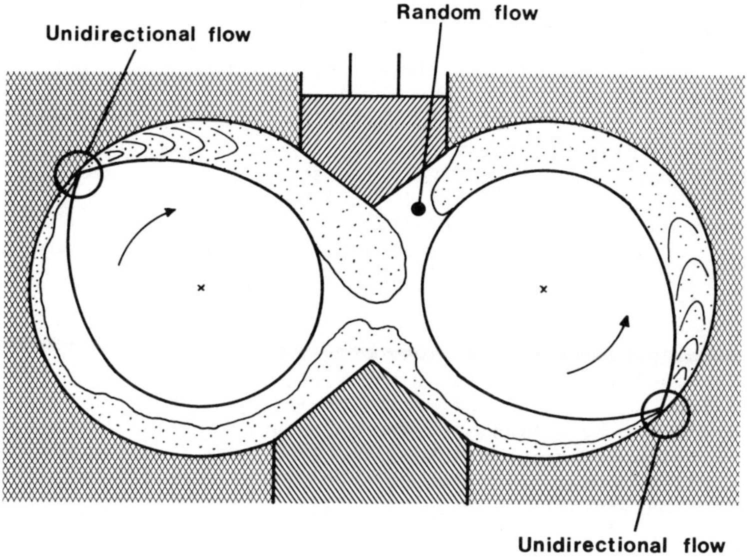

Figure 2 : Flow situations in the internal mixer

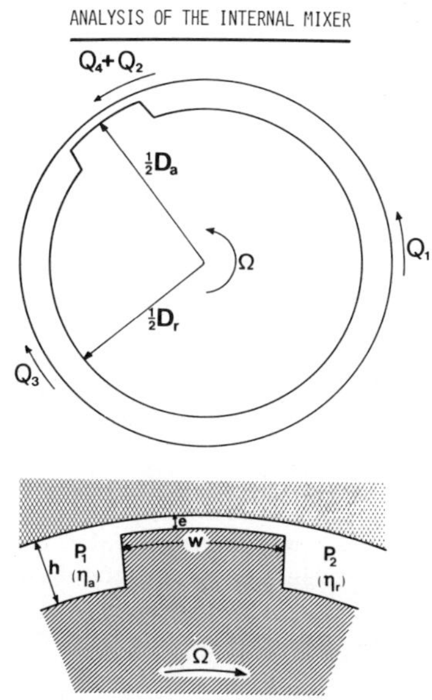

Shear rate : $\dot{\gamma}_{MAX} = \frac{\pi D_a \Omega}{e} + \frac{e \Delta P}{2 \eta_a w}$

Shear stress : $\tau_{MAX} = \eta_a \dot{\gamma}_{MAX} = \frac{\eta_a \pi D_a \Omega}{e} \left[1 + \frac{3\left(\frac{D_r h}{D_a} - e\right)}{\left(\frac{w h^3}{\pi e^2 D_r} + e\right)} \right]$

Figure 3 : Internal mixer - Simplified geometry analysis

where $(\sigma_{12})_w$ is the wall shear stress and $\Delta(N_1)_w$ is the variation along the length L of the first normal stresses difference ($N_1 = \sigma_{11} - \sigma_{22}$) in the flow direction. For Newtonian fluids, it follows that the lubrication theory is valid only if L/H >> 4.

A priori, this condition is satisfied only with intermeshing rotors. Moreover, with polymer melts, normal stresses in shear flow are very large, even at relatively low shear rate; for example, in the range of 10^5 dyn/cm^2 at $\dot{\gamma} = 1$ sec^{-1} for polystyrene at 180°C (15) (and probably higher with elastomers). Consequently the N_1/σ_{12} ratio is larger for viscoelastic fluids than for Newtonian fluids and therefore the validity condition for using the lubrication theory at the nip is surely not fulfilled.

Another consequence of the high N_1 offered by elastomers is that in the unidirectional flow regions of the mixer an elongational flow is surely combined with the shearing flow, as illustrated in Figure 4.

The importance of extensional flow in the mixing process has been pointed out by Cotten (16) and thoroughly analyzed by Nakajima (8) in the case of carbon black-filled compounds. A steady elongational flow can be developed only if the extensional rate increases exponentially (versus time) (17). Nakajima demonstrated that this type of deformation induces an anisotropy of the material, enhanced in highly filled compounds or containing oriented fibers. Therefore, the steady state is nearly impossible and, with polymers, the elongational flow is not a 'pure' deformation and necessarily involves a shear component.

The behaviour of elastomers in internal mixers reflects consequently a combination of shear and extensional responses. Any model of the mixing process has to take account of both shear flow and transitory extensional flow, since for the latter the steady state cannot be achieved. Therefore the response of the elastomers in the earlier times of an elongational process gives information about the mixing behaviour. Using a high rate extensometer, Nakajima showed that the elongational behaviour of unfilled raw rubbers can be interpreted in terms of processability (18).

Rheological properties and mixing behaviour of natural rubber

In order to study the effect of rheological properties of elastomers on their behaviour in the internal mixer, some experiments were performed using various natural rubber kindly supplied by MRPRA (x). Natural rubber rheology has not been deeply studied despite the commercial importance of this material, and only a few recent papers deal with the mixing of natural rubber (19, 20, 21) and rheological comparison between natural and synthetic cis-1,4 polyisoprenes (22).

The grades of natural rubber studied are described in Table I.

(x) Malaysian Rubber Producers Research Association - Brickendonbury, U.K.

Table I

Grade	Remarks	Intrinsic[(a)] Viscosity $[\eta]$, dl/g	Mooney Viscosity ML(1+4) $_{100°C}$
RSS 1	Standard RSS 1	9.72 ± 0.95	110.0
ICR	Initial concentration rubber	7.02 ± 0.58	95.0
SMR L	-	6.59 ± 0.24	87.0
SMR CV	Contains hydroxylamine salt	5.38 ± 0.05	57.5
SMR 10	-	5.09 ± 0.41	94.5
SMR 20	-	5.43 ± 0.56	85.5
SMR GP	Contains hydroxylamine salt	4.94 ± 0.33	64.5
HARUB/P60WF	Peptized SMR L from Harrison and Crosfield	4.06 ± 0.22	56.0

(a) in Toluene; at 25°C; gel removed from solution by centrifugation

In addition to Mooney viscosities, the rheological properties of raw natural rubber samples were determined at 100°C using the Monsanto Processability Tester (MPT) with a die of 2.01 mm diameter (L/D ratio = 16). Elongational properties were simply estimated at 22°C, using a tensile tester and dumbbell samples die-cut from compression moulded sheets.

Results are given in Table II, from the MPT, in terms of apparent viscosity at $\dot{\gamma}$ = 10 sec^{-1} and of K and n, the power law parameters; from tensile tests, in terms of stress and elongation at break and of tensile strain energy at 200% elongation.

Table II : Shear flow properties (MPT at 100°C) and extensional properties (at 22°C) of gum natural rubber samples

NR grade	Shear flow properties at 100°C			Tensile properties at 22°C		
	Apparent Viscosity at $\dot{\gamma}$ = 10s^{-1} (kPa. s)	Power law parameters K (kPa. s)	Power law parameters n	Tensile Strength (MPa)	Elong. at break (%)	Strain energy at 200% (MPa)
RSS 1	41.5	(188.1)*	(0.342)*	0.400	250	0.42
ICR	30.6	216.9	0.149	0.222	280	0.25
SMR L	28.5	210.3	0.132	0.127	580	0.22
SMR CV	18.0	133.5	0.130	0.081	320	0.14
SMR 10	29.5	198.0	0.174	0.233	350	0.26
SMR 20	25.6	(130.7)*	(0.292)*	0.184	390	0.19
SMR GP	20.4	138.6	0.169	0.083	360	0.15
P60 WF	18.6	120.2	0.189	0.054	460	0.11

(*) Results with limited significance due to non-linearity of the $\log \eta_a - \log \dot{\gamma}_a$ graph

Mixing experiments were performed in a Farrel-Bridge Banbury, BR type of 1.573 liter (rotor speed : 117 rpm; ram pressure : 0.5 MPa) using the formulation and procedure given in Table III.

Table III: Formulation and mixing procedure

Ingredients	phr	weight (g)
Natural rubber	100	800
Carbon black N330	50	400
Zinc Oxide	5	40
Processing oil	3	24
Stearic acid	2	16
		1280

Mixing operations (starting temp.: 50°C)	Time (min.)
- Introduce elastomer	0
- Add 1/2 black, ZnO	2
- Add 1/2 black, oil, stearic acid	4
- sweep	6
- dump	8

Total mixing time: 8 minutes

The total lot weight of 1280 g corresponded to a fill factor of 0.76 (for a mean density of 1.13 g/cm3). Mixing power consumption was recorded using the Monsanto Power Integrator, and kWh readings were converted in mixing energies (MJ/m^3) by taking into account the lot volume, according to:

$$\text{Mixing energy} : \frac{PI \times 3.6 \times D \times 10^6}{W} \qquad (3)$$

where PI is the power integrator reading (kWh), W the lot weight (g) and D its density (g/cm^3). All mixing experiments were repeated in order to estimate the reproducibility and mean results are given in Table IV, in terms of gum mixing energy and of complete mixing cycle energy.

Table IV : Mixing energy (MJ/m^3) of natural rubber samples

Natural Rubber Grade	Gum mastication	Complete mixing cycle
RSS1	905.0 ± 59.0	2933.5 ± 35.5
ICR	763.5 ± 27.5	2747.5 ± 7.5
SMR L	685.5 ± 12.5	2687.0 ± 2.0
SMR CV	590.5 ± 10.5	2363.5 ± 17.5
SMR 10	727.5 ± 76.5	2655.0 ± 45.0
SMR 20	702.5 ± 29.5	2605.5 ± 13.5
SMR GP	700.0 ± 19.0	2482.0 ± 3.0
P60 WF	708.5 ± 10.5	2504.0 ± 9.0

When comparing viscosity results (either Mooney ML (1+4) from Table I or apparent viscosity at $\dot{\gamma}$= 10 sec^{-1} from Table II) with mixing data given in Table IV, it can be seen that, as expected, the higher the viscosity of the gum rubber, the larger the mixing energy. Nearly linear relationship are observed between the gum mixing energy and the viscosity and between the energy for the complete mixing cycle and the viscosity. This indicates that the gum viscosity is controlling not only the mastication step but also the carbon black incorporation and dispersion stages. This experimental observation is well in line with the theoretical considerations above, and current mixing practice.

Figure 5 shows a linear relationship between the mixing energy and the tensile strain energy of gum natural sample. Since the slope of the graph is higher for the complete cycle than for the mastication stage, it can be concluded that the extensional flow resistance of the elastomer is also playing some rôle in the filler incorporation and dispersion stages. Similar graphs can be drawn with the tensile strength but obviously with a greater scatter of data.

The above theoretical considerations, the relationships between the mixing energy and the viscosity, and the tensile strain energy demonstrate in fact that mixing operations of elastomers are controlled by '<u>energy absorption processes</u>' which are dependent upon the viscoelastic properties of the polymer materials.

In a capillary rheometer, such as the Monsanto Processability Tester, the extrusion pressure is measured, within the barrel, using a pressure transducer, located just above the die entrance. Therefore, the calculated apparent shear stress is affected by an error due to entrance pressure loss which, practically can be corrected using the so-called 'Bagley method' (23). In addition, there is a relationship between the entrance pressure loss and the mean extensional stress in this region (24). Consequently, when using a capillary rheometer the apparent viscosity is determined without applying the Bagley correction, the results obtained are in fact, including the shear response (within the die) and the elongational response (at die's entrance) of the material, in the shear rate and temperature conditions of the experiment.

By definition, the instantaneous rate of energy absorption per unit volume, EA_i, is given by the product of the shear stress σ and the shear rate $\dot{\gamma}$ (23):

$$EA_i = \sigma \cdot \dot{\gamma} \tag{4}$$

where EA_i is in Pa/s (or J/m^3). When replacing the shear stress σ by $\eta_{(\dot{\gamma})} \cdot \dot{\gamma}$ where $\eta_{(\dot{\gamma})}$ is the shear rate dependent viscosity of the materials, it follows:

$$EA_i = \eta_{(\dot{\gamma})} \cdot \dot{\gamma}^2 \tag{5}$$

In a shear rate range $\dot{\gamma}_A - \dot{\gamma}_B$, the rate of energy absorption (per unit volume) is given by:

$$EA_{(\dot{\gamma}_A - \dot{\gamma}_B)} = \int_{\dot{\gamma}_B}^{\dot{\gamma}_A} \eta_{(\dot{\gamma})} \cdot \dot{\gamma}^2 \, d\dot{\gamma} \tag{6}$$

and for a material obeying the power law, it follows:

$$EA_{(\dot{\gamma}_A - \dot{\gamma}_B)} = \frac{K}{n+2} \left[\dot{\gamma}_B^{n+2} - \dot{\gamma}_A^{n+2} \right] \tag{7}$$

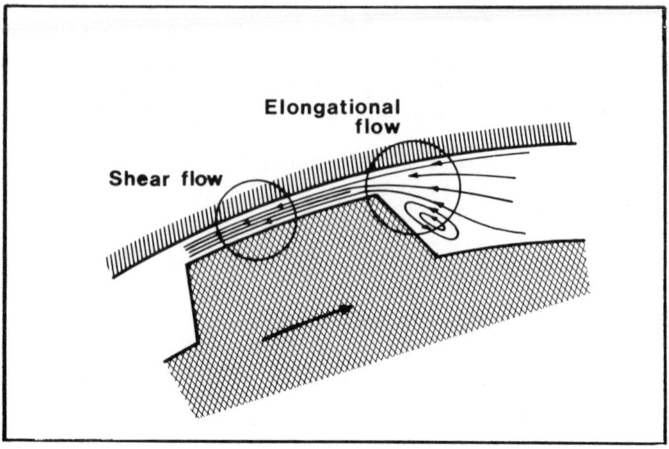

Figure 4 : Unidirectional flow region in the internal mixer

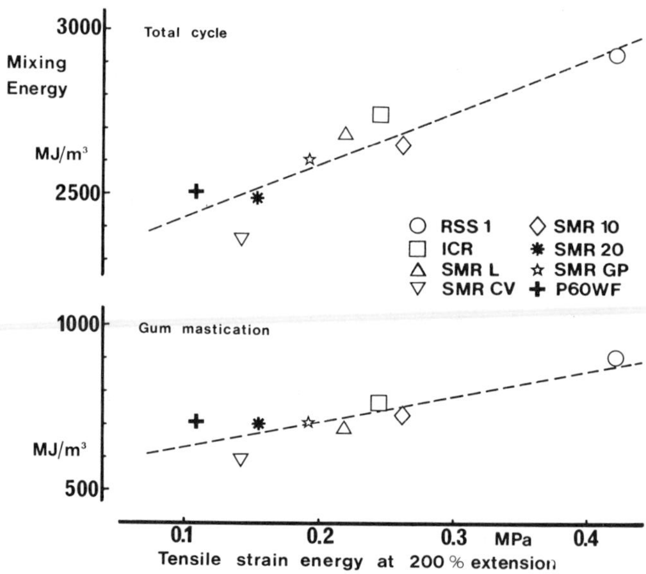

Figure 5 : Mixing energy versus tensile strain energy of gum natural rubber samples

Equation (7) allows the rate of energy absorption to be calculated, using K and n parameters of the power law. (The unit of $EA_{\dot{\gamma}_A-\dot{\gamma}_B}$ is the Pa/s^2 or the $J/m^3\ s^2$).

From the K and n values given in Table II for the gum natural rubber samples, the rates of energy absorption were calculated in the shear rate range of MPT experiments (1.52 - 15.2 sec^{-1}) and plotted versus the mixing energy. As shown in Figure 6, there is a linear relationship between the energy dissipated during the mixing process and the rate of energy absorption of the gum rubber.

This exercise demonstrates clearly that the mixing process is closely controlled by the viscoeleastic responses of the polymer undergoing shear and elongational stresses in the unidirectional flow regions of the internal mixer. Owing to their macromolecular nature, polymers exhibit an elongational flow response as important as their shear flow response but, with respect to the actual flow situations in the internal mixer, only a transitory extensional field has to be considered. Therefore, the behaviour of elastomers in the earlier stages of an elongational process is already indicative of what is happening during the mixing process. Since the true flow situation in a capillary rheometer combines shear and elongational flow fields, a suitable treatment of capillary extrusion experiments can provide information about the ability of a given elastomer to dissipate energy during a mixing process.

Rheology and extrusion of rubber

In the rubber industry, the prime function of extrusion is to preform the material and, whilst several rheological problems are common to both thermoplastics and rubber compounds, the plasticization is less critical with the latter. Therefore, the rheology of the extrusion screw will not be considered here, while it is worth noting that the flow situation in a screw is far from being completely understood.

The extrusion die remains the critical part of an extruder since its design and operational control deeply affect the output and quality of the extrudate. In rubber processing, dies used are generally short and consequently, the elastic character of the extruded compound plays a highly significant rôle, and it is still difficult to separate effects due to the die geometry and to the viscoelastic response of the material.

1. Extrusion through short dies. Let us describe the present state of knowledge about the flow situation in an extrusion die. As shown in Figure 7, in a die of sufficient length, four regions can be distinguished:

(i) the entrance region, mainly within the barrel, where a converging flow field is established. In this region, an extensional flow field is superimposed on a shear flow field.

(ii) the relaxation region of the entrance effects where the perturbations associated with the converging flow field are progressively damped down.

(iii) the viscometric flow region where there is a steady simple shear flow.

(iv) the exit region.

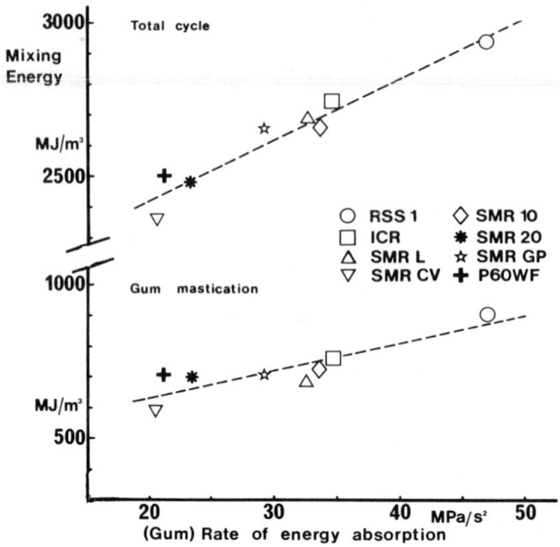

Figure 6 : Mixing energy versus rate of energy absorption of gum natural rubber samples

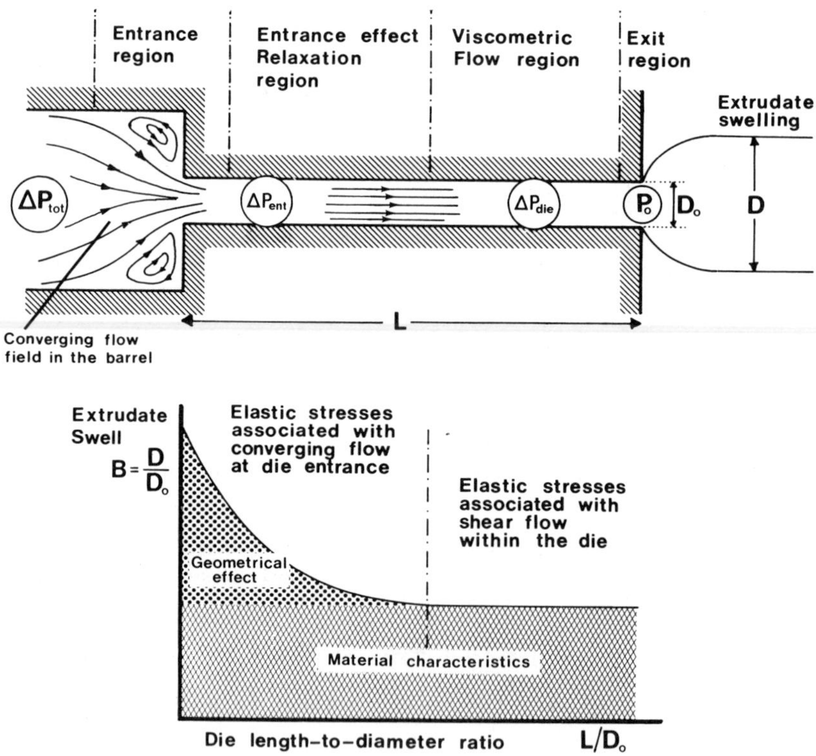

Figure 7 : Flow situations in an extrusion die

The overall extrusion pressure is divided into several components associated with those various regions (26, 27):
- the entrance pressure drop
- the pressure gradient within the die
- the exit pressure.

When considering in comparison, the dependence of the extrudate swell ratio (B) versus the length-to-diameter ratio of the die (L/D_o) it can be seen (Figure 7) that the post-extrusion swelling has two components: firstly, the recovery of the elastic strain stored at the entrance of the die and secondly, the recovery of the elastic deformation related to the shear flow within the die. The former begins to relax in the duct and if the die length is sufficient, only a steady equilibrium value remains. This constant value is practically reached with dies of $L/D_o \geq 20$ and is a material characteristic depending only upon the shear rate and temperature considered. A more complete discussion of extrusion behaviour of rubber compounds can be read in (28). With the short dies used in practice, the steady state is not obtained and therefore, the elastic response (i.e. the post-extrusion swelling) results from a combination of geometrical effects and viscoelastic behaviour of the material. It consequently consists in a transitory rheological state which, despite its great practical interest, has received little attention.

2. Experimental study. Using short dies of various L/D_o ratios (D_o= 6.35 mm; L/D_o=1.088, 2.828 and 5.704; entry angle : 90°), experiments were performed at 121°C with the Monsanto Processability Tester, on a tread band compound (*). Results are given in Table V, in terms of extrusion pressure and 60 sec. relaxed extrudate swell versus the various shear rates used.

By plotting the extrusion pressure versus the length-to-diameter ratio, at constant shear rate, the so-called Bagley correction (23) can be determined, as shown in Figure 8. The Bagley correction e allows the true shear stress σ to be calculated following

$$\sigma = \frac{\Delta P}{2(2\ L/D_o + e)} \qquad (8)$$

where ΔP is the extrusion pressure, L and D_o the die length and diameter respectively.

As can be seen, the correction e increases with the shear rate. Practically, this means that the greater the output and/or the shorter the die, the larger the (geometrical) entrance effects.

Data given in Table V demonstrate a significant effect of die dimensions on post-extrusion swelling, at constant shear rate. As pointed out above, the swelling at the exit of short dies is directly affected by viscoelastic processes in the relaxation region of the entrance effects (see Figure 7). This behaviour can be analysed by considering the post-extrusion swelling versus the

(*) Oil-extruded, solution SBR: 137.5/N375 carbon black: 70/ZnO :70/ stearic acid: 2/2,2,4-trimethyl-1,2 dihydroquinoline, polymerized 1/N-(1,3 dimethylbutyl)-N'-phenyl-p-phenylenediamine : 1.5.

Table V: Extrusion of a tread band compound through short dies at T = 121°C

Dies diameter : D_o = 6.35 mm								
L/D_o:	1.088		2.828		4.628		5.704	
Shear rate	P[a]	E.S.[b]	P	E.S.	P	E.S.	P	E.S.
(sec^{-1})	(MPa)	(%)	(MPa)	(%)	(MPa)	(%)	(MPa)	(%)
0.48	0.74	27.0	1.24	19.4	1.76	18.0	2.07	16.0
1.01	1.06	35.2	1.70	23.0	2.35	20.0	2.76	19.0
2.02	1.26	40.0	2.00	26.0	2.72	21.5	3.17	20.5
4.80	1.59	49.0	2.46	32.0	3.32	25.0	3.86	21.0
9.98	1.93	58.0	2.97	38.0	3.86	28.0	4.66	23.5
19.97	2.21	63.0	3.59	47.0	4.66	32.0	5.34	26.0
47.95	2.76	73.0	4.34	53.0	5.59	34.5	6.62	28.0

(a) : P = extrusion pressure
(b) : E.S. = 60 sec. relaxed post-extrusion swelling, at T = 121°C

'residence time' within the capillary die. By definition, the residence time is given by (29):

$$t_r = \frac{\pi D_o^2 L}{4Q} = \frac{8\ L}{D_o \dot{\gamma}} \tag{9}$$

where L and D_o are the die length and diameter respectively, Q the volumetric output and $\dot{\gamma}$ the shear rate.

Figure 9 is a plot of post-extrusion swelling versus residence time, as calculated from data given in Table V. As can be seen, the

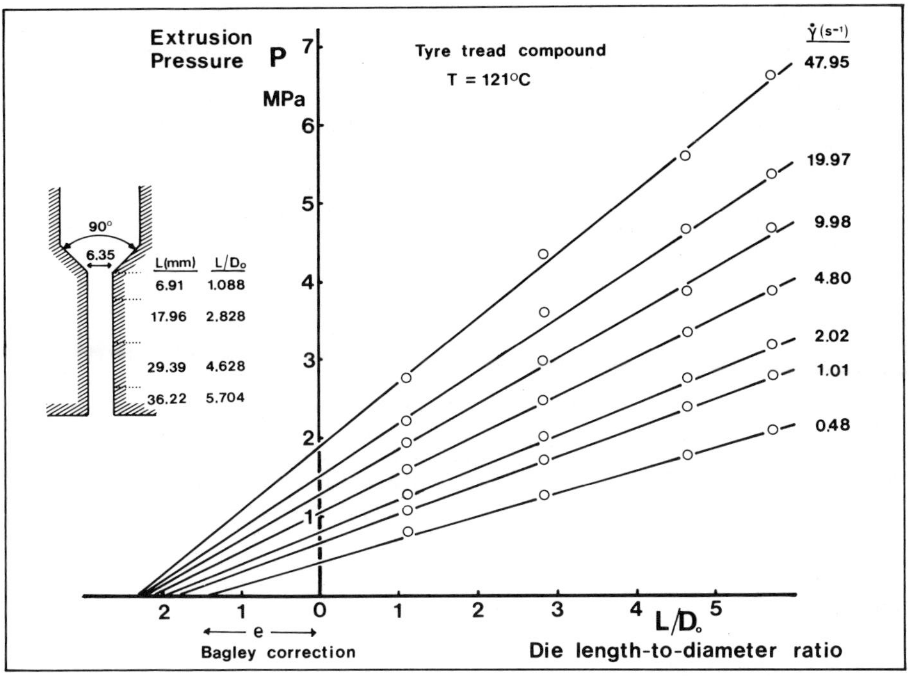

Figure 8 : Bagley correction for a tyre tread compound

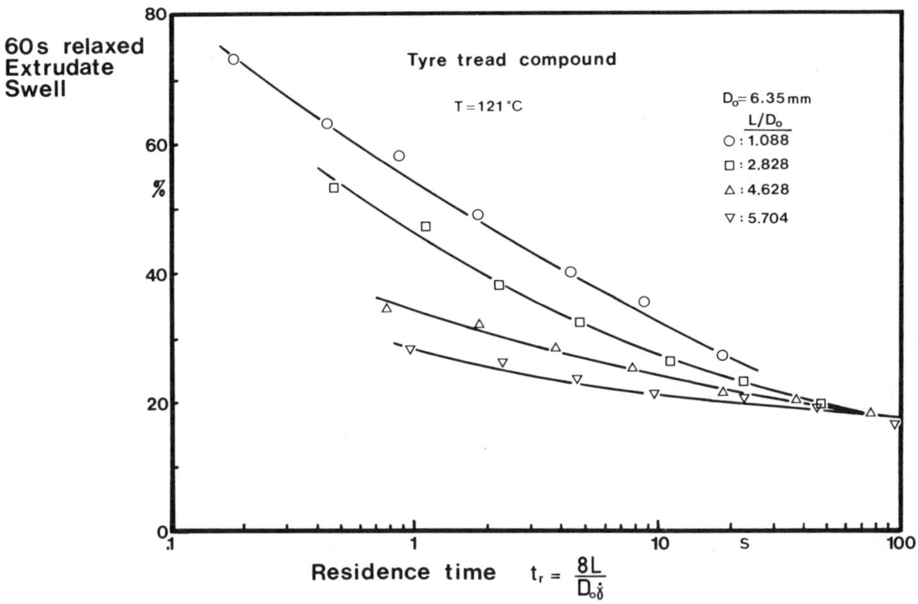

Figure 9 : Extrudate swell versus residence time within extrusion dies for a tyre tread compound

curves obtained for the four dies considered are not superimposed and the smaller the L/D_o ratio and the shorter the residence time, the larger the discrepancy.

With respect to Figure 7, it is clear that with short dies, the post-extrusion swelling includes the entrance effect contribution and the shear response in the viscometric region. Consequently, the extrudate swell not only depends upon the residence time within the die, but also upon the transit time within the entrance zone. Therefore, it is possible to define a corrected residence time, t_{rc}, according to:

$$t_{rc} = \frac{\pi D_o^2}{4Q} (L + L_c) = \frac{8}{D_o \dot{\gamma}} (L + L_c) \quad (10)$$

where L_c is a fictitious extension of the die within the barrel, equal to the length of fully developed capillary flow having a pressure drop equivalent to the excess pressure drop resulting from entrance effects. This fictitious length L_c is obviously related to the Bagley correction, according to:

$$e = -2 \frac{L_c}{D_o} \quad (11)$$

and it follows:

$$t_{rc} = \frac{8\,L}{D_o \dot{\gamma}} - \frac{4\,e}{\dot{\gamma}} = \frac{4}{\dot{\gamma}} \left(2 \frac{L}{D_o} - e\right) \quad (12)$$

When plotting the extrudate swell versus the corrected residence time t_{rc}, an excellent fit is obtained for the four dies (x), as can be seen in Figure 10. This demonstrates that with the short dies of practice, the entrance converging flow region plays a very important rôle in post-extrusion swelling.

According to equation (12), it follows that when the L/D_o ratio is not significantly larger than e, small variations of the shear rate (and hence of the output) will directly affect the corrected residence time and therefore the extrudate swell. The shorter the time, the larger the effect.

Post-extrusion swelling in fact, is not really a processing problem in itself, as long as it remains constant since, in this case, one can always manage it to control extrudate dimensions. The real problem is the extrudate swell variations arising during the process.

The above considerations explain why with the short dies of industrial practice, rubber extrudates frequently exhibit swelling variations which are actually compensated for by adjusting the take-away conveyor speed. This is however not always satisfactory and the results presented here suggest that a suitable design of the entrance geometry could reduce post-extrusion swelling variations.

(x) In fact, with the shorter die (L/D_o = 1.088), negative or zero values were obtained for t_{rc}. These values were not taken into consideration because the errors in determining the Bagley correction were probably too large.

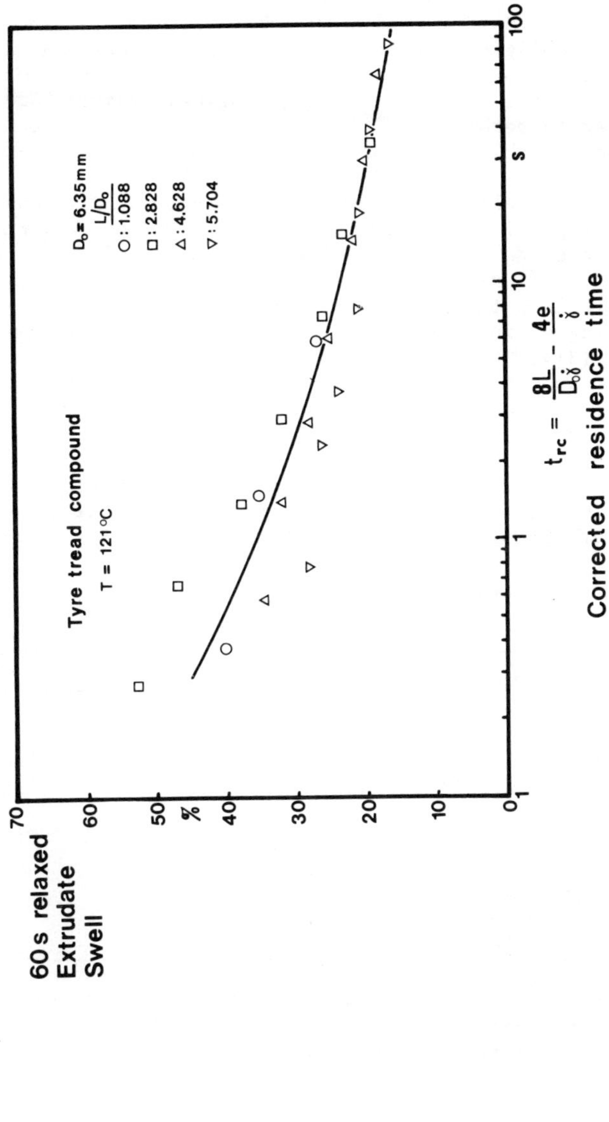

Figure 10 :Extrudate swell versus 'corrected' residence time within extrusion dies for a tyre tread compound

Rheology of heterogeneous rubbery materials

Heterophasic polymer systems are finding increasing applications because they lead to products with improved engineering performance and sometimes better and useful melt processing characteristics. Even when considering the simplest case of two phase polymer blends, the rheological behaviour and hence the processing characteristics are far from being well understood, despite considerable current efforts (30). Good examples of such materials are the so-called ' 'thermoplastic elastomers', which due to their two-phase morphology, exhibit peculiar rheological properties. The present understanding of relationships between flow properties and processing behaviour of polymer allows however the appropriate rheological evaluations to be performed in order to recommend adequate processing conditions. This can result in considerable time and cost savings when compared with the classical processing trial-and-error approach.

Santoprene, a new class of thermoplastic rubber introduced by Monsanto, is heterophasic polymer material offering good examples of peculiar and interesting flow properties. This material is prepared by in-situ dynamic vulcanization of a conventional thermosetting rubber polymer within a thermoplastic matrix undergoing shear, and therefore producing a fine dispersion of crosslinked micron-size rubber particles (31). Since these fully cured particles retain their identity at processing temperatures and flow conditions, such materials exhibit the rheology associated with dispersions of fine particulate fillers in a molten polymer (32).

Using dies of various L/D_o ratios (i.e. : D_o = 3 mm, L/D_o = 12; D_o = 1.51 mm, LD_o = 20; D_o = 0.762 mm; L/D_o = 30), experiments were performed at 204°C with the Monsanto Processability Tester, on three hardness grades of Santoprene, as described in Table VI.

Table VI : Santoprene grades tested

Material	Shore Hardness[a]	Specific Gravity[b]
Santoprene 201-73	73A[c]	0.98
Santoprene 201-87	87A	0.96
Santoprene 203-40	40D	0.95

(a) Test method: ASTM D-2240
(b) Test method: ASTM D-297
(c) Letter indicates shore hardness scale

Piston rates were selected in order to cover a shear rate range of 10-5200 sec^{-1} with the three dies used. The corrections for entrance pressure loss were determined according to the Bagley method (23), in order to calculate the corrected shear stress. Results, as given in Table VII in terms of apparent viscosity and (instantaneous) extrudate swell, are essentially similar to the ones published by Goettler, Richwine and Wille (32).

Table VII: MPT results on Santoprene grades (T = 204°C)

Santoprene grade:			201-73		201-87		203-40	
DIE D_o	DIE L/D_o	$\dot{\gamma}_a$ (s^{-1})	η_a (kPa.s)	E.S. %	η_a (kPa.s)	E.S. %	η_a (kPa.s)	E.S. %
3	12	10	4.39	4	4.59	3	4.75	6
		20	2.44	-	2.79	-	2.97	9
		40	1.30	4	1.50	4	1.60	9
		80	0.71	5	0.83	6	0.93	11
1.51	20	82	0.70	8	0.82	12	0.91	15
		160	0.41	10	0.50	17	0.56	20
		320	0.23	11.5	0.27	19	0.30	24
		640	0.13	14	0.15	24	0.17	30
0.762	30	638	0.13	13	0.15	19	0.17	29
		1305	0.072	19	0.090	29	0.10	43
		2609	0.042	22	0.052	38	0.058	53
		5190	0.024	26	0.030	47	0.034	63
Power law fit: η_a =			$28.1\dot{\gamma}_a^{0.170-1}$		$29.7\dot{\gamma}_a^{0.190-1}$		$30.7\dot{\gamma}_a^{0.202-1}$	

$\dot{\gamma}_a$: apparent shear rate; η_a : apparent viscosity; E.S. :extrudate swell

Results given in Table VII show that the viscosity versus shear rate variation of these three Santoprene grades fits the power law over the entire range from 10 to 5200 s^{-1}. Both the viscosity and the extrudate swell at constant shear rate increase with decreasing rubber particle content (or increasing hardness). Thus, these olefinic thermoplastic vulcanizates essentially behave like highly filled fluids in flow. When compared with unvulcanised rubber (see previous sections) or polypropylene - EPDM blends (32), the extrudate swell appears low and there is no Newtonian viscosity plateau.

By a suitable treatment of these MPT data, it is however possible to derive some more information on flow behaviour and morphological effects of these olefinic thermoplastic vulcanizates (OTV).

Using the extrusion pressure (ΔP) measured in the barrel just above die entrance, the apparent shear stress is calculated according to the well-known equation.

$$\sigma_a = \frac{\Delta P}{4\ L/D_o} \qquad (13)$$

where L and D_o are the die length and diameter respectively. When introducing the Bagley correction in terms of equivalent length-to-diameter ratio, $(L/D_o)_{ent}$, the true shear stress is obtained as follows

$$\sigma = \frac{\Delta P}{4\left[\frac{L}{D_o} + \left(\frac{L}{D_o}\right)_{ent}\right]} \tag{14}$$

In fact, as seen in the previous section (see Figure 6), the overall extrusion pressure ΔP consists of several components, i.e.

$$\Delta P = \Delta P_{ent} + \Delta P_{die} + P_{exit} \tag{15}$$

where ΔP_{ent}, ΔP_{die} and P_{exit} are the entrance pressure drop, the pressure gradient within the die and the exit pressure respectively. Since P_{exit} is generally very small, it can be neglected in first approximation and it follows that:

$$\Delta P_{die} = \Delta P - \Delta P_{ent} \tag{16}$$

And since the Bagley correction basically takes into account the entrance pressure drop effects (and the negligible exit effects), it can be written that:

$$\sigma = \frac{\Delta P}{4\left[\frac{L}{D_o} + \left(\frac{L}{D_o}\right)_{ent}\right]} \simeq \frac{\Delta P_{die}}{4\,\frac{L}{D_o}} \simeq \frac{\Delta P - \Delta P_{ent}}{4\,\frac{L}{D_o}} \tag{17}$$

Equation (17) allows an estimation of the entrance pressure drop, ΔP_{ent}, to be calculated, knowing the extrusion pressure and the Bagley correction, as follows:

$$\Delta P_{ent} \simeq \Delta P \left\{ \frac{\left(\frac{L}{D}\right)_{o\ ent}}{\frac{L}{D_o} + \left(\frac{L}{D}\right)_{o\ ent}} \right\} \tag{18}$$

At the entrance of a die, there is a converging field which produces a strong extensional flow. Cogswell (24) has suggested a procedure for calculating an 'apparent extensional viscosity', η_{EC}, from measurements of entrance flow. Since neither the local normal stresses nor the local extension rate can be calculated directly from measurable quantities, it is necessary to make a number of assumptions about the flow. Cogswell assumes that the entrance pressure drop, ΔP_{ent} can be represented as a sum of two terms, one related to shear and the second to extension. He further assumes that the fluid follows streamlines that result in the minimum pressure drop, and for a fluid having a power law viscosity function, he derived the following equation for an 'apparent extensional viscosity', η_{EC} :

$$\eta_{EC} = \frac{9(n+1)^2(\Delta P_{ent})^2}{32\,\eta_{(\dot{\gamma}_a)}\cdot\dot{\gamma}_a^2} \tag{19}$$

where n is the flow index, and $\eta_{(\dot{\gamma}_a)}$ the apparent viscosity at shear rate $\dot{\gamma}_a$.

Using experimental results obtained with the three Santoprene grades studied, the entrance pressure drop and the apparent extensional viscosity were calculated according to equations (18) and and (19). Results are given in graphical terms in Figures 11 and 12 respectively.

Figure 11 shows that, as expected, the entrance pressure drop increases with increasing shear rate. In addition, the ΔP_{ent} vs $\dot{\gamma}_a$ curves of the three samples are significantly different, obviously related to the rubber particle content. The lower the hardness the higher the entrance pressure loss. As far as entrance effects are indicative of the magnitude of melt elasticity, it is easily understood that the higher the rubber particle content, the higher the melt elasticity. This result associated with the fact that the extrudate swell of Santoprene decreases with increasing rubber content indicates the limits of the analogy between OTV's and highly filled polymer. Indeed, the analogy can only be correct if 'elastic fillers' are considered. Furthermore, since the curves in Figure 11 seem to converge at very high shear rate, it is clear that the rheological behaviour of OTV's is controlled by the existence of a connective structure of rubber particles, progressively destroyed when shear increases, as suggested by Goettler, Richwine and Wille (32).

As far as Cogswell's approach is correct in estimating elongational viscosity from entrance pressure drop, Figure 12 presents very interesting results. As can be seen, the 'extensional viscosity' increases with decreasing hardness of the material. This can be explained using the connective structure model pointed out above. These particle associations however are more likely to form while the melt is not under shear and therefore, the 'apparent elongational viscosity' at die entrance is decreasing when shear rate increases. As the connective structure is destroyed, the entrance flow resistance is weaker. In addition, it seems that the η_{EC} vs $\dot{\gamma}$ curves exhibit a maximum, depending upon the rubber content. The higher the rubber content, the lower the shear rate for this maximum. As can be seen in Figure 12, the 40D grade exhibits an increasing elongational viscosity below 30-40 s^{-1} and a decreasing η_{EC} at higher shear rate. Lower hardness grades probably exhibit the same behaviour but with the maximum arising at lower shear rate. Whilst not yet fully understood these results clearly demonstrate that the processing rheology of Santoprene thermoplastic rubbers is dominated by singular effects associated with the connective structure of rubber particles.

Conclusions

The rheological approach to rubber processing allows a better understanding of the flow behaviour of elastomers to be achieved. Obviously the picture is far from being complete and due to their complexity, rubber compounds are exhibiting phenomena not yet completely understood.

The specific examples presented in this paper, either in mixing rubber compounds, or in shaping through short dies or when considering the melt rheology of thermoplastic elastomers, demonstrate clearly that one has to pay as much attention to elongational flow as to shear flow in rubber processing situations.

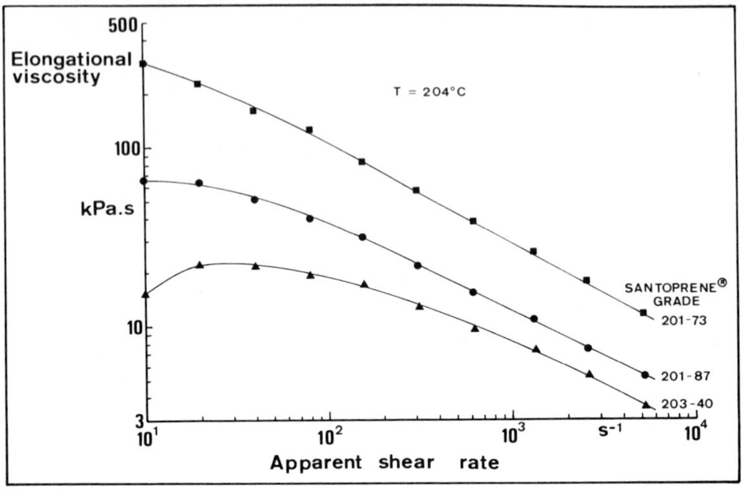

Figure 11 :Entrance pressure drop versus shear rate for Santoprene thermoplastic elastomers at 204°C

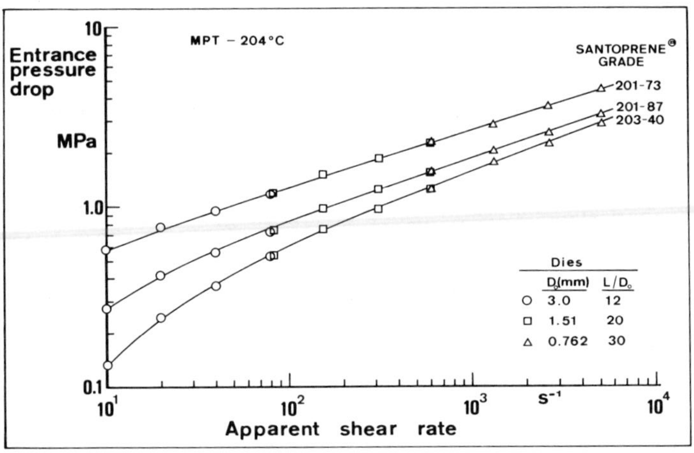

Figure 12 :Apparent elongational viscosity versus shear rate for Santoprene thermoplastic elastomers at 204°C

Correct and imaginative use of suitable rheological equipment, such as the Monsanto Processability Tester, allows pertinent information to be obtained at lower cost than the classical trial-and-error approach. Furthermore, better cost efficiency can be achieved when interpreting capillary rheometer results in terms of processing and selecting the best operating conditions with respect to material flow properties.

Acknowledgements

The author wishes to thank Dr. M. Bristow of Malaysian Rubber Producers Research Association - Brickendonbury, U.K. who kindly supplied various grades of natural rubber. He is grateful to Dr. L.A. Goettler and Mr. J. Sezna from Monsanto Polymer Products Company, Akron, Ohio, for the experimental data on Santoprene thermoplastic rubbers used in this paper.

Literature

(1) H. PALMGREN - Rubb. Chem. Technol. 48, 462 (1975)
(2) J.M. FUNT - "Mixing of Rubbers" - RAPRA, Shrewsbury, England (1977).
(3) E.C. BERNHARDT - "Processing of thermoplastic materials" van Nostrand Reinhold, New York, pp 424-446 (1959).
(4) J.M. McKELVEY - "Polymer Processing" - chap.12, Wiley, New York (1962).
(5) P.S. JOHNSON - Elastomerics, 115 (1), 9, January 1983.
(6) N. TOKITA and T. PLISKIN - Rubb. Chem. Technol.,46,1166 (1973).
(7) I. MANAS-ZLOCZOWER, A. NIR and Z. TADMOR - Rubb. Chem. Technol., 55 (5), 1250 (1982).
(8) N. NAKAJIMA - Rubb. Chem. Technol.,53, 1088 (1980).
(9) W.M. WIEDMANN and H-M. SCHMID - Rubb.Chem.Technol. 55, 363(1982)
(10) P.K. FREAKLEY and W.Y. VAN IDRIS - Rubb.Chem.Technol.,52, 134 (1979)
(11) H. LAMB - "Hydrodynamics" - chap. XI, pp 583, Dover Publ., New York (1945).
(12) W.R. BOLEN and R.E. COLWELL - SPE Antec, 14, 1004 (1958).
(13) J.L. WHITE and N. TOKITA - J. Appl. Polym. Sc., 12, 1589 (1968).
(14) J.L. WHITE - Polym. Eng. Sc., 19 (11), 818 (1979).
(15) K. ODA, J.L. WHITE and E.S. CLARK - Polym. Eng. Sc., 18, 25 (1978).
(16) G.R. COTTEN - Plastics and Rubber : Processing, 4 (3),89 (1979).
(17) J. MEISSNER - Trans. Soc. Rheol., 26, 405 (1972).
(18) N. NAKAJIMA - Polym. Eng. Sc., 19, 215 (1979).
(19) G.M. BRISTOW - NR Technology, 10 (3), 53 (1979).
(20) J.L. LEBLANC - Plastics and Rubber Processing and Applications, Vol.1, No 2, 187 (1981)
(21) G.M. BRISTOW and A.G. SEARS - NR Technology, 11 (3), 45 (1980); ibid, 11 (4), 77 (1980); ibid, 12 (2),36 (1981)
(22) S. MONTES and J.L. WHITE - Rubb. Chem. Technol, 55, 1354 (1982)
(23) E.B. BAGLEY - J. Appl. Phys., 28, 624 (1957)
(24) F.N. COGSWELL - Trans. Soc. Rheol., 16, 383 (1972)
(25) J.M. McKELVEY - "Polymer Processing" - Chap. 2, sec. 2-7, p 51, Wiley, New-York (1962)

(26) C.D. HAN and M. CHARLES - Trans. Soc. Rheol., 15, 371 (1971)
(27) J.L. WHITE and A. KONDO - J. Non-Newtonian Fluid Mech.,3,41(1978)
(28) J.L. LEBLANC - Rubb. Chem. Technol., 54, 905 (1981)
(29) E.B. BAGLEY, S.M. STOREY and D.C. WEST - J. Appl. Polym. Sc., 7, 1661 (1963)
(30) J.L. WHITE; A.P. PLOCHOCKI and h. TANAKA - Polym. Eng. Rev., Vol. 1, N° 3, 218 (1981)
(31) A.Y. CORAN and R. PATEL - Rubb. Chem. Technol.,53, 141 (1980)
(32) L.A. GOETTLER, J.R. RICHWINE and F.J. WILLE - Rubb. Chem. Technol., 55, 1448 (1982)

RECEIVED February 6, 1984

CARBON FIBERS IN POLYMERS

12

Mesophase Formation in Polynuclear Aromatic Compounds

A Route to Low Cost Carbon Fibers

R. J. DIEFENDORF

Materials Engineering Department, Rensselaer Polytechnic Institute, Troy, NY 12181

Pitch precursor carbon fibers have the potential of providing good mechanical properties at low cost. The high preferred orientation of graphite basal planes that is necessary for high modulus is obtained by forming a discotic liquid crystal pitch. Requirements for liquid crystallinity are discussed for model compounds. These considerations are generalized and expanded to describe liquid crystal formation in pitches.

The element carbon in the form of graphite has the highest specific stiffness (stiffness/kg) and highest theoretical, tensile strength. While graphite is stiff and strong in the basal plane networks, it is compliant and weak normal to the planes. Furthermore, the layer planes can easily shear over each other in perfect graphite. A problem to be solved in producing a high performance carbon fiber is how to align the graphite basal planes roughly parallel to the fiber axis, and have a high defect structure such that basal plane shear is difficult. Since graphite can not be dissolved and only melts at very high temperature under pressure, the fiber must be made from a precursor which can be converted subsequently to carbon. Desired characteristics of the precursor are that it:

1) develops an oriented structure which is carried through to the carbon fiber,
2) has a high yield of carbon,
3) is cheap, and
4) is easy to handle through the various processing steps such as spinning and carbonization.

Pitch materials appear to be able to meet all four of these criteria.

Carbon Yield

The carbon yield from pitches can be understood by studying the yields of the generic components of pitch using the method of corbett (1,2). The amounts of saturates, napthene-aromatics, polar aromatics & asphalentenes can be determined for different pitches for various

0097–6156/84/0260–0209$06.00/0

heat treatment temperatures and times, Figure 1. Saturates and napthene-aromatics are observed to evaporate during heat treatment leaving the less volatile polar aromatics and asphaltenes, Figure 2. The carbon yield at 800°C is found to be linearly related to the amount of polar aromatics and asphaltenes in the precursor pitch, Figure 3. Although yield depends upon more than molecular weight, the carbonization of model polynuclear aromatic compounds indicates that a minimum molecular weight of 400 is needed to prevent volatilization before molecular weight increasing, condensation reactions occur. Figure 4. From these results an asphaltene pitch would give a high yield and would be low cost.

Orientation

Liquid crystallinity is determined by the length to diameter ratio, L/d, for rod-like molecules, and diameter to thickness, D/h, for disc-like molecules. The results of calculations predict liquid crystallinity in rod-like molecules with L/d ratios greater than 6.3 and in disc-like molecules with D/h ratios greater than 3.5 (4,5). Liquid crystallinity is observed in the rod-like para polyphenylene series when the L/d ratio approaches or exceeds 6.3, Figure 5. While both quinque- and sexi- paraphenyl have sufficiently high L/d ratios to form rod-like liquid systems, no disc-like polynuclear aromatic compounds are available with a high enough D/h ratio to form a liquid crystal. Ovalene, which has ten fused rings, is close (D/h=3.0-3.2), but liquid crystallinity has not been observed for it or for the binary system of ovalene/coronene, Figures 5. Ovalene has a melting point of 472°C. Presumably, fused ring systems with higher D/h would have higher melting points, and thermal decomposition would occur before the crystalline/mesophase transition temperature is reached.

Brooks and Taylor (6) showed that heat treatment of coal tar and petroleum pitch forms a liquid crystal (mesophase) with discotic symmetry. Generally, mesophase formation was assumed to occur by polymerization and condensation of smaller pitch molecules to form large planar aromatics. It is true that the molecular weight increases upon heat treatment, but our studies on carbon yield indicates that volatilization might be more important than polymerization and condensation. If so, the original isotropic pitch already may contain the mesogenic species.

Mesophase can be observed in pitches which have a higher molecular weight than ovalene yet can have a softening point well below the melting point of ovalene. Several factors cause this. Firstly, a wide range of species with differing molecular weights and structures are present in a pitch. Melting point depression will occur, and mesophase regions may be uncovered. For example, in a binary eutectic system, two nonmesogenic compounds often will "uncover" a liquid crystalline region near the eutectic composition where the melting point depression is greatest, Figure 6. Secondly, the wide range of species makes crystallization sluggish, so that the system can be supercooled and mesophase may form before crystallization occurs or the glass transition temperature is reached. Thirdly, mesophase pitches often appear to consist of flexibly linked polynuclear aromatics with three to six rings with a number average molecular weight of 1000 or more. By analogy with calculated effects of adding flexible links to rigid-rod molecules, the softening point is more

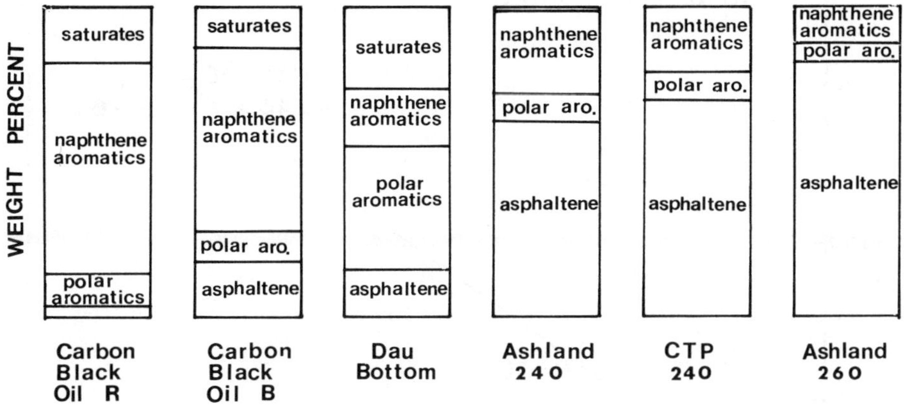

Figure 1. The saturate, napthene-aromatic, polar aromatic, and asphaltene content of valious pitches and heavy oils.

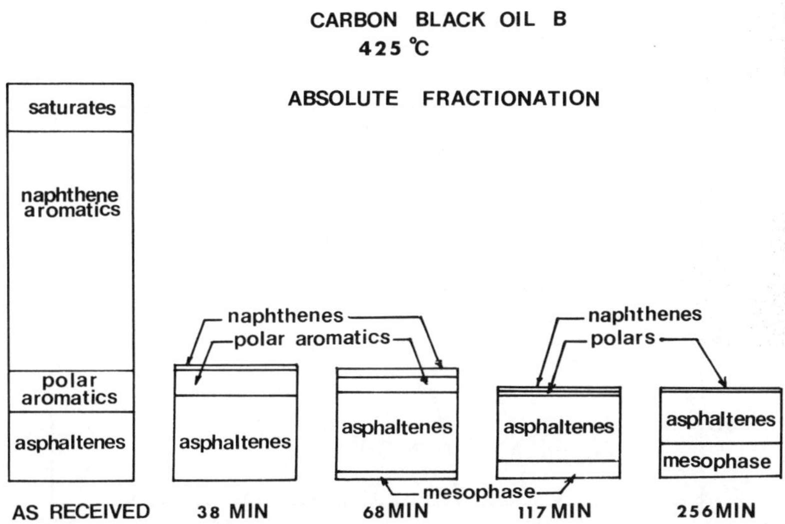

Figure 2. The generic fractions for a catalytic cracker oil upon heat treatment at 425C.

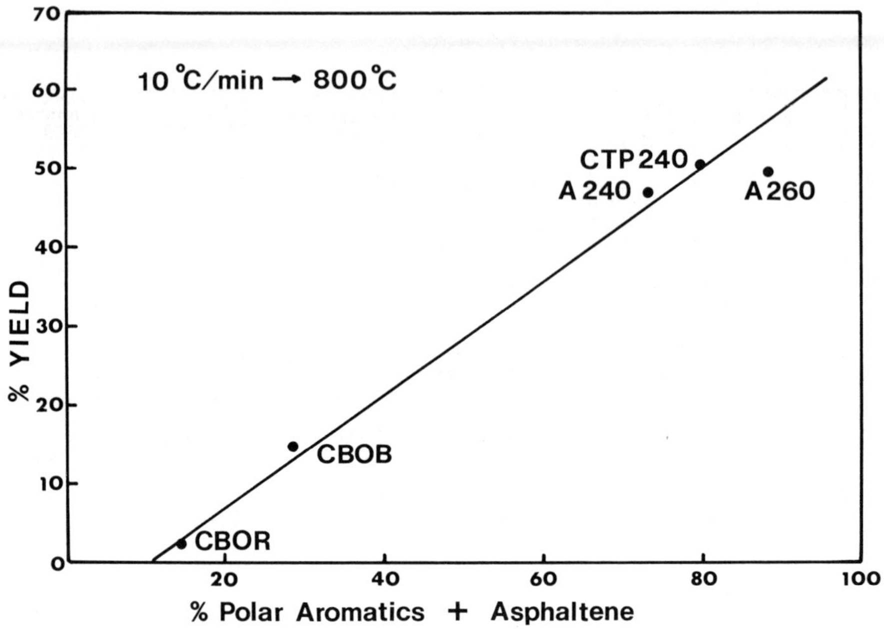

Figure 3. The carbon yield of various pitches and heavy oils versus polar aromatic and asphaltene content.

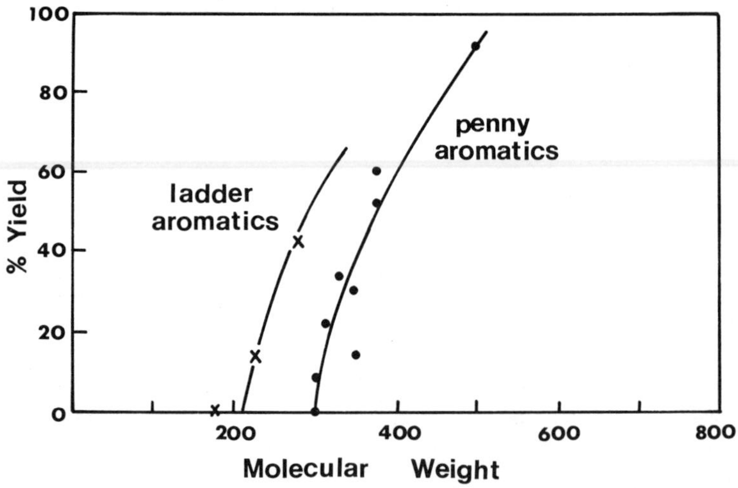

Figure 4. The carbon yield of polynucleat aromatic compounds upon heating to 800C.

depressed than the mesophase/isotropic transition temperature. Fourthly, short alkyl side arms on the polynuclear aromatic cores may enhance mesophase formation similarly to the effects of alkyl tails on rod-like liquid crystals. Finally, mesophase pitches have many of the characteristics of lyotropic liquid crystal systems. In this case, low molecular weight nonmesogenic species act as solvents or plasticizers for the high molecular weight mesogenic species. Without these low molecular weight plasticizers, the high molecular weight species would decompose before a sample could be heated hot enough to form fluid mesophase.

These five factors which enhance formation of mesophase are hard to study independently. Chemical reactions are occuring at the temperatures required to study pitch mesophase such that the system characteristics are continuously changing. Viscosity is also frequently high, and optical microstructures can be misleading as evidence for mesophase formation. Quenched and slow cooled specimens may appear isotropic and 100% mesophase respectively.) Pitches are not very ideal solutions. Hence, mesophase formation is altered by solubility considerations, which considerably clouds interpretation of experimental data. Finally, the complexity of a pitch makes the analytical characterization very difficult (10,11). Simple averages such as number average molecular weight are just insufficient, and even these values may be hard to measure properly (12). These problems make for ambiguities in the interpretation of experimental data.

A gross simplification is to classify pitch molecules into two general groups. One group consists of mesogens with real mesophase/isotropic transition temperatures (T_R), Figure 7. Generally, the higher molecular weight, more aromatic species would belong to this group. Mesogenic species where the mesophase/isotropic transition temperature is just above the softening point would be easily disrupted by small additions of non-mesogenic species. Larger mesogenic species would have higher mesophase/isotropic transition temperatures, and would require larger amounts of non-mesogenic species to destroy the mesophase. The second group consists of nonmesogens which have virtual transition temperatures (T_V), e.g., the mesophase/isotropic transition temperature is below the softening point and only an isotropic liquid is observed. Nonmesogens which have vitrual transitions temperatures close to the softening point will not be very disruptive to a mesophase, while a nonmesogen with a low virtual transition temperature would be very disruptive. Also, an isotropic pitch may contain an appreciable quantity of mesogens. The only requirement is that the transition temperature (range) be virtual, e.g., below the softening point. Similarly, a mesophase pitch may contain appreciable quantities of nonmesogenic species. In a two phased, isotropic & mesophase pitch, both phases will contain mesogenic & nonmesogenic species but with different relative concentrations.

The proof that an isotropic pitch can contain an appreciable mesogenic component can be obtained by removing a sufficient amount of the nonmesogenic species. Volatilization has been successfully used to reduce the lower molecular weight nonmesogenic species. However, appreciable chemical reaction occurs at the temperatures required to remove the less volatile nonmesogens. Hence, the chemical reactions may be the cause for mesophase formation rather than volatilization. Solvent fractionation of pitches circumvents this problem as the

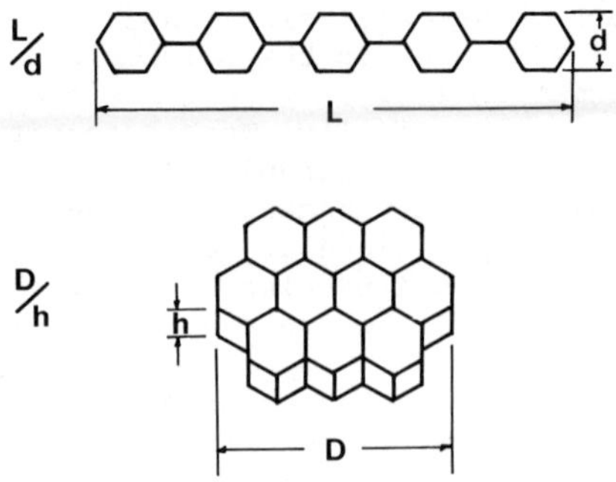

Figure 5. The chemical structures of quinque parapolyphenylene and ovalene.

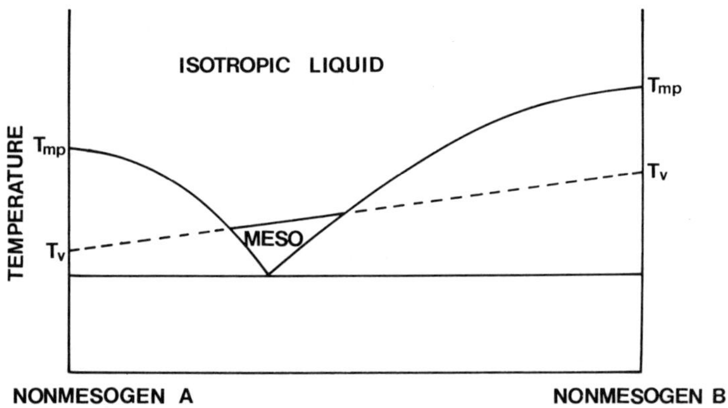

Figure 6. A schematic binary phase diagram of two nonmesogens which show mesophase formation near the eutectic composition.

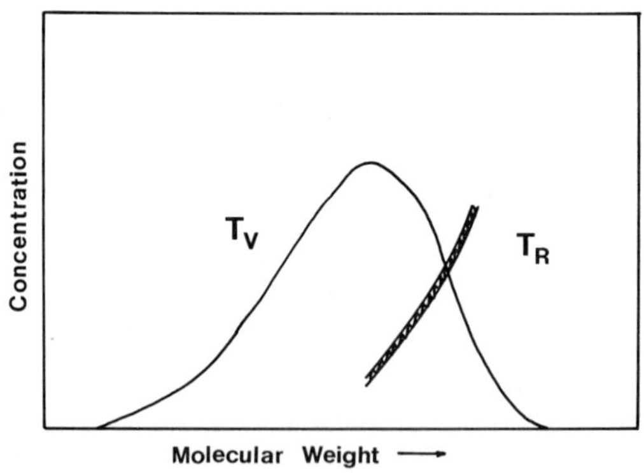

Figure 7. A schematic illustrating that an isotropic pitch can be considered to consist of mesogens and nonmesogens.

separation can be carried out at ambient temperature. The wide range of available solvents allows a tailoring of the separation process (13). This technology is a critical factor in producing low cost carbon fibers as it allows a 100% mesophase pitch to be made which can be spun at temperatures where chemical reactions occur slowly. (The 100% mesophase structure produces high modulus in the carbon fiber.) Earlier pitch precursor carbon fibers were made from pitches containing only 55% mesophase to allow a suitable, lower spinning temperature.

Solvent separations depend on solvent/solute interactions, and the mesogen/nonmesogen separation obtainable with solvents may not be very good as solvent/solute interactions are not controlled by mesogenicity. However, the toluene insoluble fraction of a commercial pitch, Ashland A240, directly forms mesophase upon heating. The gel permeation chromatogram for the whole pitch and the toluene soluble portion shows, by difference, that the toluene insoluble molecular, weight distribution is quite broad, Figure 8. Although the toluene soluble portion of the pitch still contains an appreciable fraction of mesogens, they do not appear to be separable with common solvents without including too high an amount of nonmesogens. A final proof that mesophase formation is mainly due to the physical separation of nonmesogenic species rather than chemical reactions was performed by adding 25% of Ashland A240 pitch back to the toluene insolubles of A240 mesophase pitch. Almost all the mesophase was destroyed upon heating even for this relatively minor addition of virgin pitch (14).

Grouping pitch molecules into mesogenic and nonmesogenic fractions allows construction of a "pseudo" binary phase diagram, Figure 9. Transition temperatures become broadened to transition temperature ranges because of the distributions. Conventional heat treatment of isotropic pitches would increase the temperature to the heat treatment temperature with little change in composition. As the nonmesogens evaporate, the composition moves towards the mesophase region. If the temperature is sufficiently high and the time sufficiently long, the composition will change until mesophase is formed. By contrast, solvent fractionation removes the nonmesogens and the composition will be in the mesophase region. Heat treatment need only be severe enough to cause softening and sufficient domain growth to produce microscopically observable mesophase. In general, the mesophase/isotropic transition is not observed in mesophase pitches because of molecular weight increases at higher temperatures and coking.

There are many types of natural and synthetic pitches. Some consist of large polynuclear aromatic cores with a large number of relatively long alkyl side arms which solubilize and soften the species. Others with similar softening points consist of relatively small polynuclear cores with no or at most a few methyl groups attached. The structures for a solvent fractionated, catalytic cracker mesophase pitch are illustrated in Figure 10, (15). The lowest molecular weight species are nonmesogenic but plasticizing. The highest molecular species won't soften by themselves, but stabilize the mesophase structure. The most probable molecule has a molecular weight of about 1000. Molecular structures are constructed by combining results from IR, UV/VIS, NMR, VPO, GPC, HPLC and elemental analysis (16). These structures are only based on the average analytical results. Even for averages, a wide range of structures, only one of which is illustrated, can be drawn. Since molecules can deviate from these average results, a much wider range of structures will exist with

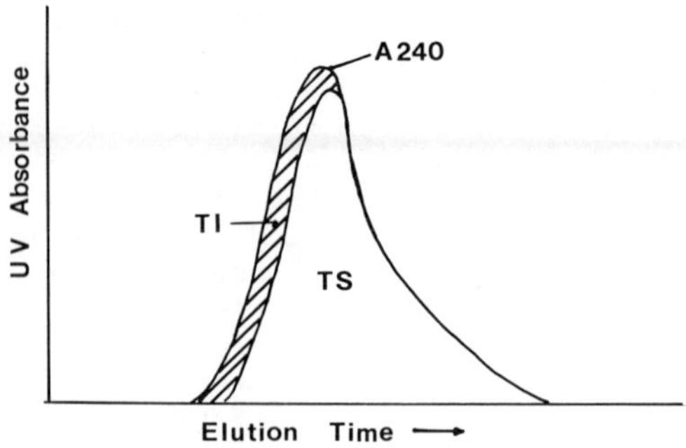

Figure 8. The gel chromatogram for Ashland A240 pitch and the toluene soluble (TS) fraction.

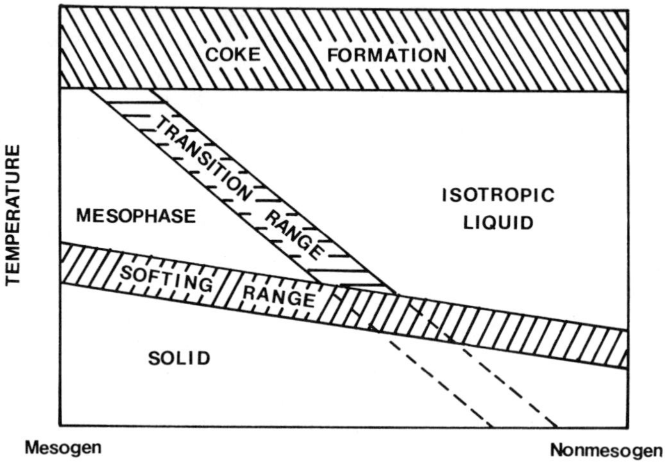

Figure 9. A pseudo binary phase diagram which explains mesophase formation in a pitch.

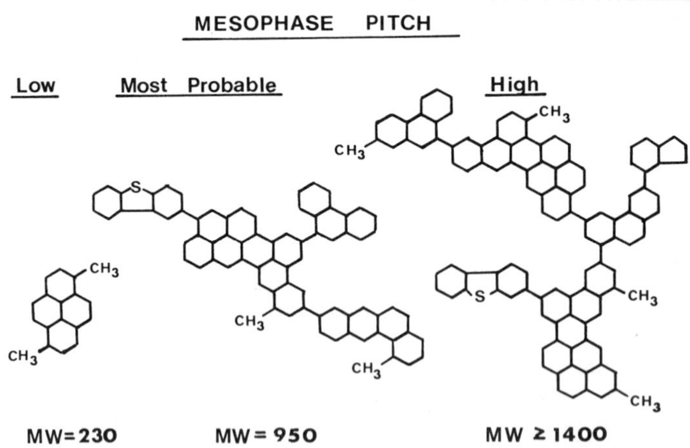

Figure 10. Typical average chemical structures for the toluene insoluble fraction of a catalytic cracker pitch.

the only requirement being that the distribution must give the average analytical results.

Processability

Mesophase pitch can be melt spun into fibers. Major requirements are that the pitch is:

1) thermally stable at spinning temperatures,
2) free from particulates, and
3) rheologically acceptable.

Solvent fractionated mesophase pitches can be made which meet these three requirements. As spun strength of pitch fibers is low but in the same range as some other commercially produced textile fibers. Subsequent processing to carbon fiber is similar to that used for polyacrylonitrile precursor carbon fiber. A more complete description of this process and fiber properties are presented in the paper of Riggs & Diefendorf in this volume (17).

Conclusions

Calculation and model compounds show that a length to diameter ratio of 6.3 is sufficient to form a rigid rod liquid crystal. Similarly, calculation indicates that a diameter to thickness ratio of at least 3.5 is required to form a discotic liquid crystal. Mesophase formation in pitches is enhanced by the broad molecular weight distributions, flexible links between polynuclear aromatic elements and alkyl side groups. Generically, pitches can be considered to consist of two species: mesogenes and nonmesogens. Solvent fractionation can be used to partially separate the two types of species. Finally, a 100% mesophase pitch can be produced which can be melt spun and processed into a carbon fiber with high mechanical properties.

Acknowledgment

The author is indebted to his students, particularly: S.H. Chen, D.M. Riggs, W.C. Stevens, and J. Venner. Special thanks should be given to S.H. Chen for the artwork and D. Ruddy for typing the manuscript.

References

1. Riggs, D. M. ; Diefendorf, R. J. "Factors Controlling the Thermal Stability and Liquid Crystal Forming Tendencies of Carbonaceous Materials", Preprint, CARBON '80, Baden-Baden, 1980, pp. 330-333.
2. Corbett, L. W. "Relationship Between Composition and Physical Properties of Asphalt" presented at the Assoc. of Asphalt Paving Technologists, Kansas City, Missouri, Feb. 1970.
3. Riggs, D. M.; Diefendorf, R. J. "Thermal Stability of Aromatic Compounds", Extended Abstracts, 14th Biennial Conf. on Carbon, Pennsylvania State Univ., June 1979, pp. 350-351.
4. Flory, P. J.; Ronca, G. "Theory of Rod-like Particles - Part II: Thermotropic Systems with Orientation-Dependent Interactions", Mol. Cryst. Liq. Cryst., 1979, Vol. 54, pp. 311-330.
5. Alben, R., Phy. Rev. Lett. 24, p. 1041, 1970.

6. Brooks, J. D.; Taylors, G. H. "Chemistry and Physics of Carbon", P. L. Walker, Jr. and P. A. Thrower, Eds., Marcel Dekker, New York, 1968, Vol. 4, pp. 243-268.
7. Smith, G. W. "Phase Behavior of Some Condensed Polycyclic Aromatics", Mol. Cryst. Liq. Cryst., Vol. 64, Letts., pp. 15-17.
8. Chen, S. H.; Diefendorf, R. J. "Mesophase Formation in Polynuclear Aromatic Compounds", Extended Abstracts, 16th Biennial Conf. on Carbon, U.C. San Diego, July, 1983, pp. 28.
9. Riggs, D. M.; Diefendorf, R. J., This volume.
10. Greinke; O'Conner, L. H. "Determination of Molecular Weight Distributions of Polymerized Petroleum Pitch by Gel Permeation Chromatography with Quinoline Eluent", Anal. Chem., 52, 1980, pp. 1877-81.
11. Chen, S. H., Stevens, W. C. and Diefendorf, R. J. "Molecular Weight and Molecular Weight Distributions in Pitches" International Symposium on Carbon, Toyohashi, Japan, 1982, pp. 59-72.
12. Chen, S. H.; Diefendorf, R. J. "Molecular Weight Determination of Pitches", Extended Abstracts, 16th Biennial Conf. on Carbon, U. C. San Diego, July, 1983, p. 433.
13. Venner, J.; Diefendorf, R. J., This volume.
14. Riggs, D. M.; Diefendorf, R. J. "A Phase Diagram for Pitches", Preprint, CARBON '80, Baden-Baden, 1980, pp. 326-329.
15. Chen, S.H.; Diefendorf, R. J. Unpublished data.
16. Chen, S. H.; Diefendorf, R. J., This volume.
17. Riggs, D. M.; Diefendorf, R. J., This volume.

RECEIVED January 17, 1984

Pitch–Solvent Interactions and Their Effects on Mesophase Formation

J. G. VENNER[1] and R. J. DIEFENDORF

Materials Engineering Department, Rensselaer Polytechnic Institute, Troy, NY 12181

High performance pitch carbon fiber can only be manufactured from mesophase forming pitches. Generally, commercially available pitches require an extensive heat-treatment to increase the molecular weight of the pitch by removing or condensing the low end species, prior to the mesophase formation. A selective solvent can also be used to remove the low molecular weight, mesophase-inhibiting species. The multidimensional solubility parameters of a solvent provide qualitative information as to the type of molecular interactions a solvent is capable of undergoing, and the suitability of a solvent for separating a mesophase-forming fraction from the pitch. The effects of varying the selective solvent and altering the pitch to solvent ratio are reported in this paper.

High strength, high modulus, pitch based carbon fiber can be made from a precursor pitch which forms an anisotropic phase, or a mesophase. The aligned pitch molecules in the mesophase provide the fundamental structural elements of graphitizable carbons (1,2). Unfortunately, commercially available pitches, such as Ashland A-240, are complex mixtures of mesophase forming species and mesophase disrupting species(3). A commercial pitch must be processed to remove the low molecular weight mesophase inhibiting species to provide a pitch fraction that will form a coalesced mesophase rapidly upon melting (Figure 1). Formerly, it was thought that a pitch could not form 100% coalesced mesophase without undergoing extensive heat-treatments to increase molecular weight and aromaticity(4). The non-mesophase forming species were thought to react to form larger, planar molecular blocks more capable of alignment. Recently, however, Riggs and Diefendorf(5) patented the use of selective solvent as an alternative to extensive heat-treatments. A selective solvent can

[1] Current address: Exxon Enterprises Materials Division, Fountain Inn, SC 29644.

0097–6156/84/0260–0219$06.00/0

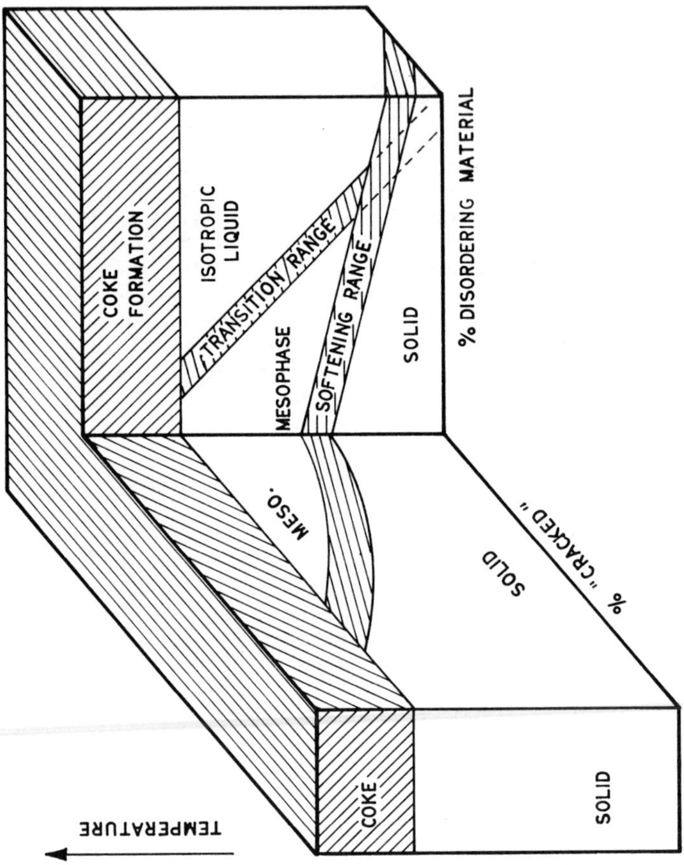

Figure 1. Three-dimensional pseudo-phase diagram of a typical pitch.

dissolve many of the lower molecular weights, disordering species leaving the larger, more orientable species, insoluble. The insoluble distributions are capable of forming a 100% coalesced structure rapidly upon melting provided enough of the low molecular weight species have been removed, and the insolubles have the proper structure to form a mesophase. The mesophase forming tendencies of a pitch fraction depend largely on the structure of the precursor pitch, and on the extraction capability of the solvent. Therefore, the selection of an extracting solvent and extraction conditions are critical.

This research focused on answering three fundamental questions:

1. What solvents can be used to extract a fraction from a commercial pitch (such as A-240) which is capable of forming a coalesced mesophase rapidly upon melting, and what criteria will predict their effectiveness?
2. What extraction conditions are critical?
3. Can the extraction be modified, or tailored to produce desirable fractions?

Solvent Selection

The regular dissolution of a solid in a non-electrolytic solvent has been described in detail by Hildebrand(6) who showed that mole fraction of solid dissolved depends on the molecular weight of the solid, (which affects its enthalpy of fusion), its melting temperature, the system temperature, and the difference between the solbility parameter of the solvent and the solute.

$$\log \frac{1}{x_2} = \frac{\Delta H^F m(Tm-T)}{4.575TmT} - \frac{\Delta Cp}{4.575} \frac{Tm-T}{(T)} + \frac{\Delta Cp}{1.987} \log \frac{Tm}{T} + \frac{V_2}{4.575} (\delta_1-\delta_2)^2 \Phi_1 2 \quad (1)$$

x_2 = mole fraction of solid soluble in solvent

ΔH^F_m = heat of fusion at the melting point

Tm = melting point of the solid

ΔCp = the difference between the heat capacities of the solid and liquid solute

V_2 = molar volume of solute

δ_1, δ_2 = solubility parameters of solvent and solute

and Φ = volume fraction of the solvent

If the solute, the concentration, and the system temperature are kept constant, then differences in solubility depend solely on the difference between the solubility parameter of the solid and the solvent.

Riggs showed the solubility of a pitch in a series of solvents with increasing solubility parameters followed a bell shaped curve, with maximum solubility occurring when the solubility parameter of the solute and solvent were equal(5). In general, non-polar solvents with total solubility parameters ranging from 8.0-9.5 were desirable for extracting mesophase forming fractions from pitches. In this work, A-240 pitch was dissolved in 38 solvents, both dispersive and polar (Table I). A plot of weight percent dissolved versus solubility parameter shows that A-240 solubility in dispersive solvents does follow the typical bell shaped curve, however, the total parameter solubility fails to describe A-240 solubility in polar solvents (Figure 2). Alcohols, aldehydes, acetonitrile, acetone, dioxane, and other non-dispersive solvents dissolve less than the predicted percentage of A-240. As Hildebrand's solubility parameter was intended for use in dispersive systems, this is not surprising.

The inadequacy of using the total solubility parameter to describe the behavior of polar and hydrogen bonded solvents in polymer systems led Charles Hansen(7) to generalize Hildebrand's parameter. The total solubility parameter was broken down into three components representing the three types of interactions possible between molecules: dispersive, polar, and hydrogen bonding.

$$\frac{\Delta E}{Vm} = \frac{\Delta Ed}{Vm} + \frac{\Delta Ep}{Vm} + \frac{\Delta E_h}{Vm} \tag{2}$$

$$\delta_T^2 = \delta_d^2 + \delta_p^2 + \delta_h^2 \tag{3}$$

Although generalizing the solubility parameter into three dimensions is not entirely rigorous (as the geometric mean expression is used to describe hydrogen bonding interactions), the three dimensions provide an improved qualitative understanding of solute-solvent interactions, and should be useful in predicting solubility of pitches. A solubility map can be created by plotting the weight percentage of pitch dissolved as a function of dispersive and polar components of the solubility parameter. Symbols or contours are used to highlight regions of reduced or enhanced solubility. A solubility plot is shown in Figure 3 where solubility is indicated by the shape of the symbol used to show location on the plot. Strong dispersive solvents with moderate polar character are most able to dissolve A-240, while solvents with small dispersive parameters or large polar parameters, are generally incapable of dissolving an appreciable percentage of the pitch. In between these two regions exists an area of partial solubility. Solvents, such as benezene, dioxane, toluene, xylene, carbon tetrachloride, DMF, or DMA, dissolve between 75 and 96% of A-240. Because many of the lower molecular weight species can be removed by these solvents, their insoluble fractions should be ideal candidates for forming mesophase.

The mesophase forming tendencies of the insoluble fractions were investigated by encapsulating samples under vacuum, then heat-soaking to 375°C for two hours. A list of the mesophase forming fractions is presented in Table II. As expected, solvents with solubility parameters located at the outer edge of the enhanced solubility region extract A-240 fractions capable of forming coalesced mesophase. These solvents range in total solubility parameter from 8.65 to 12.14 $(\frac{cal}{cc})^{1/2}$, with dispersive parameters ranging from 8.7 to 9.3. Solvents, such as DMF and acetaldehyde, show appreciable polar parameter while DMF and dioxane are strongly hydrogen bonded. MeCl,

TABLE I

Effect of Solubility Parameter on Weight % A240 Dissolved

Solvent	Abbreviation	Solubility Parameter $(cal/cc)^{1/2}$	Wt.% A240 Dissolved*
1,2,4 Trichlorobenzene	TCB	10.45	99.8
Bromobenzene	Bb	10.60	99.8
Acetophenone	A	10.60	99.8
Nitrobenzene	Nb	10.62	99.7
Tetrahydrofuran	Tf	9.52	99.7
Quinoline	Q	10.75	99.6
Pyridine	Py	10.61	99.3
Chlorobenzene	Cb	9.57	98.9
Tetramethylurea	Tmu	10.60	98.9
Ethylene bromide	Eb	11.70	98.6
Trichloroethylene	Tce	9.20	98.0
Carbon disulfide	Cd	10.00	97.5
Ethylene dichloride	Ed	9.80	97.3
Chloroform	Cf	9.2	96.7
Methylene chloride	Mc	9.93	96.0
Dimethyl acetamide	DMA	11.10	95.0
Benzene	B	9.15	88.8
Xylene	X	8.80	88.6
Dioxane	D	10.00	87.4
Toluene	T	8.91	86.8
Dimethyl formamide	DMF	12.14	86.1
Carbon tetrachloride	Ct	8.65	82.8
Decalin	Dl	8.80	66.1
Diethyl amine	Da	7.97	63.6
Acetaldehyde	Aa	9.86	49.8
Benzyl alcohol	Ba	11.64	48.0
Nonane	Nn	7.70	44.8
Decane	Dn	7.70	44.7
Acetone	Act	9.77	40.5
Dimethyl sulfoxide	DMS	12.93	36.6
Octane	On	7.60	36.3
Nitromethane	Nm	12.90	31.8
Hexane	Hn	7.30	31.1
Pentanol	Pol	10.61	22.7
Acetonitrile	Acn	11.90	19.5
Propanol	Ppl	11.97	16.0
Ethylene glycol	EG	16.30	7.1
Methanol	M	14.28	2.5

*Room temperature 1g pitch/100 ml solvent.

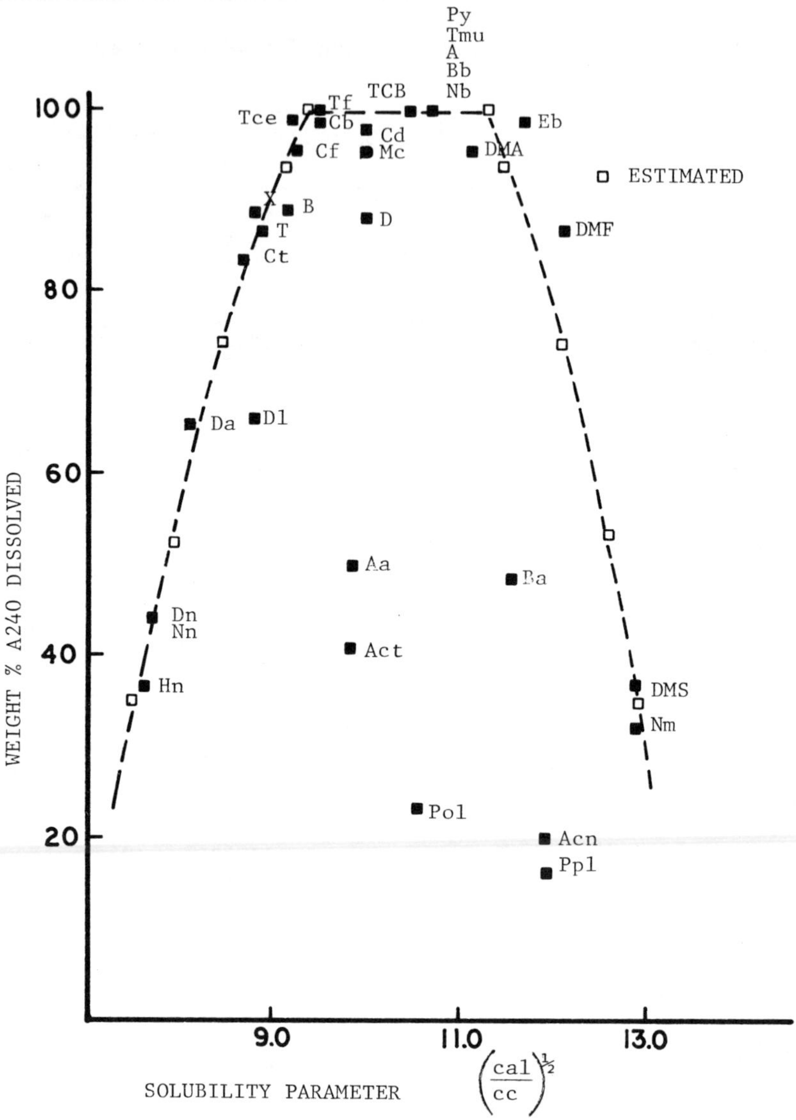

Figure 2. Effect of overall solubility parameter on A240 solubility.

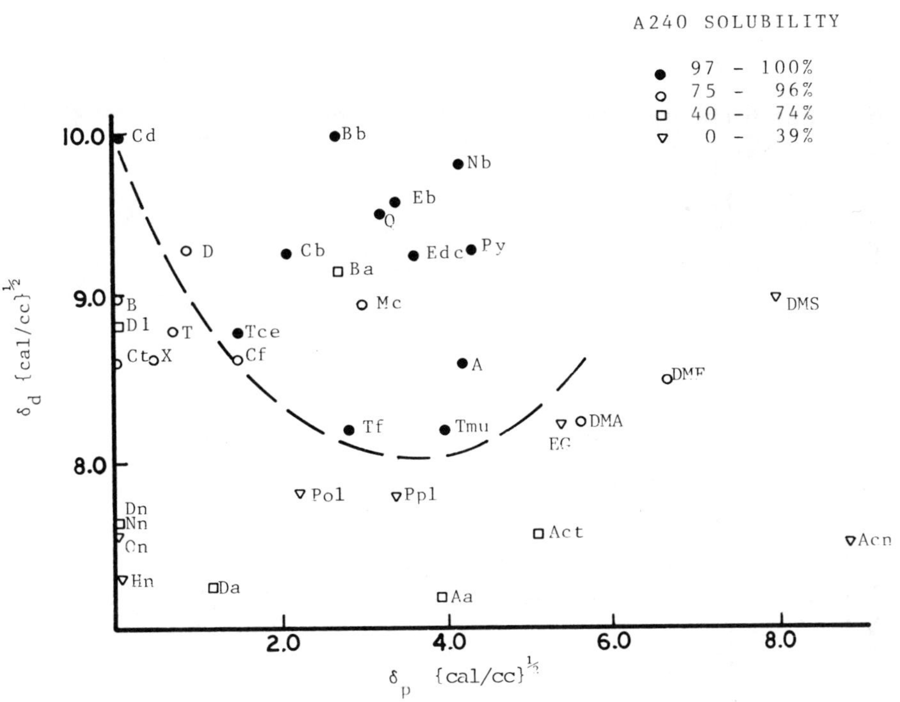

Figure 3. Two-dimensional solubility map. δ_d versus δ_d

TABLE II

Solubility and Molecular Parameters of Solvents which Separate Mesophase Forming Fractions

Solvent	δ_T $(cal/cc)^{1/2}$	δ_d $(cal/cc)^{1/2}$	δ_p $(cal/cc)^{1/2}$	δ_h $(cal/cc)^{1/2}$	Wt.% A240 Dissolved
Benzene (B)	9.15	9.0	0.0	1.0	88.8
Xylene (X)	8.80	8.65	0.5	1.5	88.6
Dioxane (D)	10.00	9.3	0.9	3.6	87.4
Toluene (T)	8.91	8.8	0.7	1.0	86.8
Dimethyl formamide (DMF)	12.14	8.5	6.7	5.5	86.1
Carbon tetrachloride (Ct)	8.65	8.7	0.0	0.3	82.8
Decalin (Dl)	8.80	8.8	0.0	0.0	66.1
Diethyl amine (Da)	7.97	7.3	1.1	3.0	63.6
Acetaldehyde (Aa)	9.86	7.2	3.9	5.5	49.8
Benzyl alcohol (Ba)	11.64	9.0	3.1	6.7	48.0

chloroform, and DMA dissolve 95% of A-240 leaving too small a quantity of precipitate to test for mesophase formation. The last four insoluble fractions, from benzyl alcohol, acetaldehyde, diethylamine, and decalin form mesophase spheres within an isotropic matrix. These fractions represent 34 to 50% by volume of the A-240, and still contain a large percentage of disordering components. However, the spheres of mesophase are promising signs which suggest that given suitable heat-treatment, a larger percentage of the pitch could form a coalesced mesophase, if a suitable extraction technique could be found.

Characterization of Select Mesophase Forming Samples

The molecular weight distribution of a pitch affects its ability to form mesophase(8, 5). Both Gel permeation chromatography (GPC) and high pressure liquid chromatography (HPLC) were used to monitor the molecular weight distributions of select soluble and insoluble fractions. In this way, it was possible to directly determine the effects of solvent fractionation on the molecular distribution of A-240 and gain insight to the distribution required for rapid mesophase conversion.

The molecular weight distributions (MWD's) of DMF and toluene insoluble fractions are shown in Figure 4. Both solvents dissolve approximately 86% of the A-240 under the conditions tested, but as toluene and DMF are located on opposite sides of the enhanced solubility region of Figure 3, they extract different distributions from A-240. Toluene dissolves some large, presumably more dispersive species insoluble in DMF, while DMF dissolves some of the smaller presumably more polar species insoluble in toluene. Both insoluble fractions possess very broad molecular weight distributions, as indicated by the large amount of overlap occurring between the soluble and the insoluble distributions. The breadth of the distributions imply that the criterion for solution is complex and that solvents separate a pitch on the basis of the molecular structure as well as the molecular weight of species. Species with identical retention times can be soluble or insoluble.

Solubility is governed not only by the solubility limits of the individual pitch species, but will also be affected by associations known to occur in pitch solutions. These may enhance or reduce the solubility of a particular pitch component. Precipitation of associated clusters or micelles broaden the MWD of the extract pitch and may provide the mechanism for direct mesophase conversion upon melting. The lower molecular weight species reduce viscosity to permit low temperature realignment of the larger, more aromatic species(1). From the MWD's, a larger fraction of pitch may be capable of mesophase formation if an optimized separation technique could be developed.

High pressure liquid chromatography can be used to supplement and confirm GPC results. In HPLC, separations depend on the relative interactions between solute, solvent, and the stationary phase. By gradually varying the carrier solvents' composition, (gradient chromatography), the relative attractions can be controlled. Figure 5 shows HPLC of A-240 and the TI fractions. A Waters C-18 Radial pak column and a MeOH-pyridine gradient were used. Results have been corrected so the areas beneath the curves accurately

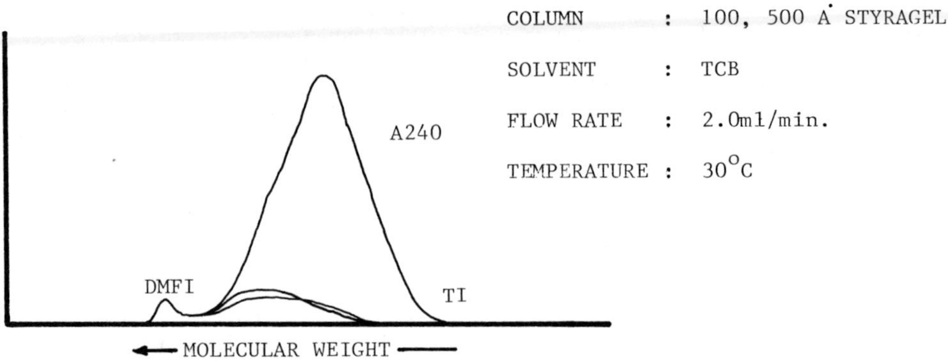

Figure 4. Relationship between DMFI, TI, and A240.

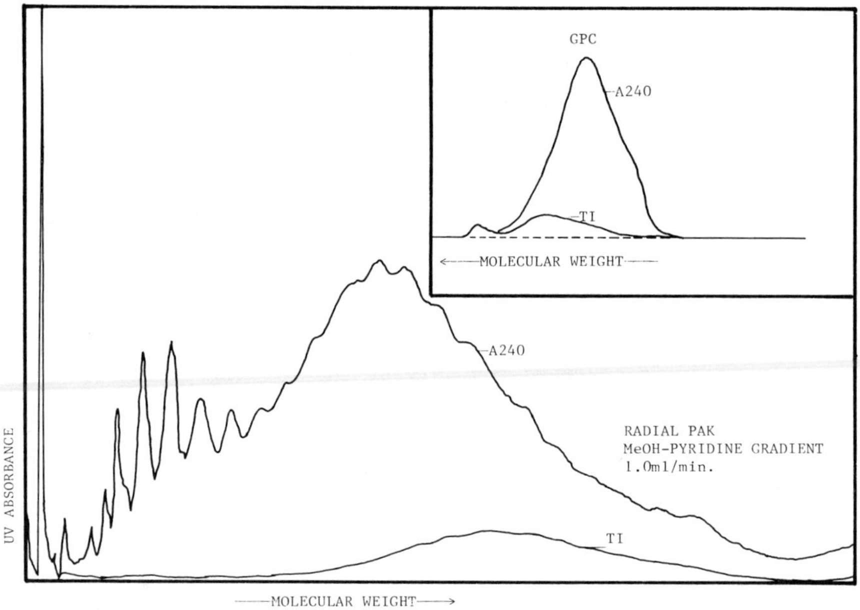

Figure 5. HPLC and GPC of relationship between A240 and TI.

represent the fraction's weight percent of A-240. The HPLC curve differs from the GPC curve in that several discrete peaks are visible early in the separation, followed by a large modulated continuous curve. The mesophase forming fractions lack the discrete peaks visible early in the A-240 separation as the smaller molecules have been stripped away by the fractionating solvent. Both GPC and HPLC results support three conclusions:

1. The mesophase forming fraction represents only a fraction of the pitch, but is made up of a variety of molecular weights and types. If a different fractioning technique could separate the pitch according to mesophase forming tendencies, yields could be improved.
2. The distributions required for mesophase formation are broad, but not specific. Both DMFI and TI distributions are different, yet they both form coalesced mesophase.
3. The solubility of high ends is strongly related to the dispersive character of the fractionating solvent.

The C/H ratio and the softening points of the meso-formers were determined to find out if other measurable characteristics are common to the mesophase forming fraction of A-240. The C/H ratio of the mesophase fractions range from 1.72 to 1.85 compared to 1.5-1.6 for the nonmesophase fractions. Unfortunately, the C/H ratios of the mesophase forming fractions do not correlate with the weight % dissolved or molecular weight distributions (HPLC or GPC results) of the extract. The softening points of the mesophase forming fractions increase markedly over the nonmesophase forming fractions. However, these softening points also fail to correlate with the C/H ratios or the chromatography results. These results do reinforce the general trends of increasing molecular weight, C/H ratio and softening points of the extract with increasing weight % dissolved, but they do not indicate specific requirements for rapid mesophase conversion upon melting. The mesophase forming fractions are similar in that many of the smaller, low molecular weight species have been removed, yet the distributions are broad. The presence of some low molecular weight components enhance rapid coalescence of mesophase by lowering viscosity enabling molecular rearrangement at temperatures where few reactions occur. Different solvents appear to extract different lower molecular weights plasticizer components. (Some of these components are able to reduce viscosity without disrupting the mesophase transition of the extract.) In this way, solvents can extract characteristic distributions with unique properties, many of which are capable of forming mesophase directly upon melting.

Effect of Altering Pitch to Solvent Ratio

Understanding the interactions occurring between a distribution of physically and chemically different species is extremely complex. When a model compound is dissolved in a solvent, composition of the precipitate will not differ from those of the solute. However, when the solute is multicomponent, the composition of any insoluble fraction will depend on the relative proportions of the individual components (governed by their individual solubility limits) and their interactions. Therefore, properties of the pitch extracts should relate to the ratio of pitch to solvent used for the separation. As the pitch to solvent ratio increases, the relative proportions of insoluble components should change, altering the properties of the extract as this simplified example will show.

A hypothetical solute is composed of equal percentages of two noninteracting components, "A" and "B". Component "A" is soluble up to 1 gram per liter while component "B" is soluble up to 3 grams per liter solvent, and the mixture up to 4 grams per liter. Initially, as this solute is added to the solvent, it is completely soluble (Figure 6). The weight % dissolved remains at 100% until the solubility limit of "A" is reached, at two grams solute per liter. As the solute to solvent ratio increases further, solubility drops rapidly to a point where if component "B" were infinitely soluble, it would asymtotically approach 50% dissolved. For finite solubility of component "B" the curve drops rapidly again and asymptotically approaches 0% dissolved. During this time the composition of any precipitate will change depending on the relative proportions of "A" and "B". Positive interactions would shift this curve to the right, negative to the left. The effect of increasing system temperature would be similar to a positive interaction. Because a pitch consists of a continuous gradient of molecular species, its behavior is expected to be much more complex. However, the properties of any extracted fraction should strongly depend on pitch to solvent ratio.

A-240 was added to a group of solvents at 10, 30, and 60g per liter concentration to examine the effect of pitch/solvent ratio on the properties of an insoluble fraction. The following plot (Figure 7) shows the solution behavior changes markedly from solvent to solvent. Aromatic solvents are much less affected by concentration than non-aromatics. DMF and toluene show identical weight percents dissolved at low concentrations, but as the concentration is increased, solubility in DMF drops much faster than in toluene.

The difference may relate back to differences between the parameters of the two solvents and the components they extract. Toulene is more dispersive and a larger dispersive parameter may be required to dissolve the larger aromatics at higher concentrations (DMF being more polar dissolves a larger percentage of the polar species). Even though the percentage of A-240 dissolved by toluene remains constant throughout the changes in pitch/solvent, the properties of the extracted pitch should change.

GPC was used to examine differences between the extracted fractions. Previous GPC results show 10g/1 DMFI possesses a larger fraction of high molecular weight species than TI at the same pitch to solvent ratio. At increased ratios, DMF dissolves a smaller percentage of the A-240 (Figure 8), leaving an insoluble fraction incapable of directly forming coalesced mesophase upon melting. Many of the high molecular weight components become insoluble, but this increase is offset by a greater increase in mid and low molecular weight components, reducing the overall molecular weight average of the fraction. The increased presence of low molecular weight species restricts mesophase formation. In contrast, toluene dissolves an almost constant weight percentage of A-240 throughout the concentration range tested. However, as the GPC plot shows, the distribution of insolubles is changed. TI at 60g/1 contains molecules from the high end of the molecular weight distribution that are soluble at low concentrations. The increase in high ends would normally improve mesophase forming tendencies but the higher concentration TI extract

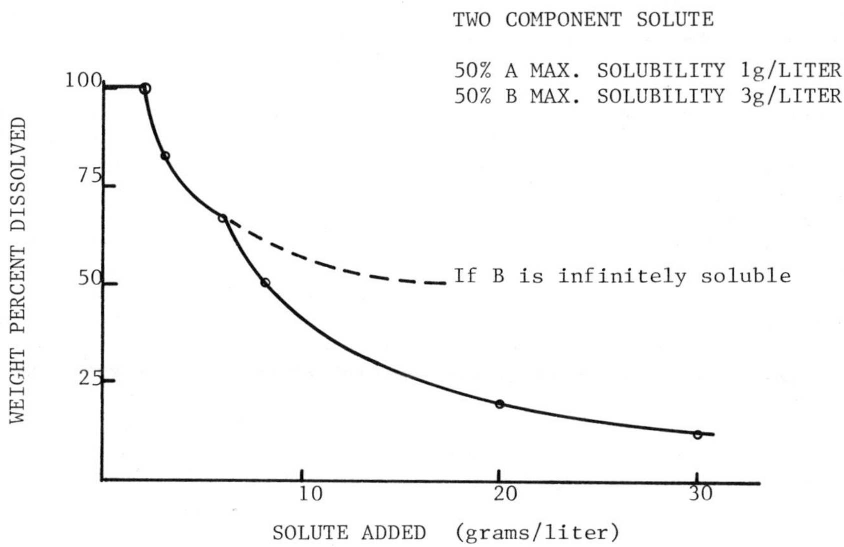

Figure 6. Effect of solute to solvent ratio on the solubility of a mixture.

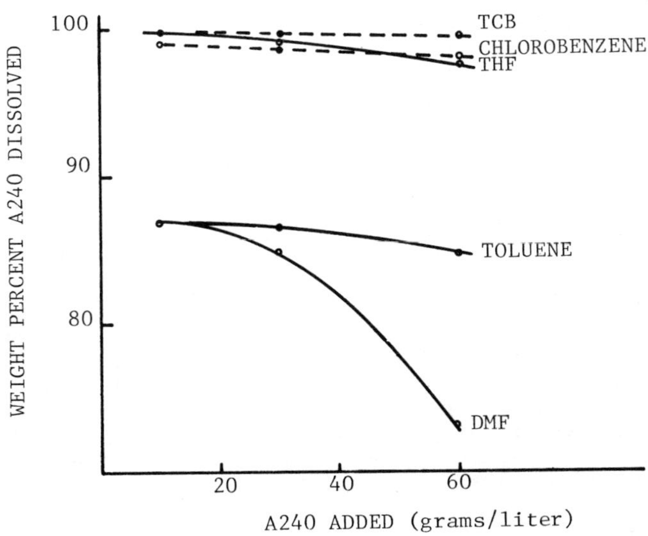

Figure 7. Effect of pitch to solvent ratio on A240 solubility.

still contains isotropic regions after being heat-soaked to 375°C for 2 hours. It appears that the increase in isotropic regions is for structural reasons. The C/H ratio of both the TI and DMFI dropped from 1.85 to 1.69 and 1.63, respectively. The drop in C/H ratio correlates with DMFI GPC results but appears to contradict the TI GPC results. The TI 60g/l may contain associated polar species which are not dispersed by the room temperature GPC, or it may contain high molecular weight aromatics that are not as condensed as the lower concentration extracts. These clusters may have precipitated from the solutions due to their increased polarity over the dispersive pitch-solvent solution or due to their increased interactions at higher concentrations. Regardless of the mechanism involved, it is clear the effect of altering pitch/solvent ratio depends greatly on the solvent. The weight % dissolved in dispersive solvents is not affected to as great an extent as in polar solvents. However, the molecular weight distribution and properties of all insoluble fractions change. An extract can be tailored because properties are sensitive to pitch/solvent ratio. Over-effective solvents can be moderated by increasing concentration, and ineffective solvents can be enhanced by reducing concentration.

Solvent Mixtures

Solvent mixtures provide an alternative tailoring process. Theoretically, the combination of two solvents located on opposite sides of a region of enhanced solubility should result in a mixture more effective than either solvent individually. A mixture of toluene and DMF was used to test this hypothesis and as expected, the blend is more effective as it dissolves 91.3% of the A-240 compared to approximately 87% by either solvent individually. The insolubles from this extraction form a fine domained mesophase when heated to 375°C.

Moderating an over-effective solvent by blending it with an ineffective solvent can also shift solubility parameters to the edge of the enhanced solubility region. A 50:50 (volume) mixture of methylene chloride, and DMSO (a mixture of an over-effective and an inaffective solvent) dissolves 82% of the A-240, and the extract verges on forming a coalesced mesophase after only moderate heat-treatment.

Both examples illustrate that solvent blends can be used to extract mesophase forming fractions from the A-240. Figure 9 suggests several other solvents may form successful blends. Solvents normally inappropriate should be blended to shift the solubility parameters of the mixture closer to the edge of the enhanced solubility region where the blend can extract a pitch fraction capable of complete coalescence. Solvent blends can also be used to moderate solvent costs without impairing solvent effectiveness.

Conclusion

A wide variety of solvents can be used to extract pitch fractions that will form mesophase upon melting. The properties of the mesophase depend on the pitch, the solvent, the insoluble fraction, and the time and temperature of the heat-treatment. A multi-dimensional

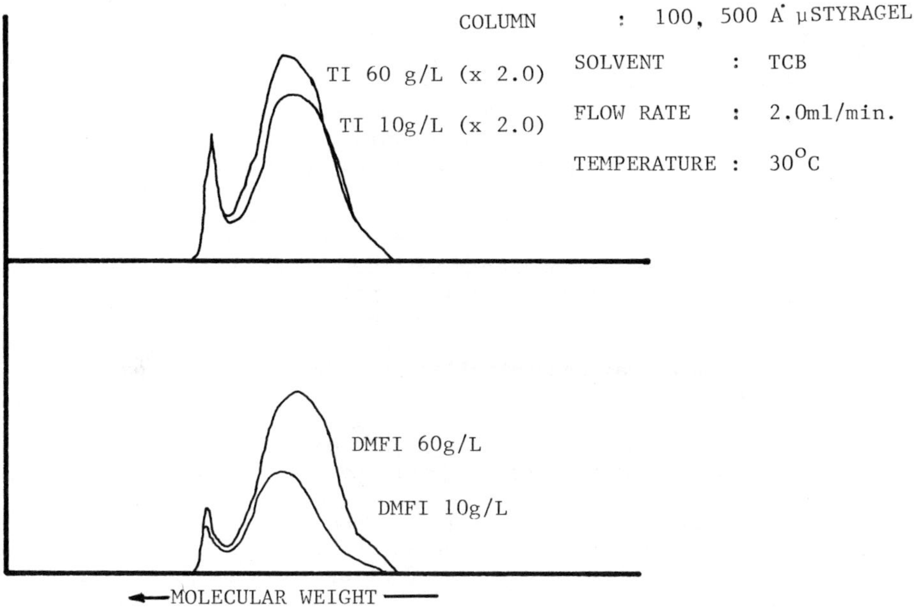

Figure 8. Effect of pitch to solvent ratio on molecular distributions.

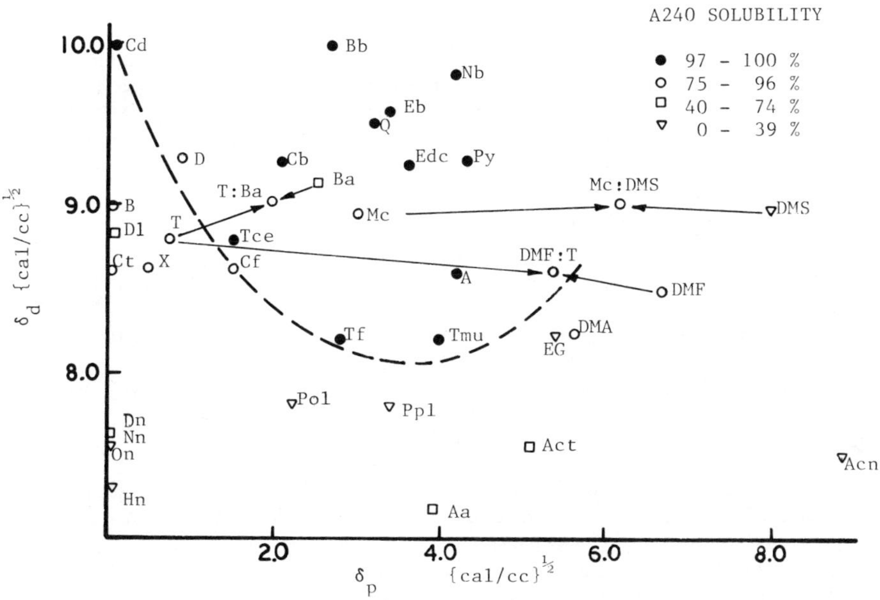

Figure 9. Two-dimensional solubility map. δ_d versus δ_d

solubility parameter, such as Hansen's qualitatively predicts the effectiveness of the fractionating solvents. Solvents, and solvent combinations, with solubility parameters located adjacent to A-240's region of enhanced solubility extract a fraction of the pitch that will form coalesced mesophase readily upon melting.

Solvents extract a pitch on the basis of molecular structure, interactions, and the molecular weight distribution of the pitch species. As a result, the insoluble pitch fractions are very broad. The lower molecular weight species act as plasticizers and reduce the pitch viscosity allowing for long range molecular order at low temperatures. Mesophase formation is not restricted to a specific percentage or distribution of the A-240 pitch as different solvents extract characteristic distributions from the A-240 capable of forming a coalesced mesophase readily when melted. Evidence suggests mesophase formation may be controlled by a small dominant percentage of the A-240, which combines with other recessive species capable of aligning along the backbone, or dissolving within, the dominant species which form the anisotropic phase. Only when the recessive species represent a large percentage of the extract is mesophase formation restricted.

Extracted pitch behavior is also dependent on the pitch/solvent ratio used for the separation. Altering ratios will affect both the weight percent dissolved, and the properties of an extract. Increasing the pitch/solvent ratio of a polar or hydrogen bonded solvent can decrease their effectiveness and shift and reduce the size of the enhanced solubility region on solubility plots. In this way an over-effective solvent can be moderated so it extracts a mesophase forming sample.

In summary, it is noted the relationships which exist between solvents and pitches are as difficult to characterize as the structure of the pitch itself. Several analytical techniques must be combined to resolve the effect of a single solvent on a pitch. The effects of several solvents on A-240 have been discussed in this paper and related to the mesophase forming tendencies of the insoluble distribution. However, as the complex pitch distribution is characterized more completely, the effect of solvents will become easier to ascertain along with the prerequisites for mesophase formation.

Literature Cited

1. Brooks, J. O. and Taylor, G. H., Carbon 3, 1965, p. 185.
2. White, J. L., Guthrie, G. L., and Gardner, J. O., Carbon 5, 1967, p. 517.
3. Riggs, D.M. Ph.D. Thesis, Rensselaer Polytechnic Institute, Troy, NY, 1979.
4. Lewis, I. C. and McHenry, E. R., U.S. Application No. 423,693, 1973.
5. Riggs, D. M. and Diefendorf, R. J., U.S. Patent No. 4,208,267, 1980.
6. Hildebrand, J. H. and Scott, R. L., "The Solubility of Non-Electrolytes", 3rd Edition, Reinhold Publishing Company, 1950.
7. Hansen, C. H., "The Three-Dimensional Solubility Parameter and Solvent Diffusion Coefficient", Danish Technical Press, 1967.
8. Didchenko, R., et.al. "Abstracts", 12th Conference on Carbon, 1975, p. 329.

RECEIVED June 5, 1984

14

Molecular Weight Determination and Distribution in Pitches

S. H. CHEN, W. C. STEVENS[1], and R. J. DIEFENDORF

Materials Engineering Department, Rensselaer Polytechnic Institute, Troy, NY 12181

Vapor pressure osmometry (VPO) and gel permeation chromatography (GPC), using 1.2.4 -trichlorobenzene (TCB) solvent, have been applied to measure molecular weight and molecular weight distribution of petroleum pitches. A high osmometry temperature is required to prevent molecular association/phase separation of high molecular weight species. VPO appears to be satisfactory for molecular weight determination with most TCB soluble fractions of petroleum pitches up to and including toluene insoluble fraction. Different pitches have different GPC calibration curves and each sample should be used to define its own molecular weight/elution volume relationship. The higher molecular weight species show a higher UV absorption coefficient. This discrete absorption coefficient makes analysis difficult.

Petroleum pitches and coal tar pitches are becoming increasingly important as raw materials for the production of low cost, high performance carbon fibers. This development has been primarily due to the ability of the pitch to form an oriented liquid crystalline mesophase. Since the 1st observation of mesophase spheres in pitch material by Brooks & Taylor (1), a number of studies on the kinetics of mesophase formation have been made by several different groups. A free radical polymerization mechanism was proposed by Singer & Lewis (2). However, studies of heat-treatment of heavy oil, by Riggs & Diefendorf (3) indicates that mesophase formation is preceded not necessarily by extensive polymerization reactions which builds up the molecular weight and aromaticity to satisfy the average structural requirements for mesophase formation. Rather, volatilization increases molecular weight and alkyl cracking reactions increase the degree of aromatic carbon, although decreasing molecular weight.

Mesophase can be thermotropic or lyotropic in nature. For thermotropic systems, the size of the molecules, or the molecular weight

[1] Current address: American Cyanamid, Stamford, CT 06904.

0097-6156/84/0260-0235$06.00/0

(MW), as well as shape is important for mesophase formation. For lyotropic systems, mesophase formation is dependent on sufficient solubility of a thermotrope in an isotropic solvent to form a mesophase, or alternatively enough plasticization of a thermotrope by a plasticizer to allow flow. Mesophase pitches appear to be of the second type and are controlled by the molecular weight/structure distribution. Obviously, molecular weight is an important parameter for mesophase formation in pitches. However, early studies indicated that mesophase was insoluble in all common solvents, and molecular weight could not be determined. Even for the more soluble asphaltenes various techniques produced values of molecular weight varying over a range of ∿1000 to greater than 50,000 (4). Further, a study of asphaltene molecular weight by Speight (5), using the vapor pressure osmometry (VPO) technique, shows that the MWs were dependent on the nature of the solvent and the VPO temperature. A higher VPO temperature or higher dielectric constant of the solvent reduced the observed MWs. The molecular association, which is responsible for the controversy regarding the determination of MW, can be removed by using a better solvent and increasing the VPO temperature. This work was performed to determine if "absolute" molecular weight of solvent fractionated mesophase pitches could be determined by VPO.

The molecular weight distribution (MWD) which is important for describing mesophase formation in pitches was determined by gel permeation chromatography (GPC). While the solvent/solute/gel interactions depend on more than just the size of the solute, smaller molecules generally are retained longer than larger molecules. The success of the technique depends on good solvent and columns, the elution volume/molecular weight calibration and the detector response/concentration calibration. Good solvent and high system temperature prevent the molecular association. Evidence of associated molecules is the observation of a high molecular weight peak at 30°C. In our studies, this peak could be removed by increasing the system temperature to 135°C.

The calibration of MW vs elution volume for linear flexible chains and rigid rod polymeric materials often uses standard samples of polystyrene. The pitch molecules are significantly different from the PS polymer and large discrepancies between measured pitch MWs and those implied from PS standards are inevitable. Aromatic compounds appear attractive as standards but are limited to 600 MW, and are significantly different from the structure of pitches. Synthetic model compound pitches provide good standards, as the individual polymer peaks are well resolved by GPC (6). Retention volumes for molecular weights up to 1000, and sometimes higher, can be determined. However, different pitches have different calibration curves (7), (Figure 1). Hence, each pitch sample should be used to define its own MW/elution volume relation.

Experimental

Since both VPO and GPC rely on the complete solution of the pitch sample, solubility study was performed. Results on a wide range of solvents showed 1.2.4- trichlorobenzene (TCB) to be the best compromize for pitch materials and was chosen for both VPO and GPC. Several solvents such as dimethylformamide (DMF), produce low insolubles.

These solvents apparently have strong associations with the pitch molecules or disperse the pitch as proton NMR does not generate any aromatic signal.

A Corona/Wescan Model 232A VPO was used for measuring the number average molecular weight of the samples. The molecular weight distributions of pitch samples were determined by a Waters 150 C GPC. An UV detector was used in this study. The 305 mμ wavelength cut off for the TCB solvent prevents the use of a more desirable lower detector wavelength. UV/Vis spectra of model compounds and pitches were used to select the compromise of 365 mμ detector wavelength. However, small aromatics such as napthalene have no absorption above 365 mμ and hence are not detected.

Results and Discussions

The following equation can be derived for the vapor pressure osmometry determination of MW from the virial theorem and the Clausius-Clapeyron equation:

$$\Delta V = K' C_w \frac{M_1}{M_2} + K'\alpha \left(\frac{\bar{V}_2}{\bar{V}_1}\right)^2 C_w^{\ 2} + K' \beta \left(\frac{\bar{V}_2}{\bar{V}_1}\right)^3 C_w^{\ 3} + \cdots\cdots \quad (1)$$

where ΔV is voltage output of the VPO
K' is a calibration constant
C_w is weight concentration of solute, gm/c.c.
M_1 & M_2 are molar mass of solvent and solute
α & β are the second and third virial coefficients and
$\bar{V}_1$ & $\bar{V}_2$ are the partial specific volume of solvent and solute

Neglecting the higher terms, Equation 1 is modified to plot of $\Delta V/C_w$ vs C_w.

$$\frac{\Delta V}{C_w} = K' \frac{M_1}{M_2} + K' \alpha \left(\frac{\bar{V}_2}{\bar{V}_1}\right)^2 C_w \quad (2)$$

The plot of voltage output of the VPO per unit concentration vs concentration was used to determine the MW. The intercept determines the MW and the slope α. Alpha in turn, provides information on the interaction between solute and solvent. The sign and magnitude of the slope determine if good solubility is obtained for a given temperature.

VPO is a relative method for MW determination and a calibration is required. Although there is some controversy in the literature regarding calibration of VPO, the calibration with a material of 200 MW appears to be valid up to 100,000 for the apparatus used (8). To clarify this point, several model compounds were used as calibrants. Table I. Although three compounds were pure unsubstituted polynuclear aromatics and the other two contained oxygen, the calibration constants agreed within a few percent.

The calibration data for coronene are shown in Figure 2. The negative slope is indicative of a poor solution at the given temperatures. The reason is that the solubility parameter for coronene, estimated from the Riggs (9) plot is about 12.5, while that for TCB is only 10.4. Therefore, coronene is sparingly soluble in TCB and even though dilute solutions are used in VPO studies, there is a

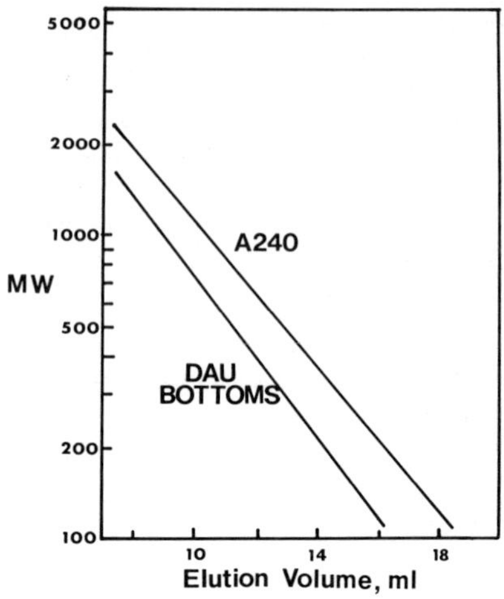

Figure 1. GPC calibration curves, using VPO, shows that different pitches have different calibration curves.

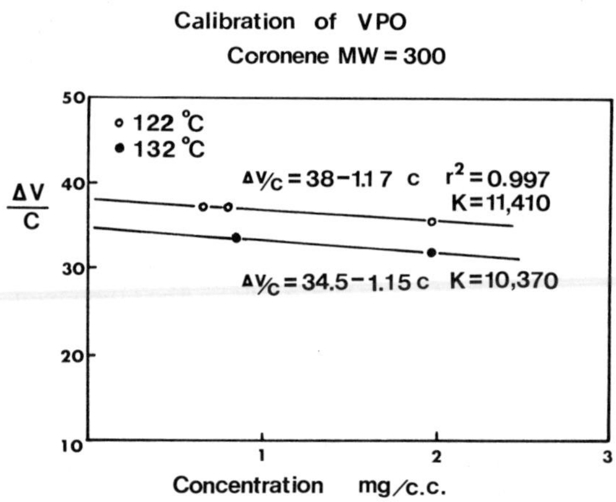

Figure 2. Calibration of VPO by coronene at 122 and 132°C.

Table I. Calibration Constants of VPO at 122°C.

Calibrant	MW	K
Perylene	252	11,700
Coronene	300	11,400
5,6,11,12 Tetraphenylnaphthacene	533	11,350
Tetraphenylcyclopentadienone	384	11,280
Cholesteryl cinnamate	516	11,430

significant contribution from the second virial coefficient. The 5,6,11,12 tetraphenylnaphthacene, which has a solubility parameter about the same as that of TCB and a melting point about 100°C less than that of coronene, yields a ∿ zero slope in the $^{\Delta V}/C_w$ plot, (Figure 3).

Results for a low MW pitch fraction, the toluene soluble/heptane insoluble (TS/HI) fraction of A 240 pitch are shown in Figure 4. The solubility parameters of pitch and solvent are reasonably well matched, and as can be seen, there is zero slope to the extrapolation line. Results for A 240 pitch, which by comparison to the $^{TS}/HI$ fraction has an overall lower MW, but also a high MW tail, is also shown in Figure 4. At the higher temperatures of 83°C & 122°C, there is again zero slope, but the lower temperature of 47°C has a slight positive slope, and a higher intercept. (The plot of MW vs C_w is inverted from the earlier $^{\Delta V}/C_w$ vs C_w plot.) The positive slope indicates a "poor" solution, one that is closer to phase separation, and the higher intercept indicates either association or phase separation of the highest MW species. Association or phase separation causes an apparent increase in MW. However, if only the highest MW species are involved, the removal of a relatively large MW but relatively small number of species does not change the number average MW much. Figure 5 shows the data for a toluene insoluble fraction of a heat-treated A 240 (CPTI). The slopes decrease with increasing temperature indicating more ideal solution behavior with increasing temperature.

The intercept for MWs tends to be constant while the slope increases with decreasing temperature. It might be expected that the intercept would increase when phase separation or association occurs at a sufficiently low temperature. Figure 6 shows the effect of temperature on molecular weight for different fractions of A 240 pitch. A 240 and its acetone insoluble (AI) fraction show higher MW at lower temperature but a temperature of 83°C is sufficient to prevent association or phase separation. Also, the differences between the MW determined at different temperatures were much less than that of CPTI. Studies of the MW determination of the $^{TS}/HI$ fraction of A 240 pitch shows no molecular association or phase separation at temperatures as low as 47°C. These results indicate that the association or phase separation primarily arises from the TI molecules. Although a similar behavior was observed with A 240 pitch and its AI fraction, (molecular association/phase separation occurs below 83°C), the AI fraction may have a slightly higher critical temperature. The level-off of the MW of CPTI at temperatures less than 83°C indicates a phase separation between 83°C and 122°C.

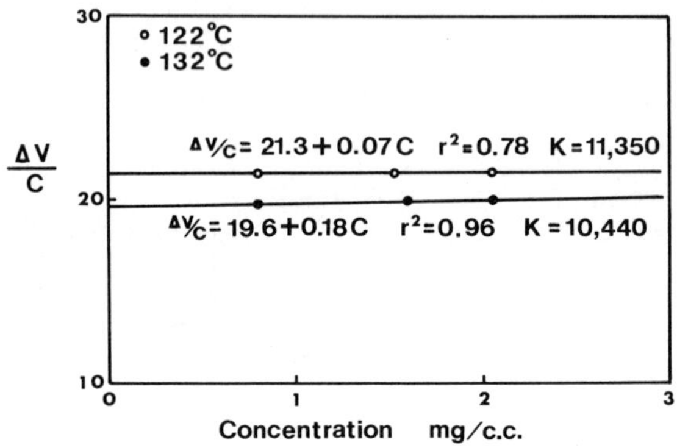

Figure 3. Calibration of VPO by 5,6,11,12 - Tetraphenylnaphthacene.

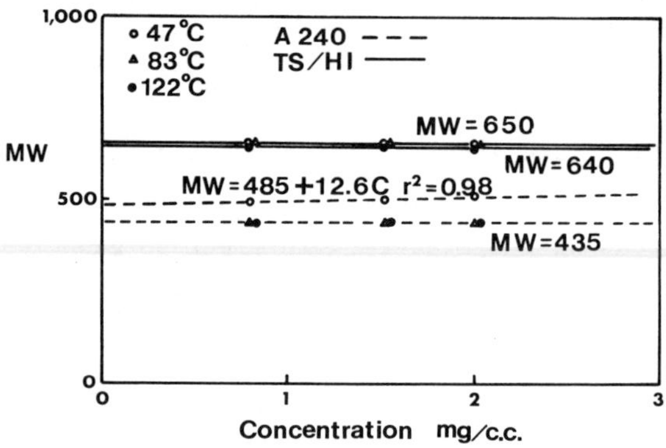

Figure 4. MW's of A240 pitch and TS/HI fraction at different temperatures.

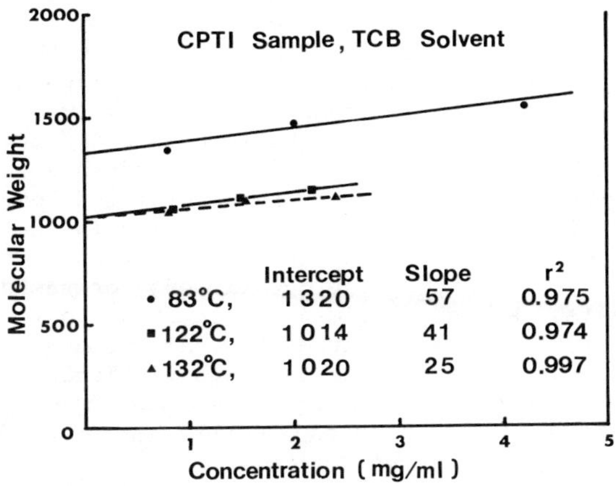

Figure 5. MW determination of CPTI at different temperatures.

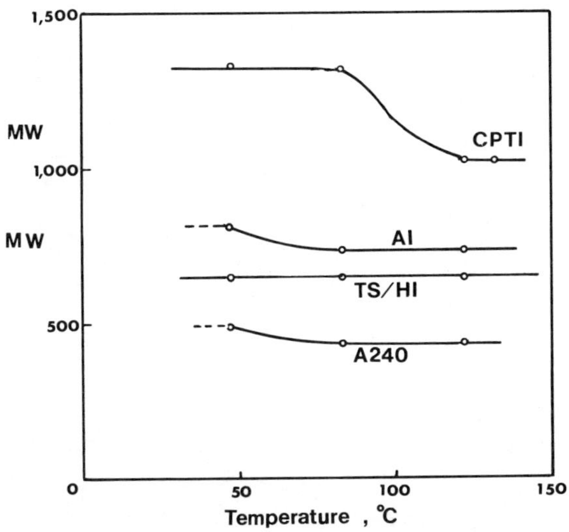

Figure 6. MW's of different fractions of A240 pitch at different temperatures.

A CPTI sample was fractionated into seven cuts by GPC and the procedure of Greinke and O'Connor (10) was used for calibration. The weights of the GPC cuts, corrected for solvent impurities, were plotted vs the integrated detector output during the time the cut was taken, Figure 7. The absorption was higher for early cuts (higher MW) than later cuts for the same sample weight. When these cuts were rerun on GPC, the fraction did not show a peak continuously shifting upwards in MW, but rather combinations of mainly two MWDs, Figure 8. Hence, Sample 3 was mostly the high MWD, samples 6 & 7 were predominantely the low MWD, while sample 4 & 5 were mixtures. Even for a commercial pitch produced from a broad range of molecules, multiples of the molecular weight of the more common species in the feedstock tend to dominate in the molecular weight distribution of the pitch. Our results show that the absorption coefficient for the higher molecular weight species is several times higher than the lower. The discrete absorption coefficients rather than a continuum of absorption coefficient vs molecular weight make analysis difficult. The poor resolution of GPC compounds the problem. However, the chromatogram could be deconvoluted into components and the discrete absorption coefficients used to yield concentrations.

The validity of MW/retention volume and the UV absorption coefficient/concentration relationships were checked by comparing the number average MW of CPTI as determined by VPO of the overall cut and by GPC. The value of 1034 from GPC compared favorably with the value of 1020 by VPO.

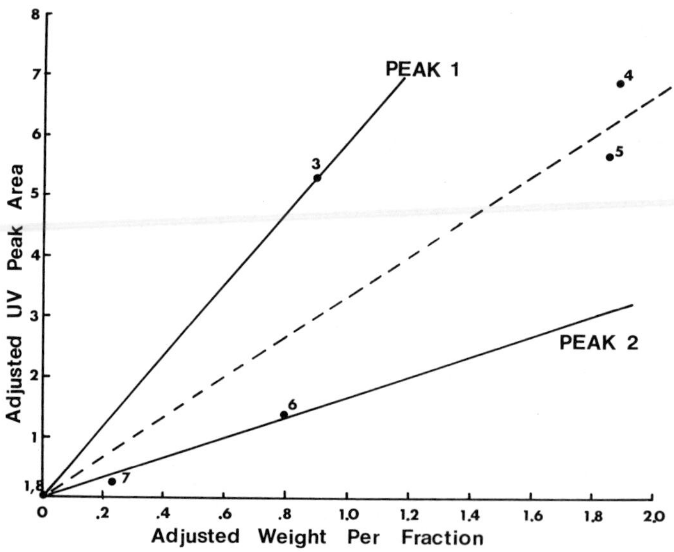

Figure 7. Relation between integrated UV absorption vs sample weight for seven GPC cuts with decreasing MW.

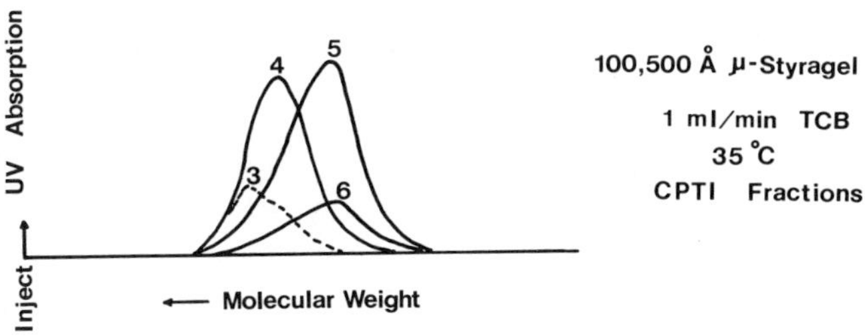

Figure 8. MWD's of the GPC cuts.

Conclusions

VPO appears to be satisfactory for MW determination for many fractions of petroleum pitches. Practically, one must operate VPO at a sufficiently high osometer temperature to prevent molecular association. Data collected which generates a relatively high slope should be accepted with some reservation.

Different pitches have different GPC calibration curves and each sample should be used to define its own MW/elution volume relationship. A quantitative evaluation performed on the CPTI sample shows a higher UV absorption coefficient for high MW species. This quantum nature of the absorption coefficient makes the definition of a continuous relationship between absorption coefficient and MW difficult.

References

1. Brooks, J. D.; Taylors, G. H. "Chemistry and Physics pf Carbon"; P.L. Walker, Jr. and P.A. Thrower, Eds.; Marcel Dekker: New York, 1968, Vol. 4, pp. 243-268.
2. Singer, L. S.: Lewis, I.C., "ESR Study of the Kinetics of Carbonization"; Carbon, Vol. 16, 1978, pp. 417-423.
3. Riggs, D. M.; Diefendorf, R. J., "Effects of Heat-Treatment on the Molecular Characteristics of Mesophase Forming Materials", Extended Abstracts, 14th Biennial Conference on Carbon, 1979, pp. 413-414.
4. Dickie, J. P. and Yen, T. F., "Macrostructures of the Asphaltic Fractions by Various Instrumental Methods", Anal. Chem., Vol. 39, No. 14, Dec. 1967, pp. 1847-52.
5. Speight, J. G., "Solvent Effects in the Molecular Weights of Petroleum Asphaltenes", PREPRINTS, ACS, Div. of Pet. Chem., Vol. 26, No. 4, pp. 825-31, 1981.
6. Chen, S. H. Ph.D. Thesis, Rensselaer Polytechnic Institute, 1982.
7. Sorensen, I. W. Ph.D. Thesis, Rensselaer Polytechnic Institute, 1982.
8. Burge, D. E. "Molecular Weight Measurements by Osometry", American Laboratory, June 1977.
9. Riggs, D. M. Ph.D. Thesis, Rensselaer Polytechnic Institute, 1980.
10. Greinke, R. A.: O'Conner, L. H., "Determination of Molecular Weight Distributions of Polymerized Petroleum Pitch by Gel Permeation Chromatography with Quinoline Fluent", Anal. Chem., 1980, 52, pp. 1877-81.

RECEIVED January 17, 1984

15

Solvent-Extracted Pitch Precursors for Carbon Fiber

D. M. RIGGS

Exxon Enterprises Materials Division, Fountain Inn, SC 29644

Pitch is a "pseudo solution" of a wide variety of different generic classes of hydrocarbons ranging from paraffins at one extreme to very highly aromatic species at the other. By using the Theory of Solubility for Non-Electrolytes, specific fractions can be isolated from a pitch by properly selecting a solvent system and extraction conditions. "Tailored" precursors for carbon fiber and other carbon products, such as carbon/carbon matrices and bulk graphites, can thus be obtained. The technique of extraction, the characteristics of different precursors, and the structure and properties of carbon fiber and composites made from solvent extracted precursors will be discussed.

Pitches are composed of a wide variety of different generic classes of compounds ranging from low molecular weight paraffins at one extreme to very highly condensed aromatic species at the other. The relative proportion of each of the molecular types and the subtleties of their molecular structures are very much a function of the pitch source (e.g. coal, oil, synthetic, etc.) and its previous thermal history. When processing carbon products from pitch precursors, the distribution of molecular types and the inherent characteristics of the the parent pitch acquire varying degrees of importance depending upon the actual carbon product being manufactured. In the case of carbon fiber processing, control over the pitch precursor composition is very important since this composition ultimately controls the pitch thermal stability, its carbon yield, the liquid crystal (or mesophase) forming ability, the pitch rheological properties, and most importantly, the properties (e.g. strength and modulus) of the fibers themselves.

Since it is very difficult to find a source of pitch with a consistent batch to batch distribution of molecular types and molecular structural characteristics, then the initial steps of a pitch-based carbon fiber process should "tailor" both the distribution and

0097-6156/84/0260-0245$06.00/0

the characteristics to a predetermined point in order to achieve the consistency desired for spinnability and fiber properties. There are various ways of accomplishing this "tailoring" including extended thermal treatments(1), sparging(2), and solvent extraction(3). Of these methods, solvent extraction may be the most effective since the pitch is really an "intrinsic" solution itself and hence, solubility relationships can be manipulated to force the extraction of desirable fiber feed components.

A "desirable" carbon fiber feed from a pitch precursor should satisfy at least four main criteria. These criteria include the following:

a. the precursor fraction should be fully liquid crystalline
b. the thermal stability and carbon yield should be high
c. the rheological characteristics and spinnability should be "manageable"
d. the softening point and glass transition temperature should be high

A fully liquid crystalline precursor is desirable for two reasons. First, it is inherently easier to spin a single phase material compared to a multiphase material, and second, the orientation that is imparted to the mesophase during spinning gives rise to the graphitic orientation developed in the fiber during the carbonization step. The graphitic orientation in turn governs the moduli of the carbonized fiber. A thermally stable precursor with a high carbon yield is also desirable for reasons related mainly to volatiles and gas evolution. Lower evolved levels of gas and volatiles significantly reduce the chances for the generation of voids or bubbles in the fiber. Thus, the chances of obtaining low strength values are reduced. The rheological characteristics and spinnability considerations are important for more or less self-evident reasons. Essentially, if the precursor is non-spinnable, then good carbon fiber most assuredly cannot be obtained. High softening points and glass transition temperatures are required because the second stage of fiber processing involves an oxidative thermosetting step at temperatures usually above 240°C. The fiber should not relax, soften, or melt during this process step, and hence, the precursor fraction should have a glass transition temperature at least in this range.

In this paper, a process for producing a carbon fiber from a solvent extracted pitch feed will be discussed.

General Characteristics of Pitches

A typical pitch is essentially a solution of hydrocarbons. As such, the interelationships among the hydrocarbons comprising the pitch are governed by the solubility relationships for non-electrolytes developed by Hildebrand and Scott(4) many years ago. The formation of a regular solution requires that the free energy of mixing among the components be negative. In order for this to occur, the heat of mixing of the components must be small relative to the product of the temperature and entropy of mixing (Equation 1).

$$\Delta G_{mix} = \Delta H_{mix} - T\Delta S_{mix} \quad (1)$$

ΔG_{mix} = free energy of mixing
ΔH_{mix} = heat of mixing
ΔS_{mix} = entropy of mixing

The entropy of mixing for a hydrocarbon solution is primarily a function of the temperature and the melting points of the individual components. The heat of mixing on the other hand is primarily dependent upon the difference in solubility parameters (or cohesive energy densities) of the components and, to a lesser degree, volume fractions of the components and the temperature. A large difference in solubility parameter between two components will result in a high heat of mixing, and hence, poor mutual solubility. If one examines a typical petroleum pitch and plots the relative proportions of the different molecular species by their respective solubility parameters, a distribution similar to that shown in Figure 1 will be obtained. It can be seen from Figure 1 that the solubility parameters for components in a typical petroleum pitch range from low values of about 7-7.5 $(\frac{cal}{cc})^{1/2}$ for the paraffinic material to over 11 $(\frac{cal}{cc})^{1/2}$ for the more highly condensed aromatics. The mesophase forming components of the pitch, which are of utmost importance to carbon fiber manufacturers, possess solubility parameters at the upper end of this distribution. The mesophase forming components have these higher solubility parameters because they are predominantly composed of large aromatic species which possess the necessary molecular structural features for liquid crystallinity (or mesomorphism) to occur. (It is important to realize that mesophase is a phase composed of components.) The distribution in solubility parameters for the pitch will change for different pitch types as well as for different thermal treatments, etc.

The range of differences depicted in Figure 1 for the component solubility parameters suggests, from a thermodynamic standpoint, that an "intrinsic" pitch solution is actually unfavorable because the heat of mixing between the large aromatics and the more paraffinic and naphthenic material is large. Since many pitches appear to be more or less homogeneous solutions, however, (based on measurements of glass transition, temperature, C/H, solubility, optical isotropy, etc.), then in order to reconcile the thermodynamics, a three dimensional localized micellar structure (or "gradient" solution) must exist throughout the pitch. A two-dimensional cut through one of these postulated micelles is shown in Figure 2. The heat of mixing for the overall pitch would be reduced if the low solubility parameter paraffins solubilized the naphthenic material which in turn solubilized the aromatics (and hence, mesophase forming components). In order for large scale coalescence of the highly condensed aromatic mesophase forming components to occur, the non-mesophase forming fractions on the outer surfaces of the micelles must be removed. As discussed previously, there are various techniques for accomplishing this removal including vacuum stripping and heat-treatment, sparging, and solvent extraction. Extensive thermal treatments remove the majority of the paraffinic and naphthenic components from the pitch (by distillation and cracking reactions), and as a result disrupt the micelle, shift the average pitch solubility parameter to higher levels, and allow for the coalescence of the mesophase forming fractions which are so desirable for carbon fiber manufacture. Additions of large amounts of a given solvent, on the other hand, result in an oversaturation of that part of the micelle most similar to the solvent in terms of solubility parameter. Those components of the micelle which are most dissimilar to the solvent

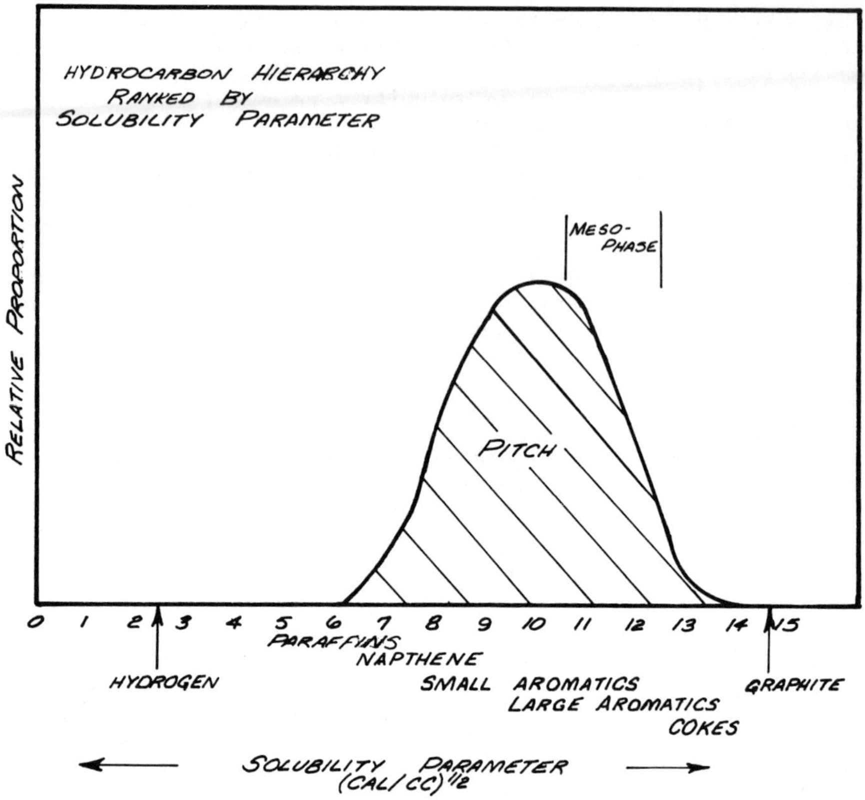

Figure 1 SCHEMATIC REPRESENTATION OF THE DISTRIBUTION OF MOLECULAR TYPES IN A TYPICAL PETROLEUM PITCH (RANKED BY SOLUBILITY PARAMETER)

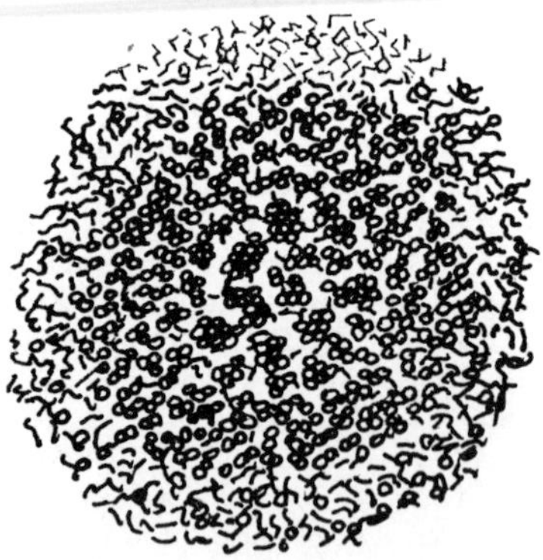

Figure 2 HYPOTHETICAL "MICELLE" FOUND IN PITCH

will precipitate from both the "intrinsic" pitch solution and the "extrinsic" pitch/solvent solution. By properly selecting the solvent, the pitch/solvent ratio and the extraction temperature, it becomes possible to remove the mesophase forming fractions from the pitch, and hence, to tailor the pitch for specific applications.

Characteristics of Solvent Extracted Petroleum Pitch

An Exxon cat cracker bottom pitch was extracted with solvents of varying solubility parameter in order to evaluate the relationships which exist between extraction conditions and the characteristics of the extracted product. The precursor pitch possessed a carbon/hydrogen ratio of 1.49 and a calculated(5, 6) overall average solubility parameter of 10.48 $(\frac{cal}{cc})^{\frac{1}{2}}$. The reflux quinoline insolubles content of the pitch was 2.3% and the refluxing toluene insolubles were at a level of 20.3%. The pitch was not optically anisotropic. Upon heating in N_2 to 530°C in a TGA, a coke yield of 37% was obtained.

The precursor pitch was extracted in solvents with solubility parameters ranging from 7.40 $(\frac{cal}{cc})^{\frac{1}{2}}$ to 8.93 $(\frac{cal}{cc})^{\frac{1}{2}}$. The results of the characterizations of the extracted fractions are shown in Table I. Samples 2 and 7 were extracted by first grinding the precursor pitch followed by both a slurry and filtration step. The slurry of sample 1 was done in heptane at room temperature, while 7 was slurried in refluxing toluene. Both of these samples represent the recovered insolubles after filtration. Samples 3, 4, 5 and 6 were produced by first grinding the pitch, slurrying the pitch in reflux ing toluene, filtering the ash and other insolubles at refluxing temperatures and then adding heptane in amounts ranging from 0 to 30% depending upon the sample. The ratio of pitch to solvent was 8:1. The toluene/heptane solution was then cooled to approximately 30°C, and the insolubles were filtered. Samples 3-6 represent the filtered insolubles.

An examination of Table 1 reveals some interesting trends. As the solubility parameter of the extraction solvent increases, the mesophase content of the extracted fraction goes to >99%. Samples 3, 4, 5 and 6 are "neomesophases"(3) in that upon melting, they instantly transform to a liquid crystalline state. Their quinoline insolubles contents were less than 1%. The extraction process breaks down the micelle and removes the non-mesomorphic species in the pitch which had inhibited the coalescence of the mesophase forming species. Sample 2, the heptane extracted fraction, did not exhibit neomesophase behavior. Extended heat treatment at 400+°C was required to convert this sample to a mesomorphic state. The heptane did not remove a sufficient amount of the non-mesophase formers.

Shown in Figure 3 is a plot of the coke yields for the different fractions as a function of the solubility parameter of the extraction solvent. The coke yields were measured by heating the fraction in a Perkin-Elmer TGS II under N_2 to a temperature of 530°C at a rate of 10°C/minute. As can be seen from both Figure 3 and Table I, the coke yields increase more or less linearly as the solubility parameter of the extraction solvent increases. Coke yields of over 90% are obtainable for samples extracted with solvents having solubility parameters over about 8.8 $(\frac{cal}{cc})^{\frac{1}{2}}$. Weight

Table I: Fraction Characteristics from Extracted Exxon CCB Pitch

SAMPLE	δ SOLVENT	COKE YIELD	C/H	MW	Tg	%rTI	%MESO	δ PROD	%QI
RAW PITCH	-	37	1.49	-	-	20.3	0	10.48	2.3
2	7.40	54	1.58	695	-	-	0	10.62	-
3	8.46	85.5	1.74	1000	232	73.9	99+	10.86	< 1
4	8.62	86.3	1.76	1120	246	78.5	99+	10.88	< 1
5	8.79	87.7	1.73	1310	266	81.1	99+	10.84	< 1
6	8.93	91.2	1.79	1600	285	85.0	99+	10.92	< 1
7	-	95.0	1.88	-	344	100	-	11.04	-

losses for the neomesophase fractions started at about 350°C. The increase in coke yield is apparently attributable to an increase in overall aromaticity of the extracted fractions. Evidence for this can be seen by an examination of the trends in C/H as a function of extraction conditions. Table I and Figure 4 both show that the C/H ratio, and hence, average aromaticity, increase with increasing extraction solvent solubility parameter. This makes sense in that solvents with relatively lower solubility parameters tend to dissolve only those pitch components having similar low solubility parameters (e.g. paraffins and naphthenes). (Refer to Figure 1.) Based on the measured C/H ratio of 1.49 for the pitch, an extrapolation of the curve in Figure 4 says that the pitch should be more or less insoluble in a solvent with a solubility parameter of 6.7 $(\frac{cal}{cc})^{\frac{1}{2}}$ or less. Shown in Figure 5 is a plot of the calculated solubility parameter of the extracted product vs. the solubility parameter of the extraction solvent. This curve is interesting in that it shows that large changes in the solubility parameter of the solvent cause only small changes in the overall solubility parameter of the pitch. This is significant because it suggests that solvents with large solubility parameters precipitate insolubles from the pitch at both ends of the distribution shown previously in Figure 1. This is reasonable in that a large difference in solubility parameter between any two components will result in poor mutual solubility. Evidence that precipitation does indeed occur at both ends of the distibution has been recently obtained(7). Under the right conditions, some pitches could conceivably yield two phase extracted fractions if the solvent system had a solubility parameter significantly different than the tails of the pitch solubility parameter distribution curve. Two phases were not observed in these samples although undoubtedly, some lower solubility parameter species were probably incorporated into the liquid crystalline fractions.

Shown in Figure 6 is a plot of the viscosity of samples 3 and 5 plotted as a function of temperature. The viscosities were measured using an Instron Capillary Rheometer at a shear rate of 450 sec^{-1}. The extracted solvent solubility parameters for samples 3 and 5 were 8.46 $(\frac{cal}{cc})^{\frac{1}{2}}$ and 8.79 $(\frac{cal}{cc})^{\frac{1}{2}}$, respectively. Figure 6 shows that the fraction viscosity increases markedly at a given temperature when the solubility parameter of the extraction solvent is increased. This again is primarily a function of the increased overall aromaticity of the fractions. Sample #6, the fraction extracted with a solvent having a solubility parameter of 8.93 $(\frac{cal}{cc})^{\frac{1}{2}}$ exhibited a very erratic viscosity vs. temperature curve. This fraction, as seen from Table I, had a relatively high C/H ratio of 1.79 and a correspondingly high fraction solubility parameter of 10.92 $(\frac{cal}{cc})^{\frac{1}{2}}$.

Shown in Figure 7 is a plot of the glass transition temperature of the different fractions plotted as a function of the extraction solvent solubility parameter.

The Tg was measured using an Ommitherm Differential Scanning Calorimeter. Figure 7 shows the glass transition temperature increases more or less linearly as the solvent solubility parameter increases. Fractions obtained by extracting the pitch with solvents having solubility parameters greater than about 8.6 $(\frac{cal}{cc})^{\frac{1}{2}}$ result in fractions with Tg's over about 245°C which as stated previously, are desirable for subsequent processing steps for pitch-base fiber. The

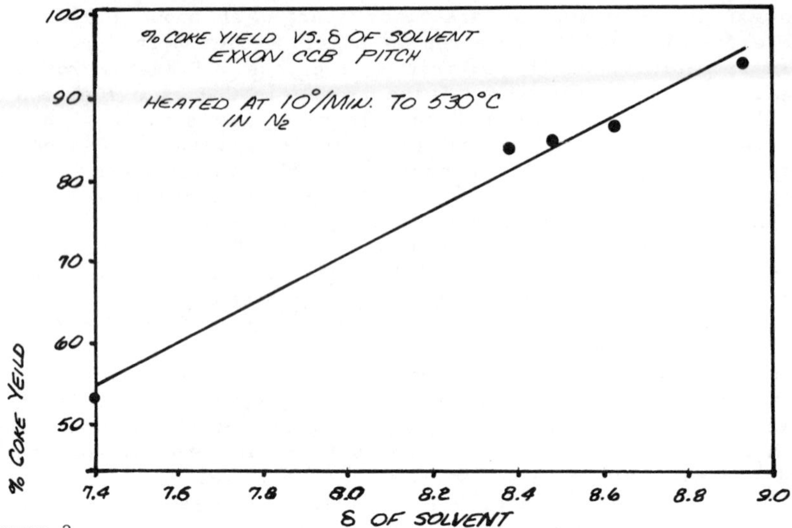

Figure 3

VARIATION IN COKE YIELD OF EXTRACTION CAT CRACKER BOTTOM PITCHES AS A FUNCTION OF THE SOLUBILITY PARAMETER OF THE EXTRACTION SOLVENT

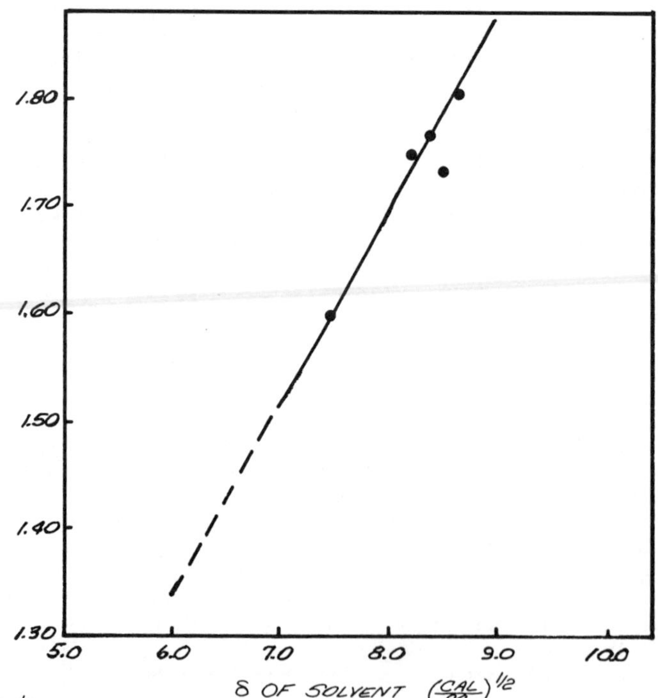

Figure 4

VARIATION IN C/H VS THE SOLVENT SOLUBILITY PARAMETER FOR VARIOUS CAT CRACKER BOTTOM PITCH FRACTIONS

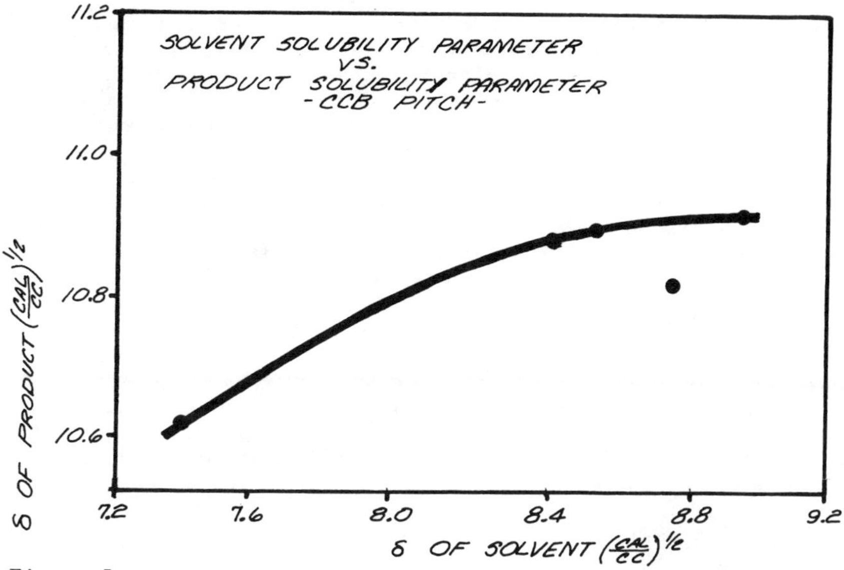

Figure 5

CALCULATED SOLUBILITY PARAMETER OF A SOLVENT EXTRACTED CAT CRACKER BOTTOM PITCH FRACTION AS A FUNCTION OF THE EXTRACTION SOLVENT SOLUBILITY PARAMETER

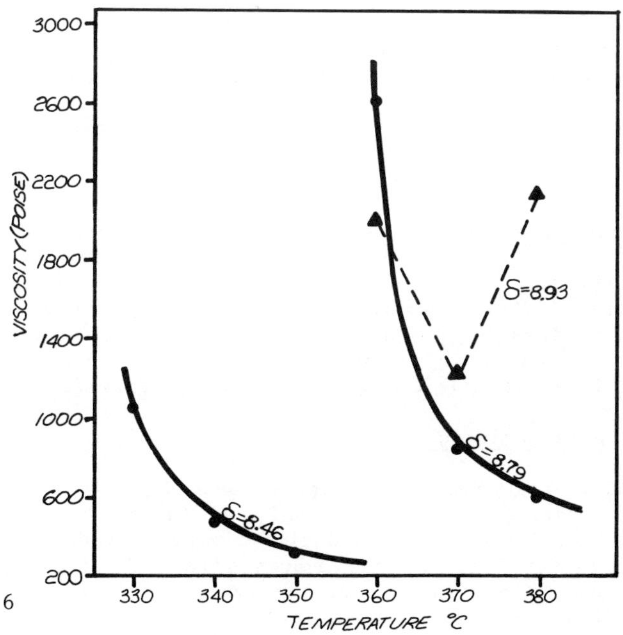

Figure 6

VISCOSITY VS TEMPERATURE RELATIONSHIP FOR 3 DIFFERENT CAT CRACKER BOTTOM EXTRACTED FRACTIONS

variation in the molecular weight of the extracted fraction as a function of the solubility parameter of the extraction solvent is shown in Figure 8. The molecular weights were measured by vapor phase osmometry (VPO) using trichlorobenzene as the solvent at 132°C. Each of the fractions was fully soluble in the as-extracted state. Figure 8 shows that the overall molecular weight of the insoluble fraction increases as the solvent solubility parameter increases. The molecular weight of the fractions exceeds 1000 when the solubility parameter of the extraction solvent exceeds $8.46(\frac{cal}{cc})^{\frac{1}{2}}$.

When one analyzes the results of the experiments just described and compares these results to the desirable criteria for carbon fiber feed, it becomes apparent that the best overall extraction solvent for the cat cracker bottom pitch is one having a solubility parameter of about $8.6\ (\frac{cal}{cc})^{\frac{1}{2}}$. The insoluble fraction resulting from an extraction with such a solvent system possesses a neomesophase character, a good coke yield (approximately 86%), a reasonable viscosity at temperatures where volatiles evolution is not too severe (approximately 360-370°C), a high C/H ratio, a Tg of about 245°C, and a molecular weight over 1100.

Carbon Fiber Processing

Based on the results discussed in the previous section, a process for producing a carbon fiber from the extracted pitch product was developed. A modified Ashland-240 petroleum pitch was extracted using the same general procedure outlined previously for samples 3-6 discussed in Table 1. The pitch was first slurried in refluxing toluene and filtered to remove catalyst fines and coke particles. The filtered refluxing toluene soluble components were then pumped to a second vessel where an anti-solvent having a solubility parameter of about $8.63\ (\frac{cal}{cc})^{\frac{1}{2}}$ was added in a ratio of 8 parts solvent to 1 part pitch. The temperature of the mixture was also lowered to 30°C. After a period of time, the insolubles from this step were removed, washed, and dried under N_2. The dried insoluble powder was then pelletized using a conventional screw extruder. The pelletization process was used mainly to remove any residual solvent still in the extracted product and also, to aid in the "feedability" of the extracted product to the fiber spinning equipment. These pellets were next spun into filament form using conventional melt extrusion equipment. The as-spun fiber was oxidized at relatively low temperatures (<300°C) in order to render it infusible for the carbonization process. Carbonization was accomplished at various temperatures under N_2 using a horizontal graphite tube resistance furnace. The fiber was run continuously through this furnace.

Carbon Fiber Characteristics

The carbonized fibers from the solvent extracted pitch precursor described in the previous section generally exhibited textures similar to those shown in Figures 9a-b. Radial textures predominated, although random textured fibers and hybrids (random/radial) were occasionally found. Fiber splits (or missing wedges) were rarely observed. The final texture obtained in the fibers was found to be a

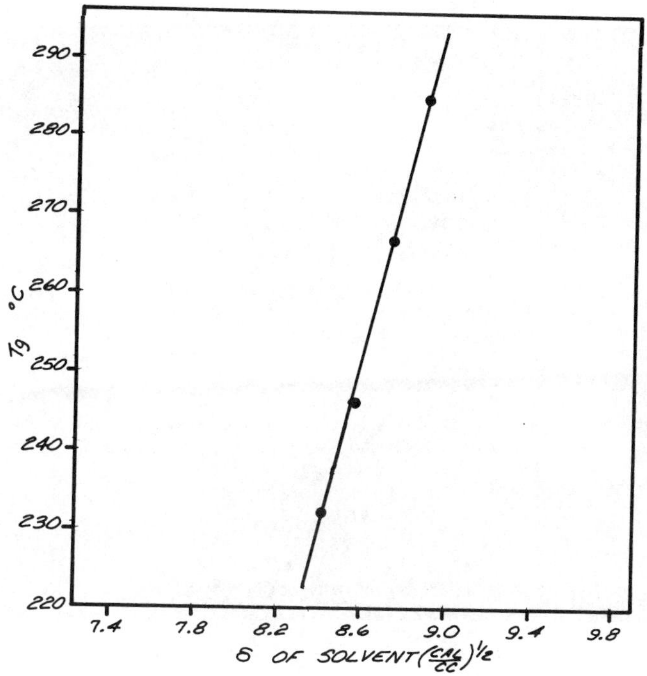

Figure 7

VARIATION IN GLASS TRANSITION TEMPERATURE FOR EXTRACTED CAT CRACKER BOTTOM PITCH FRACTIONS AS A FUNCTION OF THE EXTRACTION SOLVENT SOLUBILITY PARAMETER

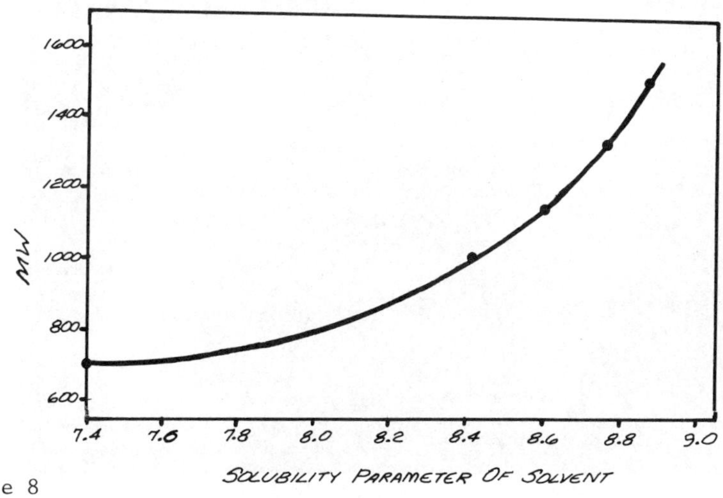

Figure 8

MOLECULAR WEIGHT VARIATION IN EXTRACTED CAT CRACKER BOTTOM PITCH FRACTIONS AS A FUNCTION OF THE SOLVENT SOLUBILITY PARAMETER

a

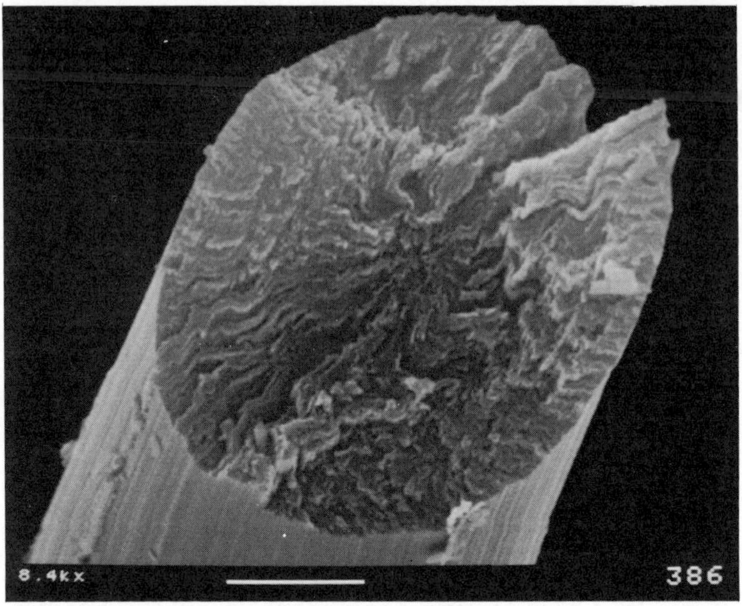

b

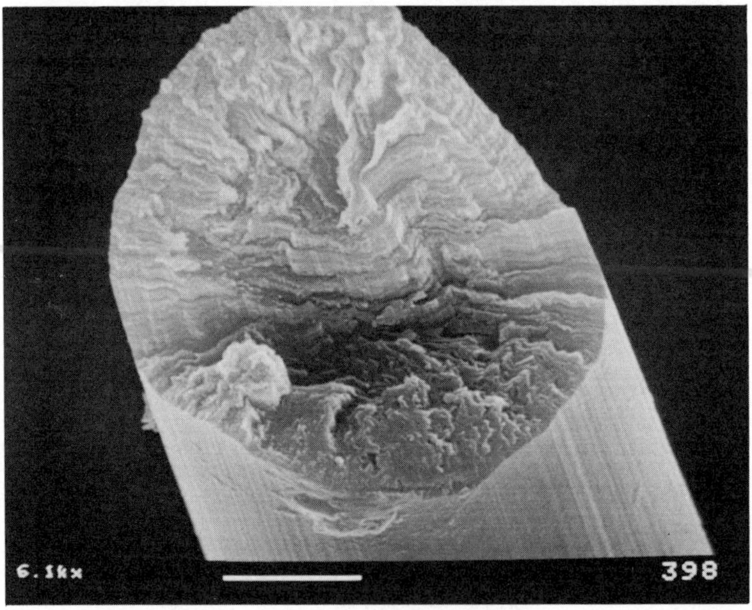

Figures 9a - 9b SEM'S OF TYPICAL SOLVENT EXTRACTED PITCH BASED CARBON FIBER

function of spinning and quenching conditions, and in addition, was a factor in determining the final fiber properties. In general, the fiber cross-sections tended to be slightly out of round.

Shown in Figure 10 is a plot of the fiber densities as a function of the carbonization temperature. The densities were measured using a liquid gradient column technique. At 800°C, the fiber density was about 1.47g/cc. The density increased to 2.00g/cc at 1500°C and to 2.10g/cc at 2000°C. The fiber diameters were progressively shrinking over this temperature range from values of about 12.0μ at 800°C to approximately 10.0μ and 9.5μ at 1500°C and 2000°C, respectively. The change in the crystallite sizes of the neomesophase based fibers, as a function of carbonizing temperature, are also shown in Figure 10. The crystallite sizes were calculated using the Scherrer equation with Cu Kα radiation.

Apparently, the graphite-like crystallites do not begin to grow in these fibers until the carbonizing temperatures exceed 800-900°C. This is probably related to the fact that the majority of material evolved during the carbonization of the fiber is removed below 800-900°C. The crystallite sizes increased from values of 13A° at 800°C to approximately 70 A° at 2000°C. The "d" spacing for fiber processed at 2000°C was 3.43A°.

The average moduli of the neomesophase based carbon fiber increased rapidly as the carbonizing temperature exceeded 1000°C. When carbonized at 1500°C, for example, the average modulus for a 10μ diameter fiber was about 32MSI. When carbonized at 2000°C, the average modulus increased to approximately 67MSI. This is shown in Figure 11. The modulus was measured in an Instron using a single filament testing technique. The variation in mean ultimate tensile strength as a function of carbonization temperature is shown plotted in Figure 12. The strength was measured using a dry strand technique. From Figure 12, it can be seen that the strength increases from a value of about 345KSI at 1300°C to 498KSI at 2000°C carbonization temperatures. At this point, it is unknown whether the mean strength will continue to rise as carbonization temperatures are increased.

Shown in Figure 13a and 13b are Weibull plots of typical strength data obtained from the neomesophase based carbon fibers heat-treated at temperatures of 1500°C and 2000°C, respectively. Both sets of fibers have strengths apparently governed by a single flaw population, though not necessarily the same type flaw. Typical flaws observed in these fibers included some surface damage, as well as very small internal pinholes.

The electrical resistivity of the fibers as a function of carbonizing temperature are shown in Figure 14. The resistivity drops from a value of several hundred thousand micro ohm centimeters at low carbonization temperatures to about 675μ ohm cm at 2200°C carbonization temperatures.

Summary and Conclusions

It has been shown that a knowledge of the thermodynamics of various hydrocarbon mixtures can be used to speculate on the nature of pitch as well as to develop extraction schemes for isolating specific components from a pitch for use as precursors for various carbon pro-

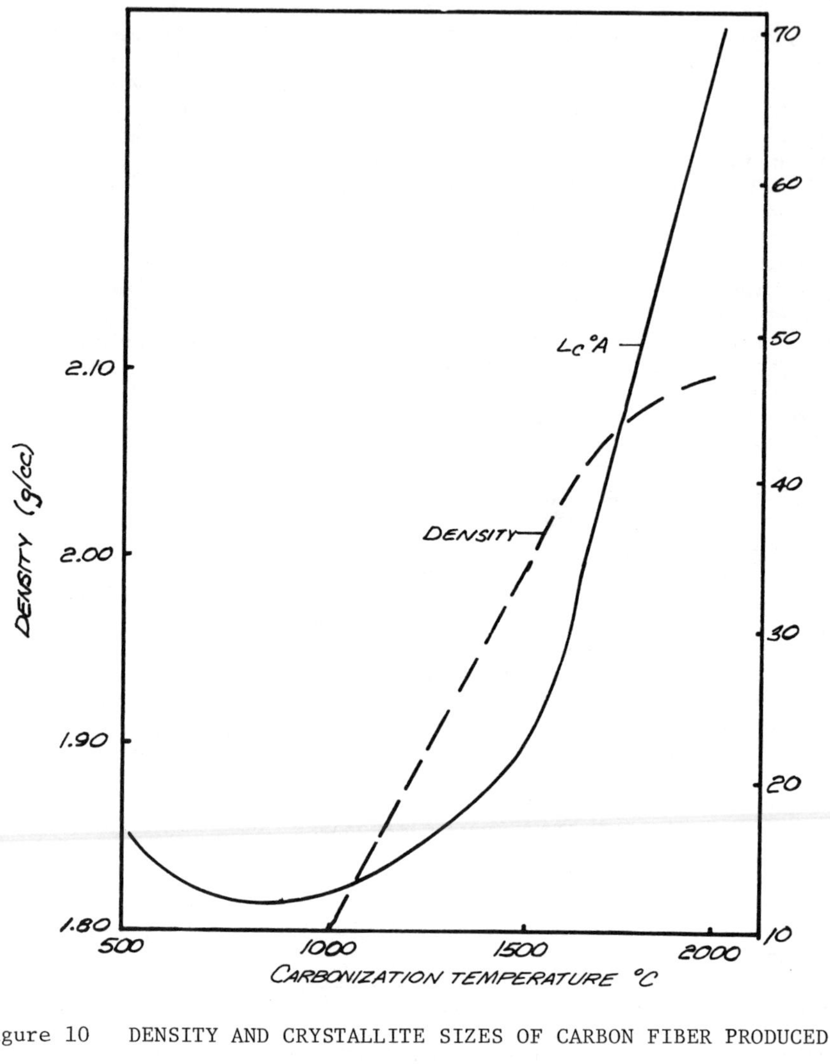

Figure 10 DENSITY AND CRYSTALLITE SIZES OF CARBON FIBER PRODUCED FROM A CAT CRACKER BOTTOM PITCH EXTRACTED WITH A SOLVENT HAVING A SOLUBILITY OF 8.64 $\left(\frac{\mathrm{cal}}{\mathrm{cc}}\right)^{\frac{1}{2}}$

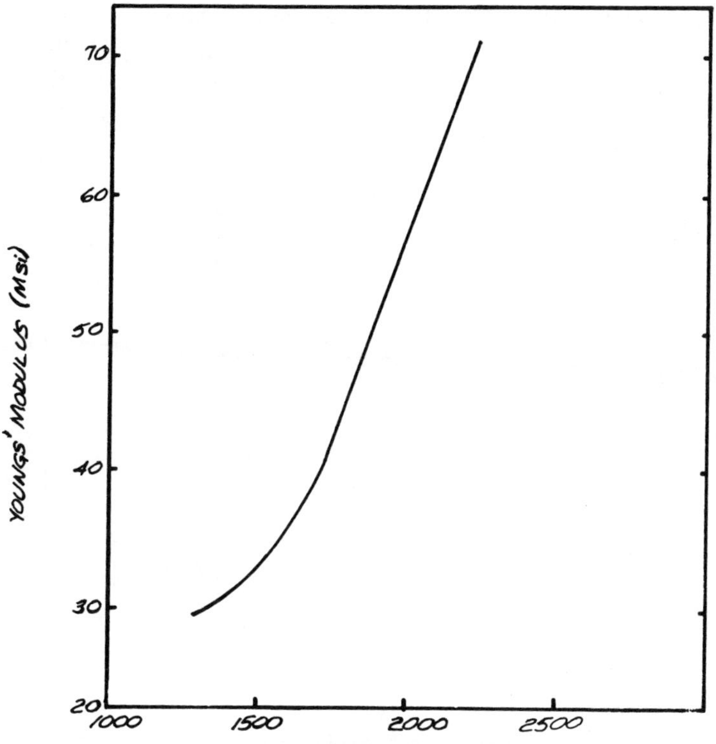

Figure 11

VARIATION IN CARBON FIBER MODULUS AS A FUNCTION OF CARBONIZING TEMPERATURE. PRECURSOR WAS A CAT CRACKER BOTTOM PITCH EXTRACTED WITH A SOLVENT HAVING A SOLUBILITY PARAMETER OF 8.64 $\left(\frac{\text{cal}}{\text{cc}}\right)^{1/2}$

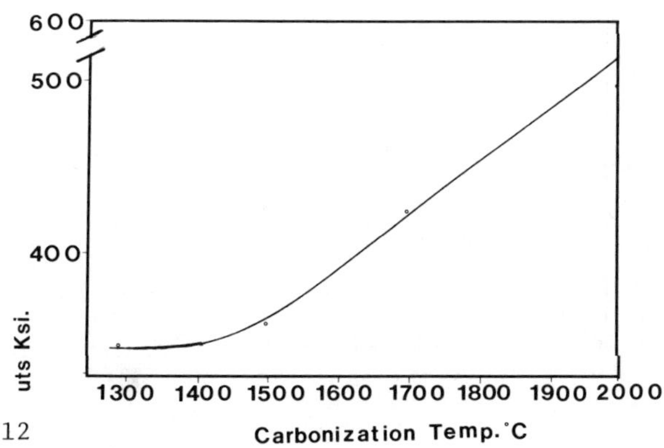

Figure 12

MEAN STRENGTH VS. CARBONIZATION TEMPERATURE EXTRACTED WITH SOLVENT OF SOLUBILITY PARAMETER = 8.64 $\left(\frac{\text{cal}}{\text{cc}}\right)^{1/2}$

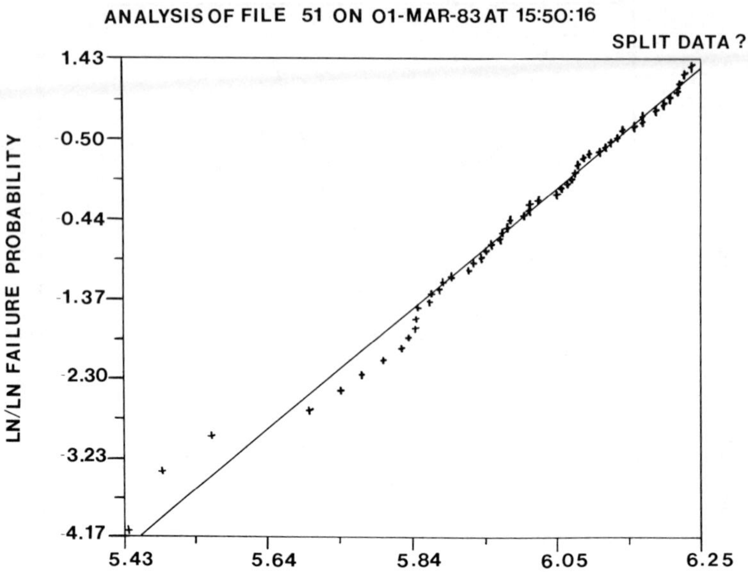

Figure 13a

WEIBULL STRENGTH DISTRIBUTION PLOTS FOR FIBER PROCESSED AT 1500°C

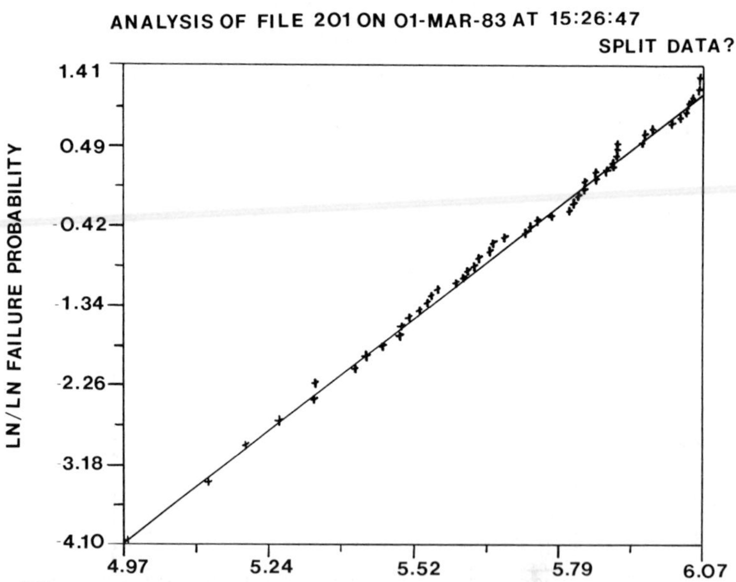

Figure 13b

WEIBULL STRENGTH DISTRIBUTION PLOTS FOR FIBER PROCESSED AT 2000°C

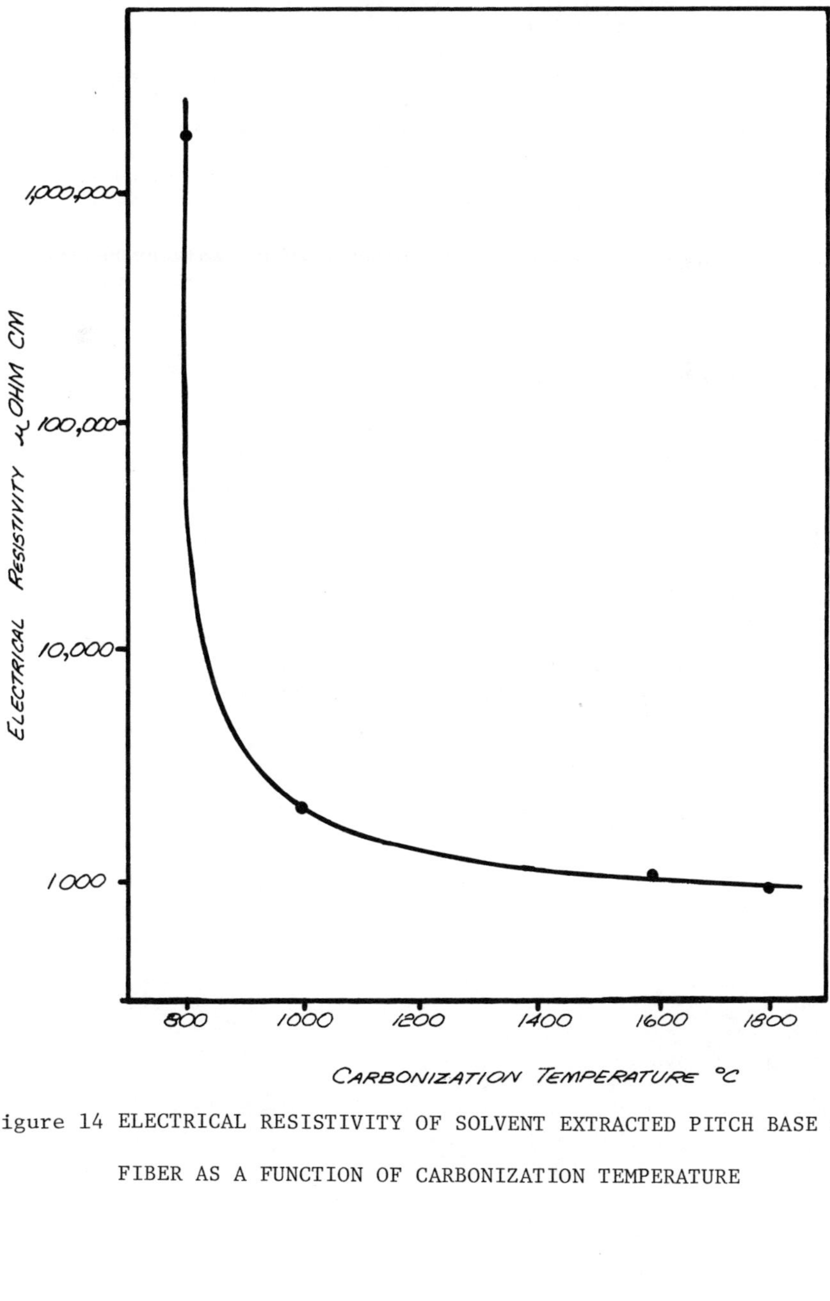

Figure 14 ELECTRICAL RESISTIVITY OF SOLVENT EXTRACTED PITCH BASE CARBON FIBER AS A FUNCTION OF CARBONIZATION TEMPERATURE

ducts. From a thermodynamic standpoint, it would seem that the pitch would form a micellar structure in order to minimize heats of mixing among the various components of different solubility parameter. The formation of the micelle would inhibit large scale coalescense of mesophase forming material and thus removal of the outer non-mesophase forming material from the micelle is necessary in order to form a fully anisotropic pitch.

By solvent extracting the pitch with a solvent system having a solubility parameter of 8.63 $(\frac{cal}{cc})^{1/2}$ a more or less ideal fraction for carbon fiber production is obtained. Extraction with solvent systems of different solubility parameter results in uniform changes in extracted product characteristics which can be tailored to other specific carbon product applications, such as carbon/carbon matrix materials, anodes, etc.

With regard to pitch-based carbon fiber, solvent extracted pitches show excellent potential as a precursor for high quality fiber.

Literature Cited

1. Singer, L., US Patent 4,005,183, January, 1977.
2. Chwastiak, S., US Patent 4,209,500, June, 1980.
3. Riggs, D. and R. Diefendorf, US Patent 4,208,267, June, 1980.
4. Hildebrand, J.H. and R.L. Scott, The Solubility of Non-Electrolytes, 3rd ed., Reinhold Publishing Corp., 1950.
5. Riggs, D.M., Ph.D. Dissertation, Rensselaer Polytechnic Institute, Troy, NY, 1979.
6. Riggs, D.M. and R.J. Diefendorf, "The Solubility of Aromatic Compounds", Extended Abstracts, 14 Biennial Conference on Carbon, Penn State University, pp. 407-408, 1979.
7. Unpublished data, G. Dickakian, Exxon Enterprises.

RECEIVED February 22, 1984

16

Effects of Sulfur and Metals on Mesophase Formation in Coal Liquid Asphaltene and Petroleum Pitch

YONG-DA GU[1], VICTORIA A. WEINBERG, MOHAMMAD-ALI SADEGHI, and TEH FU YEN

Environmental and Civil Engineering Department, School of Engineering, University of Southern California, Los Angeles, CA 90089–0242

The effect of elemental sulfur and organic sulfur compounds and organometallics on the mesophase microstructure was investigated. Elemental and organic sulfur compounds and organometallic compounds were added to Catalytic Incorporated coal liquid asphaltene and A240 petroleum pitch. The mesophase microstructure was examined using polarized light microscopy. It was found that organic sulfur compounds have less effect on mesophase microstructure than elemental sulfur but some organometallic compounds greatly inhibit mesophase formation.

The properties of manufactured graphites are determined by the microstructure of the carbonaceous mesophase which is formed during pyrolysis, usually between the temperatures of 370 °C and 500 °C. The characteristics of the final product can be measured in a qualitative way by examination of the mesophase microstructure. Several coal liquid asphaltene and petroleum pitch samples have been screened in this way to determine their suitability as precursors for graphite materials. The physical and chemical properties of the mesophase formed from the samples and their pyrolyses residues were studied(1). It was found that the phenolic oxygen present either in the precursor or by addition during heat treatment suppresses mesophase formation by crosslinking and preventing the development of fluidity(2).

The effects of other additives on petroleum pitch and PVC pitch such as metals and sulfur have also been studied(3). The aim of the present work is to investigate the effects of

[1] Current address: Institute of Coal Chemistry, Chinese Academy of Sciences, Taiyuan, Shansi, PRC.

0097–6156/84/0260–0263$06.00/0

elemental sulfur, organic sulfur and organometallic compounds on the mesophase transformation on coal liquid asphaltene. Some work was also done on A240 petroleum pitch.

Experimental

The effect of additives on the asphaltene from the Catalytic Incorporated (Cat. Inc.) coal liquid product was studied. Asphaltene is defined as the pentane insoluble but benzene soluble part of the coal liquid. The fractionation procedure has been described in detail elsewhere(1) and is shown schematically in Figure 1. Some work was also done with A240 petroleum pitch. Elemental analysis for the Wyoming sub-bituminous coal, Cat. Inc. coal liquid product, and Cat. Inc. asphaltene and A240 petroleum pitch are shown in Table I. Measured amounts of the additive compounds to be studied were added to the Cat. Inc. asphaltene and petroleum pitch. The samples were pyrolyzed and the pyrolysis residues examined by cross polarized light microscopy. Elemental analyses of the residues were done.

Table I. Elemental Analysis (%) and Aromaticity (f_a)* of Coal**, Catalytic Incorporated (Cat. Inc.) Coal Liquid, and Cat. Inc. Asphaltene and A240 Petroleum Pitch

	C	H	N	S	O	Ash	f_a
Coal	67.8	4.8	1.1	0.6	17.9	7.3	0.76
Coal Liquid	88.5	5.5	1.3	0.2	4.3	0.3	0.84
Coal Liquid Asphaltene	87.25	5.69	1.66	0.01	4.85	0.46	0.82
A240 Petroleum Pitch	92.11	5.6	0.1	1.8	0.13	0.17	0.88

*f_a estimated from H/C atomic ratio: $f_a = -0.79\ H/C + 1.43$
**Wyoming sub-bituminous

1. Additives

Sublimed elemental sulfur and dibenzothiophene (DBT) were used to study the effects of sulfur on the pyrolysis product. $AlCl_3$, Nickel(II) phthalocyanine $(C_8H_4N_2)_4Ni$, copper(II) phthalocyanine $(C_8H_4N_2)_4Cu$, vanadyl(IV) tetraphenylporphyin $(C_{44}H_{28}N_4VO)$ and ferrocene $(C_{10}H_{10}Fe)$ were used to introduce metals into the samples. The additives were ground with the samples using an agate mortar.

2. Pyrolysis

Pyrolysis experiments were carried out in a programmed temperature controlled furnace under nitrogen gas at atmospheric

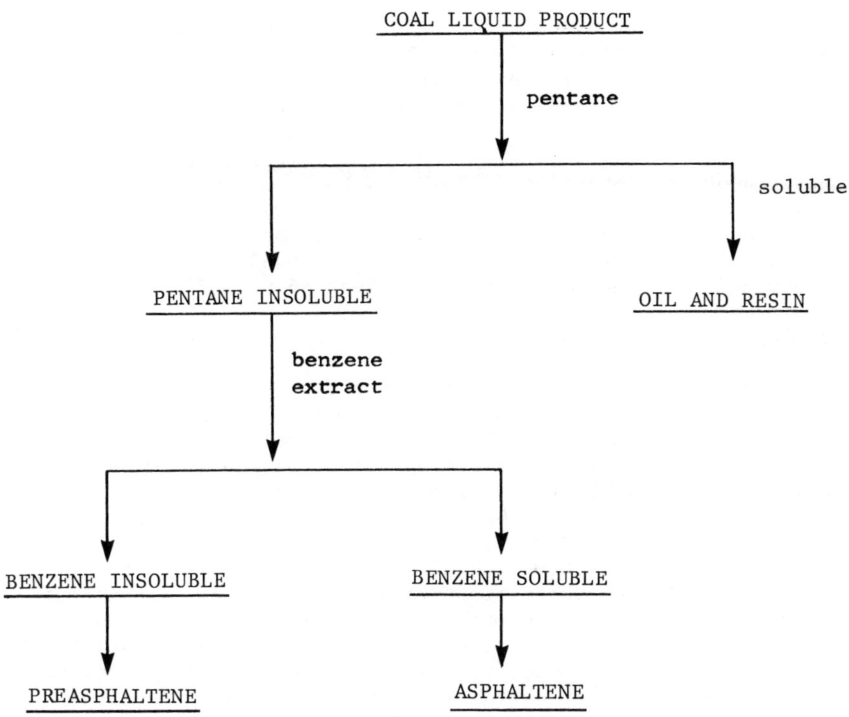

Figure 1. Flow sheet of fractionation of coal liquid.

pressure. The mixtures of Cat. Inc. coal liquid asphaltene and a measured amount of additive were added to the pyrolysis cell. This cell is an aluminum tube with a 0.5 mm pinhole in the cap for free evaporation of the pyrolysis gas. The tubes were placed vertically into the furnace and purged with nitrogen gas at 10 ml/min. The temperature was programmed at 13 ^{0}C/hr to 360 ^{0}C and then 5 ^{0}C/hr to the final temperature of 460 ^{0}C.

3. Microscopy

The micrographs of the pyrolyzed residues were obtained using an SM-LUX-POL polarized light microscope with a camera attachment. Magnification was 200X in this work.

4. Elemental Analysis of the Pyrolyzed Residues

Elemental analyses were done at Huffman Laboratories, Wheatridge, Colorado. The hydrogen and carbon contents were measured using combustion followed by CO_2 and H_2O determination (±0.3%). Oxygen was determined using pyrolysis followed by oxidation over carbon (±0.5%). Sulfur was measured using combustion to SO_2 followed by titration (±0.3%). And ash was determined gravimetrically after combustion to 750 ^{0}C (±0.3%).

Results and Discussion

The cross polarized micrographs of the pyrolyzed residues of the Cat. Inc. original coal liquid and Cat. Inc. asphaltene are shown in Figure 2. The microstructures are primarily coarse mosaic textures. The following subsections describe the different additives of elemental and organic sulfur, organometallics and their combined effects on Cat. Inc. asphaltene during pyrolysis.

1. Effect of Elemental Sulfur

Figure 2 also shows the effect of increased amounts of elemental sulfur on the microstructure of Cat. Inc. asphaltene. The microstructure changes from a coarse mosaic for 0% and 2% by weight sulfur to a very fine mosaic structure at 8% by weight sulfur. Figure 3 for the pyrolysis residues of A240 petroleum pitch and 8% by weight sulfur shows that the sulfur has less effect on the microstructure of the pitch. The microstructure of the pure pitch is flow type while that of the pitch with 8% by weight sulfur is coarse mosaic.

Table II summarizes the results of elemental analysis for the pyrolyzed Cat. Inc. asphaltene and shows how the residual pyrolysis yield increases with sulfur addition. Sulfur acts as a dehydrogenation agent in Cat. Inc. asphaltene. Loss of hydrogen leaves reactive free radical sites open resulting in increased cross linkage, aromaticity and condensation reactions. These condensation reactions decrease the fluidity and inhibit the mesophase formation as well as increasing the pyrolysis yield. Petroleum pitch is less affected by the sulfur, possibly because of its higher initial aromaticity and lower oxygen content.

Figure 2. Micrographs of pyrolyses residues of Catalytic Incorporated (Cat. Inc.) coal liquid, Cat. Inc. asphaltene and the effect of elemental sulfur on mesophase formation in Cat. Inc. asphaltene.

Table II. Elemental Analysis of Pyrolyzed Residues of Samples Modified with Elemental Sulfur (Elements in wt.%)

Samples*	C	H	O	N	S	Ash	H/C	Yield (%)**
0	91.82	3.84	2.08	1.28	0.15	0.12	0.50	52.7
4	91.25	3.88	2.71	1.69	0.22	0.63	0.51	59.5
6	91.61	3.73	2.47	1.55	0.04	0.24	0.49	58.6
8	90.61	3.83	3.18	1.24	0.35	0.71	0.51	64.7

*Each sample used is Cat. Inc. coal liquid asphaltene; numerals indicated are wt.% S added.
**Residual yields after pyrolysis (Wt.%).

2. Effect of Dibenzothiophene (DBT)

Dibenzothiophene(DBT) was added to Cat. Inc. asphaltene to determine the effect of sulfur when added as an organic molecule. Table III shows the amount in weight percent of DBT that was added to obtain the desired weight percent of sulfur. For 8% by weight organic sulfur, the sample contains almost half DBT. When DBT is pyrolyzed alone to 460°C, the pyrolysis yield is zero. As shown in Figure 4 the DBT sulfur has no effect on the microstructure of the pyrolysis residues. It is difficult to tell if this is because most of the DBT vaporizes before mesophase

Table III. Effect of Organic Sulfur Exemplified by Dibenzothiophene (DBT) Added to Cat. Inc. Coal Liquid Asphaltene

%S	Wt.% (DBT)	Yield (%)*	Microstructure**
0	0	52.7	CM
4	24	45.2	CM
6	35	43.8	CM
8	47	39.3	CM

*Residua yields after pyrolysis (Wt.%)
**CM - Coarse Mosaic

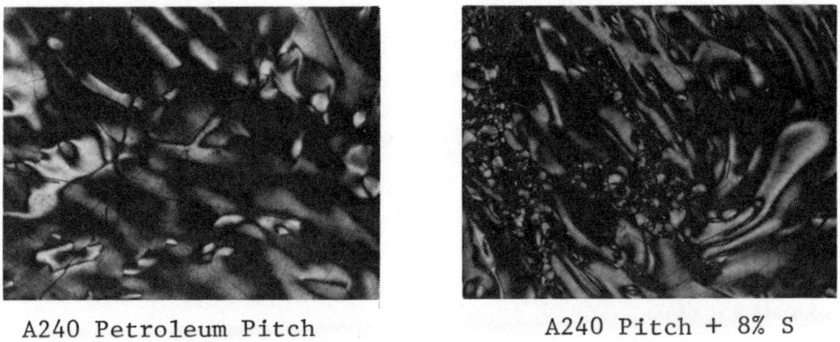

Figure 3. Micrographs of pyrolyses residues of A240 Petroleum pitch and with 8% by weight elemental sulfur.

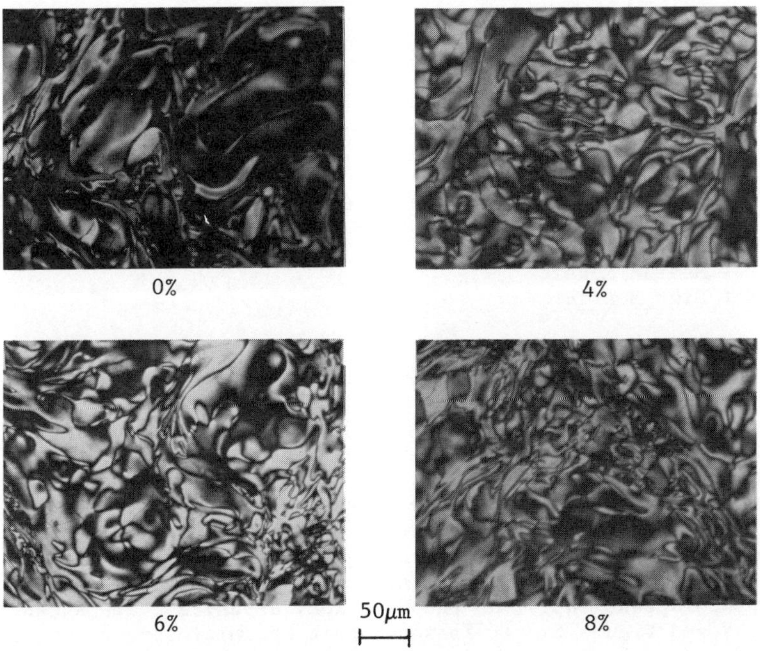

Figure 4. The effect of organic sulfur on mesophase formation in Cat. Inc. asphaltene.

formation begins so that it does not interact with the Cat. Inc. asphaltene or if the sulfur in the DBT molecule is tied up in the ring so that it cannot act as a dehydrogenation agent.

3. Effect of Organometallics

The effect of organometallics at 1600 ppm and 800 ppm on the pyrolysis residues of the Cat. Inc. asphaltene are shown in Figure 5 and Table IV. The additives used were mentioned in the experimental part. Table IV shows that organic Ni and V inhibit mesophase formation because they promote dehydrogenation and condensation reactions to a greater extent than the other listed additives. Pyrolysis yields are increased because of increased

Table IV. Effect of Organometallics* Added to Cat. Inc. Coal Liquid Asphaltene

Metal	Yield (%)**		Microstructure***	
	1600 ppm	800 ppm	1600 ppm	800 ppm
Fe	64.5	62.7	CM	CM
Al	63.0	64.4	CM	CM
Cu	--	64.8	--	Flow
Ni	--	62.3	MM	MM
V	--	68.0	--	FM

*Organometallics used are mentioned in the experimental part
**Residual yields after pyrolysis (wt.%); also with no additives the yield is 52.7% and the microstructure is coarse mosaic
***CM = Coarse Mosaic
MM = Medium Mosaic
FM = Fine Mosaic

condensation. The Fe, Al and Cu additives are all able to increase the yield without changing the microstructure. Mochida, et al. (4) have found that Friedel-Crafts catalysts such as $AlCl_3$ promote mesophase formation by removing the alkyl groups that interfere with mesophase order. Our data show that these additives also can be used to increase the pyrolysis yields.

4. Effect of Organometallics with 6% by Wt. Organic Sulfur

The combined effects of organometallic additives at 1600 ppm and 800 ppm with 6% by weight organic sulfur, exemplified by dibenzothiophene (DBT), on the Cat. Inc. asphaltene are shown in Table V and Figure 6. It is shown that the microstructures stay unchanged except for the Ni addition. Also the pyrolysis yield is increased over that of Cat. Inc. asphaltene and 6% organic sulfur without the organometallic additives. The effect of

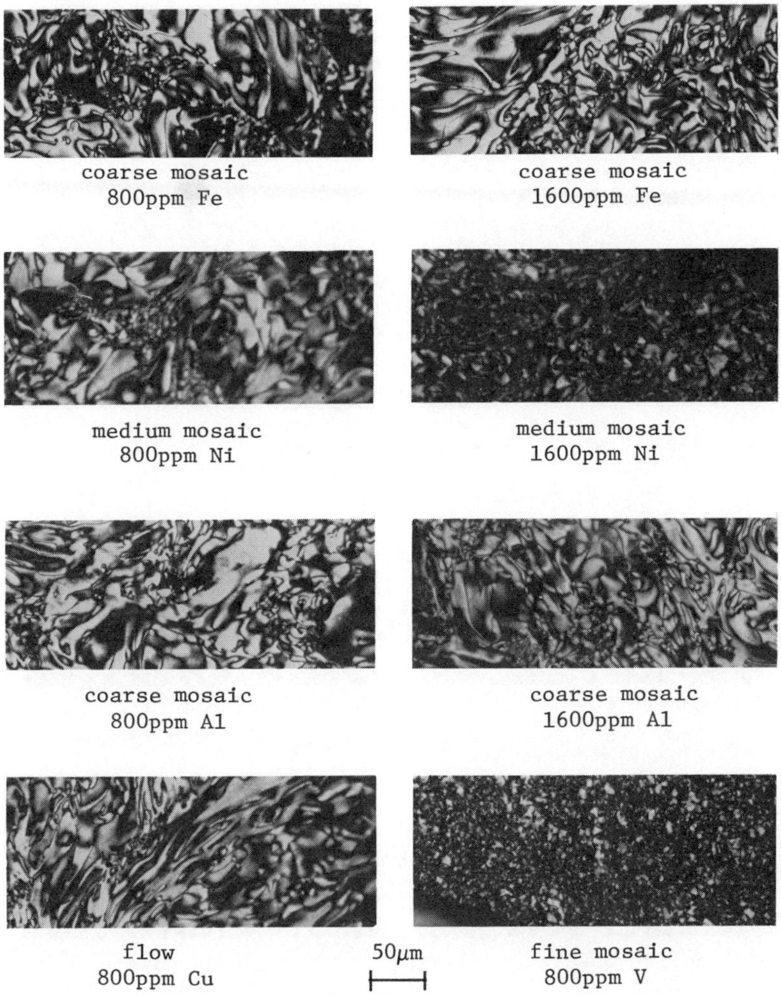

Figure 5. The effect of metals on mesophase formation in Cat. Inc. asphaltene

Figure 6. The influence of organic sulfur (6% by weight) with organometallic compounds on mesophase formation in Cat. Inc. asphaltene.

Table V. Effect of Organometallics* and 6% by weight Organic Sulfur, Exemplified by Dibenzothiophene (DBT), Added to Cat. Inc. Coal Liquid Asphaltene

Metal	Yield (%)**		Microstructure***	
	1600 ppm	800 ppm	1600 ppm	800ppm
Fe	49.1	47.8	CM	CM
Ni	50.7	50.4	FM	FM
Cu	--	--	CM	CM
V	--	--	CM	CM

*Organometallics used are mentioned in the experimental part
**Residual yields after pyrolysis (Wt.%); also with 6% organic sulfur (DBT) and no metals the yield is 43.8% and the microstructure is coarse mosaic
***CM = Coarse Mosaic
FM = Fine Mosaic

heat treatment temperature (HTT) on microstructure for the Cat. Inc. asphaltene alone and with the 6% organic sulfur and 800 ppm Ni additive is compared in Figures 7 and 8 which show that the Ni additive suppresses mesophase nucleation. Table VI shows the residual yields at different pyrolysis temperatures plus their elemental analysis and aromaticities for the combined sample of Cat. Inc. asphaltene, 6% organic sulfur and 800 ppm nickel.

Conclusions

From this work we can draw three tentative conclusions:

1. The results show that addition of elemental sulfur inhibits mesophase formation. This effect is not as pronounced on materials of higher aromaticity and lower oxygen content such as petroleum pitch. We think that this is because the sulfur acts as a dehydrogenation agent and increases the condensation reaction rate to a point where it reduces fluidity and thus suppresses mesophase growth.

2. The results with organic sulfur addition using DBT are not clear. Either the DBT molecule is very stable in the temperature range of mesophase formation or it vaporizes before the mesophase is formed. The addition of DBT does not change the microstructure but it does decrease the overall pyrolysis yield. The decrease in yield is probably due to the loss of DBT during the heat treatment.

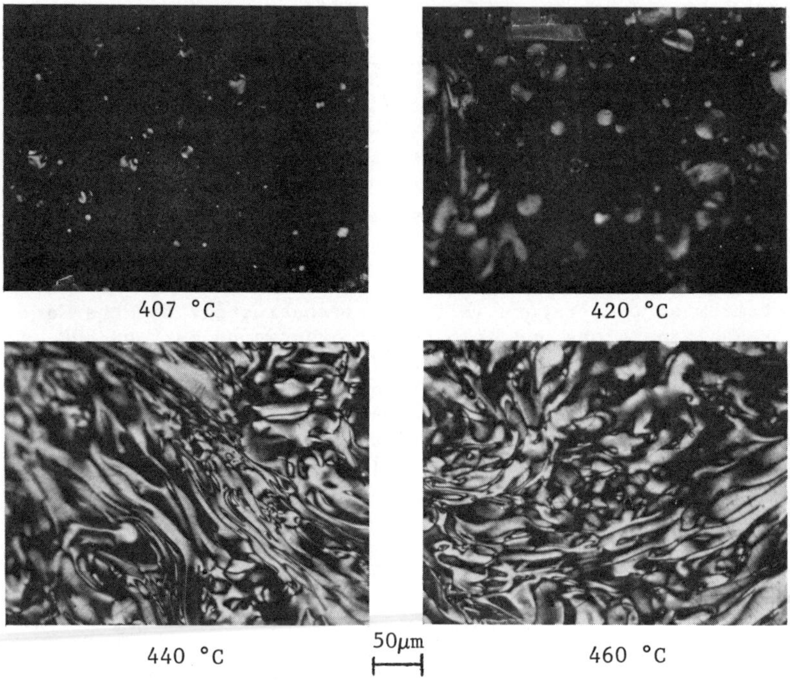

Figure 7. Micrographs of pyrolyses residues of Cat. Inc. asphaltene at different final temperatures.

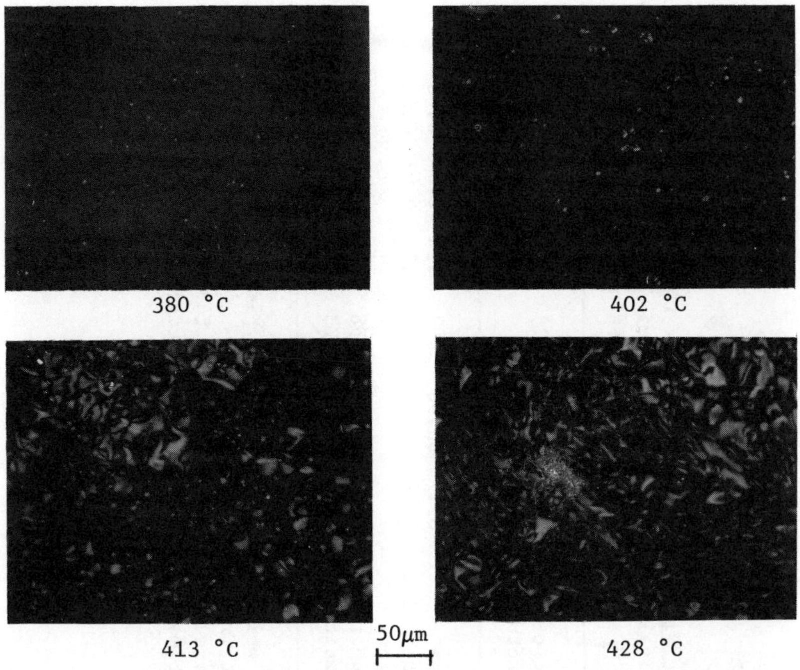

Figure 8. Micrographs of pyrolyses residues at different final temperatures for Cat. Inc. asphaltene with 6% by weight organic Sulfur and 800 ppm Nickel.

Table VI. Elemental Analysis of the Sample Residues at Different Pyrolysis Temperatures. Sample: Cat. Inc. Coal Liquid Asphaltene + 6% Organic Sulfur* + 800 ppm Nickel

HTT (^{0}C)**	Micro-structure	Yield (%)***	C(%)	H(%)	O(%)	S(%)	H/C	O/C	f_a****
380	None	81.7	88.56	5.07	4.16	0.07	0.68	0.077	0.89
402	Spherrules	74.4	89.76	4.83	3.37	0.07	0.64	0.035	0.92
413	Larger Spherrules	68.1	89.99	4.58	3.02	0.09	0.60	0.028	0.96
428	Mozaic	64.3	91.19	4.23	2.65	0.09	0.55	0.025	1.00

*Percent by weight organic sulfur exemplified by dibenzothiophene (DBT)
**HTT - Heat Treatment Temperature
***Residua yields after pyrolysis (Wt.%)
****f_a - estimated from H/C atomic ratio: f_a = -0.79 H/C + 1.43

3. The major conclusion of this work so far is that the trace inorganics Al, Fe and Cu organometallic additives will increase the pyrolysis yield without affecting the mesophase microstructure. Addition of organometallics can either have inhibitory or promotional effects. The promotional effect (e.g. Cu) is perhaps due to the termination of radical formation, whereas the inhibitory effect (e.g. V and Ni) is due to the initiation of radical formation. This will be studied further using the temperature dependence of electron spin resonance (ESR) studies.

Acknowledgment

We are grateful for financial assistance from National Science Foundation under grant no. DAR-8008755.

Literature Cited

1. Weinberg, V.L., Ph.D. Thesis, University of Southern California, Los Angeles, 1981.
2. Weinberg, V.L. and Yen, T.F., unpublished data.
3. Oi, S. and Onishi, N., "Influence of Organic Sulfur Compounds and Metals on Mesophase Formation"; Carbon 1978, 16, 455.
4. Mochida, I., Ando, T., Maeda, K., Fujitsu, K. and Takeshita, K., Carbon 1980, 18, 319.

RECEIVED May 21, 1984

FORMATION AND MODIFICATION OF FIBERS

17

Reaction of Formaldehyde Vapor with Water-Wetted Wool

S. M. AL-KHAYATT[1], H. L. NEEDLES, S. A. SIDDIQUI, and S. H. ZERONIAN

Division of Textiles and Clothing, University of California, Davis, CA 95616

Wool wetted with water at 80% wet pick up was reacted with gaseous formaldehyde for 0.25 to 2 hours, at 23, 60 and 80°C. In the absence of catalysts formaldehyde reacted rapidly with wool at all temperatures with the most rapid reaction occurring at 60°C. Although the reaction proceeded at a slower rate at 23°C, maximum improvement in wet wrinkle recovery and felting shrinkage with a minimum reduction in tensile properties occurred at this temperature. The color of the wool was unaffected by formaldehyde treatment. The formaldehyde-treated wools were yellowed by a filtered xenon plasma arc at the same rate as untreated wool and the face and back of the wool fabrics yellowed to similar degrees. Nevertheless, dyeing studies and scanning electron microscopy demonstrated that even brief formaldehyde treatment protected the wool from light-induced weathering.

The reaction of aqueous solutions of formaldehyde with wool keratin has been investigated by a number of workers (1-7). Introduction of crosslinks into wool by formaldehyde treatment was postulated in these studies, but little direct evidence for crosslinking was found. Mason (5) has presented physical evidence for formation of crosslinks as methylene groups in wool. Such crosslinks would be expected to increase the internal viscosity within the fiber, thereby decreasing the rate at which setting is lost in hot water. This hypothesis was tested by Caldwell et al. (8) who compared the crease recoveries of wool fabrics treated with aqueous formaldehyde or formaldehyde vapors with those of untreated fabrics. They found that the treatments enhanced the set through introduction of crosslinks that inhibited the rearrangement of disulphide bonds in the wool. Kann (9,10) noted that formaldedye treatment protected wool against alkali

[1]Current address: Department of Mechanical Engineering, University of Baghdad, Baghdad, Iraq.

0097-6156/84/0260-0281$06.00/0

damage. Such treatment subsequently was used in scouring, bleaching and dyeing processes to protect wool against chemical attack.

Reaction of wool with formaldehyde vapor rather than with aqueous formaldehyde would be preferred since treatment in the vapor state would have less effect on the desirable properties of wool. Previous studies (8-13) have shown that formaldehyde vapor treatment improved selected properties of wool. Morris and Molvig (11) exposed wool yarns to formaldehyde vapor at 25 to 80°C under yarn extensions of 0 to 40% at moderate humidities. They found that maximum formalization occurred at 50°C with little further effect occurring at temperatures higher than this. Tension on the yarns had very little effect on the rate or degree of formalization of the wool. The wool yarns treated with formaldehyde vapor showed slightly decreased breaking strength, major decreases in elongation at break, and greatly reduced alkali solubilities compared to control wool yarns.

Trezl and coworkers (12,13) studied vapor phase formaldehyde treatment of wool under vacuum. Treatments were conducted at 60 to 100°C using no catalyst or formic acid, trimethylamine, triethylamine, 15-crown-5-ether and 18-crown-6-ether as catalysts. In their system, the presence of water vapor was found to inhibit the rate of formaldehyde uptake. They found that more sites were attacked by formaldehyde vapor than by aqueous formaldehyde. Optimum reaction rates were observed at 70 to 80°C, and formic acid was found to be the most effective catalyst of those used. Scanning electron microscopy (SEM) did not reveal any scale damage to the wool. The treated wool was more thermally resistant, and no change in hand or dyeability of the wool was found. The treated wool had improved tensile strength and initial modulus with little change in elongation at break.

In a comprehensive study (14), formaldehyde vapor treatment of wool was very carefully examined by controlling treatment temperature, water content of the wool, and formaldehyde vapor concentration. Even short periods of formaldehyde treatment produced large decreases in alkali solubility and supercontraction. Addition of acid or base catalysts or swelling agents and thorough drying of the wool reduced the effect, whereas treatment temperatures of 80-100°C, high humidities, and high formaldehyde concentration increased the rate of reaction. The amino groups were shown to be involved in the crosslinking reaction with formaldehyde, whereas the tyrosine groups were not. Also the properties of reduced wool could be partially restored through formaldehyde vapor-treatment. The tensile properties of formaldehyde vapor-treated wools were less affected than control wool by treatment with acids, alkalis, or swelling agents. Prolonged formaldehyde treatment of the wool led to reduced abrasion resistance compared to untreated wool. Formaldehyde crosslinking of wool was effective in setting reduced or annealed wools, and alkali, heat and sunlight-induced yellowing of wool was reduced by formaldehyde treatment, presumably through blocking of amino and tryptophan groups. The dyeing properties of the wools and the lightfastness of the dyed wools were only slightly affected by formaldehyde treatment. There are no reports in the literature examining the rapid reaction of high levels of formaldehyde vapor with water-wetted wool substrate.

Recently we have treated prewetted wool fabrics with formaldehyde vapor under nitrogen in the temperature range from 23 to 80°C using very high concentrations of formaldehyde vapor. The effect of rapid reaction of high levels of formaldehyde with water-wetted wool on the properties of the treated wools were examined including tensile properties (dry and wet), wrinkle recovery (dry and wet), alkali solubility, felting shrinkage, color, dyeing characteristics, and weathering due to simulated sunlight.

Experimental

Materials and Reagents. The fabric used was a 1 x 1 plain weave worsted wool fabric (style #6561) in the ready to dye state obtained from Burlington Industries. Other chemicals and dyes used in the study were reagent grade materials from Aldrich Chemical Company and used without further purification.

Formaldehyde Treatment. Wool samples (12 x 5 cm) were prewetted in distilled water containing 0.1% sodium lauryl sulfate for 10 minutes. The samples were passed through a laboratory pad to remove excess water leaving 80% water uptake and then fixed to a fiberglass screen designed to conform to the inside contour of a 2 liter resin kettle with vertical walls. After the mounted wool sample was placed in the reaction vessel, the top of the reaction vessel was covered with a fitted glass top containing gas inlet and exit tubes and a humidity sensing device. Formaldehyde vapor was introduced into the reactor by passing nitrogen gas through a solution of 30% formaldehyde in a bubbler heated at 70°C at a rate of 5 cc/sec and into the reactor for a period of 15 to 120 minutes. The reactor was maintained at 23°, 60° or 80°C. After formaldehyde treatment, nitrogen was passed through the reactor for 15 minutes to remove unreacted formaldehyde vapor. The fabrics were removed from the apparatus and conditioned (21°C, 65 RH) prior to testing and analysis.

Testing Procedures. The tensile properties of dry and wet warp yarns from control and treated samples were measured by ASTM D2256-66T on an Instron Tensile Tester, Model TM using a gauge length of 3 inches and a rate of extension of 33% per minute (Table I). Wrinkle recovery (dry and wet) values were determined by AATCC test method 66-1978 (Table II). The felting shrinkage and alkali solubilities were determined by the methods outlined by Haron (14) (Table II).

Weathering Procedures. Samples (7 x 12 cm) of untreated and formladehyde-treated wools were exposed to simulated sunlight weathering in an Atlas Weather-ometer, Model Ci35W equipped with a filtered xenon arc. The samples were irradiated continuously for 168 hours total with 18 minute water spray following every 102 minutes of irradiation throughout the cycle. The fabric surface received 0.67 w/m^2 irradiation at a black panel temperature of 63°C and 45% RH during the irradiation.

TABLE I. Relative Tensile Properties of Formaldehyde and Untreated Wool Warp Yarns

Treatment		Rel. Breaking Strength		Rel. Elong. at Break		Rel. Energy to Break	
Temp. (°C)	Time (min)	Dry	Wet	Dry	Wet	Dry	Wet
23	15	1.01	0.77	0.83	0.86	0.81	0.73
23	30	0.99	0.81	0.83	0.84	0.75	0.77
23	60	1.01	0.76	0.83	0.84	0.84	0.71
23	120	0.97	0.81	0.87	0.88	0.72	0.79
60	15	0.97	0.77	0.80	0.82	0.73	0.73
60	30	0.95	0.70	0.73	0.75	0.64	0.62
60	60	0.98	0.70	0.73	0.74	0.68	0.63
60	120	1.03	0.67	0.77	0.74	0.70	0.59
80	15	1.02	0.82	0.80	0.89	0.79	0.78
80	30	1.03	0.78	0.83	0.82	0.86	0.73
80	60	1.05	0.72	0.77	0.74	0.78	0.66
80	120	0.98	0.74	0.77	0.81	0.72	0.70

TABLE II. Change in Properties of Formaldehyde-Treated Wool Compared to Untreated Wool

Treatment		Increase in Wrinkle Recovery (%)		Decrease in Area Felting Shrinkage (%)	Decrease in Alkalai Solubility (%)
Temp. (°C)	Time (min)	Dry	Wet		
23	15	5	16	12	29
23	30	6	13	29	32
23	60	5	16	43	32
23	120	5	9	51	56
60	15	4	10	59	49
60	30	5	11	49	48
60	60	7	10	38	60
60	120	6	10	33	60
80	15	7	11	13	30
80	30	2	12	33	49
80	60	4	8	24	50
80	120	3	6	19	48

Dyeing Procedure. Untreated and formaldehyde-treated unweathered and weathered wool samples were dyed with Acid Violet 7, Acid Green 41 and Acid Red 97 using a Renigal Laboratory Dyeing Machine, Model ST. Dyeings were carried out on 1.0 g wool samples using 2% owf dye and a liquor ratio of 100:1 at 100°C for 1 hour. The samples were then thoroughly rinsed in hot water, followed by washing in cold water, and dried prior to color measurement.

Color Measurement. The color (expresssed in x, y, Y co-ordinates) and color differences (CIELAB) between dyed and undyed untreated, formaldehyde-treated samples before and after weathering were measured using a MacBeth MS2000 color spectrophotometer using illuminant C. The Yellowness Index for undyed samples and the color of and color differences between face and back of weathered samples were also determined. The color properties of the fabrics are found in Tables III-VI.

Scanning Electron Microscopy. Samples of unweathered and weathered untreated and formaldehyde-treated wools were mounted on specimen stubbs using conducting silver paint and coated with two thin layers of silver. Scanning electron micrographs of the samples were prepared and examined for changes in the fiber surface (Fig. 1).

Results and Discussion

The autocatalyzed reaction of formaldehyde vapor with water-wetted wool at three reaction temperatures (23°, 60° and 80°C) was examined in detail. Rapid reaction of formaldehyde with the prewetted wool was observed. The physical and chemical properties of the formaldehyde-treated wool were found to change significantly, particularly within the first 15 minutes reaction time (Tables I and II). The dyeing properties of unweathered and light-weathered untreated and formaldehyde-treated wool were examined at 2.0% (owf) dyeing and compared using three acid dyes (Acid Violet 7, Acid Green 41 and Acid Red 97). The color of the unweathered and weathered wools was determined before and after dyeing (Tables III-VI). Formaldehyde treatment and weathering were found to significantly affect the color of the dyed wools.

The relative breaking strengths, elongations to break, and energies to break of warp yarns from formaldehyde-treated prewetted wools were compared with those of untreated wool (Table I). While the relative dry breaking strengths of the wool were unaffected by treatment, the wet tensile strengths showed significant losses even at the shortest treatment times. Both dry and wet elongations at break and energies to break values decreased for the treated wools with the greatest change occurring during the initial 15 minute treatment time. Treatment carried out at 60°C tended to cause a greater decrease in tensile properties of the wool than treatments at 23° and 80°.

The treated wools showed increases in dry and wet wrinkle recoveries as well as decreases in felting shrinkage and alkali

TABLE III. The Effect of Formaldehyde Treatment and Sunlight Weathering on the Color of Wool

Treatment Temperature* °C	Exposure	Color x	Color y	Color Y	Color Difference (ΔE)	Yellowness Index
Control	No	0.336	0.349	73.7	-	27.2
Control	Yes-Face	0.345	0.358	65.9	5.6	35.7
Control	Yes-Back	0.343	0.354	65.6	4.8	33.6
23	No	0.337	0.352	74.0	-	29.1
23	Yes-Face	0.346	0.359	65.9	5.2	36.4
23	Yes-Back	0.343	0.354	65.9	4.4	33.7
60	No	0.337	0.352	75.1	-	28.9
60	Yes-Face	0.347	0.360	64.3	6.4	37.2
60	Yes-Back	0.343	0.355	65.2	5.2	33.9
80	No	0.337	0.352	74.9	-	29.0
80	Yes-Face	0.347	0.360	67.3	5.3	36.9
80	Yes-Back	0.344	0.356	67.1	4.4	33.6

*Treated for 15 minutes

TABLE IV. Color Characteristics of Wools Dyed with Acid Violet 7

Treatment Temperature °C	Exposure	x	Color y	y	Color Difference (ΔE)
Control	No	0.242	0.391	8.7	-
Control	Yes-Face	0.267	0.378	11.5	7.8
Control	Yes-Back	0.246	0.393	7.0	4.6
23	No	0.242	0.390	9.0	-
23	Yes-Face	0.270	0.375	12.4	9.4
23	Yes-Back	0.246	0.392	7.4	4.4
60	No	0.242	0.390	9.6	-
60	Yes-Face	0.268	0.375	11.8	8.8
60	Yes-Back	0.247	0.390	7.5	5.8
80	No	0.242	0.389	9.6	-
80	Yes-Face	0.261	0.382	10.6	5.6
80	Yes-Back	0.244	0.394	6.7	6.8

TABLE V. Color Characteristics of Wools Dyed with Acid Green 41

Treatment Temperature °C	Exposure	x	Color y	y	Color Difference (ΔE)
Control	No	0.431	0.240	5.6	-
Control	Yes-Face	0.378	0.279	13.0	20.0
Control	Yes-Back	0.415	0.241	6.7	3.2
23	No	0.426	0.236	6.2	-
23	Yes-Face	0.378	0.277	13.2	20.1
23	Yes-Back	0.411	0.239	7.2	7.1
60	No	0.444	0.243	5.3	-
60	Yes-Face	0.390	0.256	10.1	12.6
60	Yes-Back	0.430	0.240	5.7	2.2
80	No	0.446	0.248	4.8	-
80	Yes-Face	0.384	0.261	10.4	14.4
80	Yes-Back	0.426	0.240	5.6	4.0

TABLE VI. Color Characteristics of Wools Dyed with Acid Red 97

Treatment Temperature °C	Exposure	Color x	Color y	Color Y	Color Difference (ΔE)
Control	No	0.577	0.343	13.7	-
Control	Yes-Face	0.505	0.347	18.0	16.4
Control	Yes-Back	0.564	0.345	14.5	3.1
23	No	0.570	0.345	14.9	-
23	Yes-Face	0.535	0.347	16.4	8.6
23	Yes-Back	0.572	0.344	14.1	1.3
60	No	0.570	0.345	15.0	-
60	Yes-Face	0.517	0.346	17.4	12.5
60	Yes-Back	0.565	0.344	14.6	1.9
80	No	0.574	0.345	14.6	-
80	Yes-Face	0.526	0.346	17.0	11.5
80	Yes-Back	0.568	0.344	14.4	2.1

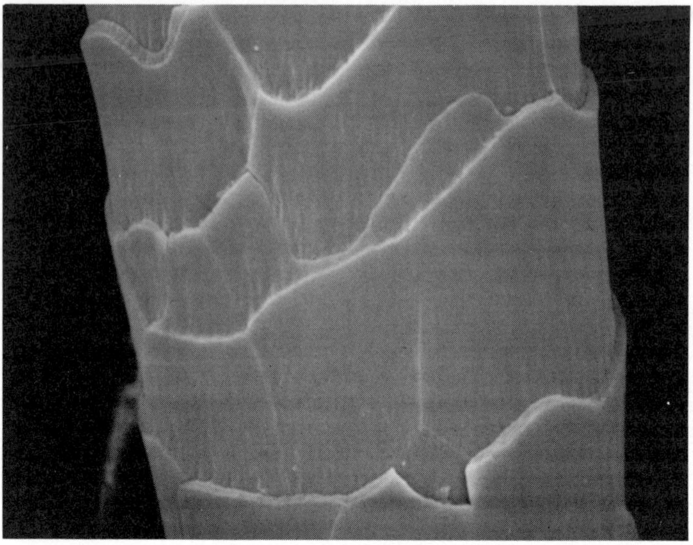

Figure 1a. Control wool at 2500X magnification.

Figure 1b. Weathered control wool at 2500X magnification.

Figure 1c. Wool reacted with formaldehyde for 15 minutes at 60 °C at 2500X magnification.

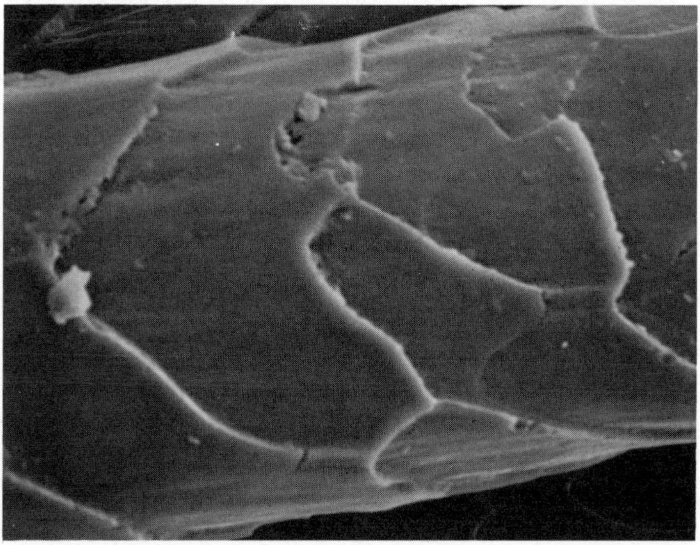

Figure 1d. Weathered formaldehyde-treated wool at 2500X magnification.

solubility (Table II). Slight improvements in dry wrinkle recoveries occurred at all temperatures with the greatest and most consistent improvement occurring at 23° and 60°C. Maximum improvements in wet wrinkle recoveries occurred on treatment at 23°C, even at short reaction times. Formaldehyde-treated wools exhibited large decreases in felting shrinkage. The decrease in felting shrinkage was progressive with reaction time for treatment at 23°C, while at 60°C the decrease in felting shrinkage was largest at short reaction times and dropped as reaction with formaldehyde continued. At 80°C, no consistent change in felting shrinkage was observed. The alkali solubilities of the wool decreased proportionally with increasing reaction time. At 60°C the alkali solubility values decreased at a much higher rate than at 23°C. At 80°C the alkali solubility value was near that of wool treated at 23° for 15 minutes reaction time and did not decrease to the degree that occurred on treatment at 60°C.

The above findings suggest that slow progressive and controlled reaction of formaldehyde with wool occurred at 23°C and that even at shorter reaction times maximum improvement in the wet wrinkle recovery of the wool was observed. At this lower temperature, more and more crosslinks were formed in the cuticle resulting in a progressive decrease in felting shrinkage. At 60°C and 80°C, the reaction of formaldehyde was less discriminate and more complex. Less improvement in wet wrinkle recoveries was observed and the decrease in felting shrinkage tended to become less with time, although the number of crosslinks increased as shown by decreasing alkali solubility as the reaction progressed. These findings differ from those of Varon (14) who carried out the reaction on wool at lower concentrations of formaldehyde vapor held at various humidities rather than on fully wetted wool. In our studies at higher formaldehyde concentrations and using wet wool, greater improvement in the properties of the wool was favoured at the lower reaction temperature (23°C) presumably due to greater control of the reaction.

Reaction of the wool with formaldehyde at the various reaction temperatures has little effect on the color of the wool (Table III). On exposure to simulated sunlight, the front and back of the control and formaldehyde-treated fabrics tended to yellow to a similar degree. Therefore, formaldehyde treatment did not protect the wool from light-induced yellowing. This finding differs from that of Varon (15) who found that formaldehyde treatment under his conditions reduced sunlight-induced yellowing as well as alkali and heat-induced yellowing.

Dyeing of unweathered and weathered control and formaldehyde-treated wools with three representative acid dyes (Acid Violet 7, Green 41 and Red 97) showed that the face and back of the weathered wools dyed very differently than the corresponding unweathered wools (Tables IV-VI). Unweathered control wool dyed to somewhat deeper shades than unweathered formaldehyde-treated wools. When the weathered wool samples were dyed, the face and back of the wool dyed to very different shades and depths of shade compared to the unweathered fabrics. The face of the weathered wools dyed to much lighter shades than the back side of the wool fabric demonstrating that weathering had markedly reduced the affinity of the wool for these acid dyes. In some cases, formaldehyde treatment tended to moderate the difference in dyeability between the face and back of the weathered fabric. These dyeing studies demonstrate that weathering of the face of the wool fabric was much more severe than noted by degree of yellowing alone and that formaldehyde treatment tended to moderate the degree of weathering.

The moderating effect of formaldehyde treatment on light-induced weathering was more readily seen when scanning electron micrographs of the samples were examined (Fig. 1). The scales of weathered control wool (Fig. 1b) showed much more damage than unweathered control wool (Fig. 1a). Formaldehyde-treated wool had a somewhat modified surface (Fig. 1c) compared to control wool (Fig. 1a) and on weathering the formaldehyde-treated wool showed much less damage to the surface scales (Fig. 1d).

Summary

These studies show that high concentrations of formaldehyde vapor can be effectively used to rapidly crosslink water-wetted wool to improve wet wrinkle recovery, felting shrinkage characteristics, and light-induced weathering properties of the wool. Future studies will concentrate on formaldehyde vapor crosslinking of reduced wetted wools to achieve maximum improvement in wrinkle recovery characteristics of the wool.

Literature Cited

1. Asquith, R.S., Parkinson, D.C., J. Textile Inst., 1967, 58, 83.
2. Frazer, L.A., Leach, S.J., Milligan, B., J. Appl. Polym. Sci., 1968, 12, 1992.
3. Griffith, J.C., Textile Res. J., 1965, 35, 1046.
4. Griffith, J.C., Mason, P., Textile Res. J., 1966, 36, 1021.
5. Mason, P., Griffith, J.C., Nature, 1964, 203, 484.
6. Watt, I.C., Nature, 1964, 201, 923.
7. Watt, I.C., Morris, R. J. Textile Inst., 1967, 58, 84.
8. Caldwell, J.B., Leach, S.J., Milligan, B., Delmenico, J. Textile Res. J., 1968, 38, 877.
9. A. Kann, German Patent 144, 485, 1901.
10. A. Kann, U.S. Patent 787, 923, 1905.
11. Morris, M.A., Molvig, T.A.W., Textile Res. J., 1966, 36, 483.
12. Trezl, L., Heizman, J., Tyihak, E., Proc., 5th Int. Wool Res. Conf., 1975; Vol. II, p. 488.
13. Crighton, J.S., Trezl, L., Rusznak, I., Proc., 6th Int. Wool Res. Conf., 1980; Vol. II, 339.
14. Haron, G.F., Ph.D. Thesis, The University of New South Wales, Australia, June, 1977.

RECEIVED January 10, 1984

18

Accelerated UV and Radiation-Induced Grafting of Monomers to Cellulose in the Presence of Additives

Applications of These Copolymers in Art Restoration and Preservation

CHYE H. ANG, JOHN L. GARNETT, STAN V. JANKIEWICZ, RON LEVOT, and MERVYN A. LONG

Department of Chemistry, The University of New South Wales, Kensington, NSW 2033, Australia

The value of ultraviolet light and ionizing radiation as initiators for the grafting of styrene in solvents to cellulose is discussed. Concepts common to both systems are compared. Thus the effect of structure of solvent and concentration of monomer on the efficiency of grafting are treated. The use of acid and polyfunctional monomers, also the synergistic effect of these two additives in enhancing grafting in both radiation systems is considered. The use of cellulose as a backbone polymer in these additive grafting studies is compared with analogous studies with wool, polyethylene, polypropylene and PVC. The significance of these additive effects in UV and EB radiation rapid cure (RRC) processes is discussed. Novel applications of both radiation and photografted celluloses in the field of art restoration and preservation are outlined.

UV and ionizing radiation have been used extensively in the grafting of monomers to cellulose (1-10). Parameters affecting the yield of copolymerization and the efficiency of the process include structure of monomer, type of solvent used and the radiation dose and dose-rate. For preparative purposes, methods for increasing the grafting yields by the inclusion of additives are important. Most of the recent work using additives for enhancement purposes have been performed with synthetic backbone polymers (11-13) especially the polyolefins (11,12). Mineral acids have been utilized successfully as additives for increasing grafting yields in ionizing radiation initiated copolymerization of monomers to the polyolefins and PVC (13). Analogous acid effects have also been reported for the cellulose grafting system involving gamma ray initiation.

By contrast with the ionizing radiation grafting studies, little additive enhancement work has been reported for UV initiated grafting reactions. It is the purpose of this paper to compare UV with ionizing radiation as initiators for grafting to cellulose, especially with respect to possible common parameters including additives which may determine optimum yields in both systems. The copolymerization

0097-6156/84/0260-0295$06.00/0

of styrene to cellulose is selected as model system. The use of acid as an additive for enhancing grafting with both UV and ionizing radiation is examined and the factors which determine the optimum yields of copolymerization with acid are considered. The acid effect as a general phenomenon in UV and gamma ray grafting is evaluated by comparing the cellulose data with analogous results from other representative trunk polymers including wool, the polyolefins and PVC. In addition to acid, new additives for enhancing copolymerization to the polyolefins are reported. The value of using these new additives with the cellulose grafting system is considered. The possible relevance of these grafting studies to the related field of UV and radiation rapid cure polymerization work is examined. Finally novel applications of both radiation and photografted celluloses in the field of art restoration and preservation will be outlined.

Experimental

Styrene was supplied by Monsanto Chemicals (Aust.) Ltd. and was purified by column chromatography on alumina. The cellulose used was Whatman 41 filter paper; wool was plain weave fabric made from 21 μ diameter Merino wool fibres; polyethylene (low density) was supplied as film (0.005") from Union Carbide; polypropylene was isotactic, doubly oriented film (0.002") ex-Shell; PVC film (0.04") was commercially available GVan material.

Grafting Procedures. The methods used for grafting were modifications of those previously reported (3,6,11,14). In the ionizing radiation work, grafting experiments were performed in stoppered pyrex tubes (15 x 2.5 cm) containing styrene/solvent solutions (20 ml) at 20 ± 1°C. Backbone polymer films or strips of cellulose of the appropriate size were fully immersed in the monomer solutions and the tubes irradiated in a 1200 Ci cobalt-60 source. At the completion of the grafting, the films were removed from the solution, solvent washed and soxhlet extracted for 72 hours. With the acid solutions, especially for the cellulose runs, the films and paper strips were washed with methanol:dioxan (1:1) prior to soxhlet treatment. After removal of homopolymer, grafted polymers were dried to constant weight at 65% RH.

In the UV work, monomer solutions (20 ml) were prepared in stoppered pyrex tubes in a manner similar to that described previously for the gamma irradiation system. The tubes were positioned on a motor driven ventilated circulating drum at distances of 12 to 30 cm from the UV source (90 W, high pressure, Hg type 93110E E_2, Phillips) at 24 ± 1°C for the appropriate time as shown on the relevant tables. The polymer films were so positioned that, during irradiation, the surfaces of the films were perpendicular to the incident radiation. After irradiation, films and paper strips were treated as for the gamma system.

Results and Discussion

Solvent Effects in Radiation and Photografting of Styrene to Cellulose. The data in Table I show that when the alcohols are used as solvents for the grafting of styrene to cellulose, similar trends in reactivity are observed for both UV and ionizing radiation systems. Thus the

Table I. Comparison of UV with Ionizing Radiation for Grafting Styrene in Alcohol Solvents to Cellulose

Styrene (% v/v)	Graft (%)							
	Gamma Ray[a]				UV[b]			
	Methanol	Ethanol	n-Propanol	n-Butanol	Methanol	Ethanol	n-Propanol	n-Butanol
20	20	30	14	0	13	9	5	5
40	43	82	67	0	28	17	12	5
60	51	105	51	0	34	22	18	5
80	62	132	15	0	53	50	30	5

[a] Dose rate of 2.99×10^4 rad/hr to total dose of 2.0×10^3 rad in stoppered tubes

[b] Solutions contained uranyl nitrate (1% w/v) as sensitizer and irradiated for 24 hr at 24 cm from 90 W high pressure lamp

lowest molecular weight alcohols, methanol and ethanol, give the highest grafting yields consistent with their wetting and swelling properties (5-8). The UV results were obtained in the presence of uranyl nitrate as sensitizer. Sensitized grafting is much more efficient than copolymerization without sensitizer. Under the UV conditions reported in Table I grafting is extremely low if no uranyl nitrate is present.

The solvent results in Table II are consistent with the above

Table II. Comparison of UV with Ionizing Radiation for Grafting Styrene in Miscellaneous Solvents to Cellulose

Styrene (% v/v)	Gamma Ray[a]				UV[b]				
	Diox.	DMF	DMSO	Hex.	No Sens. (Diox.)	Diox.	DMF	DMSO	Hex.
20	3	0	0	0	3	4	5	6	1
40	9	5	4	0	3	12	10	14	1
60	21	23	20	0	2	20	16	38	1
80	42	34	29	0	4	8	9	88	1

[a] Irradiation in stoppered vessel at 6.7×10^5 rad/hr to total dose of 1.0×10^6 rad except hexane (7.7×10^4 rad/h).

[b] Where sensitizer used, solutions contained uranyl nitrate (1% w/v) and irradiated in stoppered tubes for 24 hr at 24 cm from 90 W high pressure UV lamp, except for hexane (benzoin ethyl ether, 1% w/v).

alcohol data. Those solvents such as dimethyl formamide, dimethyl sulfoxide and dioxan which wet and swell the cellulose also give relatively high grafting yields with both radiation initiators. Grafting in DMSO is particularly efficient at high monomer concentrations using sensitized UV. Grafting yields in the absence of uranyl nitrate are very low (< 4%). Although wetting and swelling are important with this group of solvents, it is known that their swelling characteristics are significantly different to those of the alcohols. Thus in comparing the data from the two groups of solvents, it is obvious, particularly with the alcohols, that chemical processes also participate in grafting, presumably involving radical reactions from the radiolysis of the alcohol as previously proposed (6).

Mechanistically the processes involved in grafting styrene in a polar solvent to cellulose with ionizing radiation and sensitized UV are analogous. Each process is predominantly free radical in nature with a possible contribution from energy transfer processes. In the gamma ray system grafting sites in the cellulose (CelloH) are formed by the sequence of processes shown in Equations 1 to 3 with methanol as representative solvent.

$$\text{CelloH} \rightarrow \text{Cello}^{\bullet} + \text{H}^{\bullet} \quad (1)$$

$$\text{CH}_3\text{OH} \rightarrow \text{CH}_3\text{O}^{\bullet} + \text{H}^{\bullet} \quad (2)$$

$$\text{CelloH} + \text{CH}_3\text{O}^{\bullet} \rightarrow \text{CH}_3\text{OH} + \text{Cello}^{\bullet} \quad (3)$$

At temperatures of the copolymerization, participation of ionic process is minimal.

With the uranyl nitrate sensitized UV system, radical sites in the cellulose are formed from reaction with excited uranyl ion species (Equations 4-6).

$$UO_2^{2+} + h\nu \longrightarrow [UO_2^{2+}]^* \quad (4)$$

$$[UO_2^{2+}]^* + CelloH \longrightarrow Cello^{\cdot} + H^+ + UO_2^{2+} \quad (5)$$

$$[UO_2^{2+}]^* + CH_3OH \longrightarrow CH_3O^{\cdot} + H^+ + UO_2^{2+} \quad (6)$$

Thus in both radiation systems, monomer can diffuse to a site and graft by the charge-transfer mechanism previously proposed (3,5,15).

Effect of Acid as an Additive on UV and Radiation Grafting. Previous studies have shown that mineral acids can enhance the radiation grafting of styrene to cellulose under certain radiation conditions. (6). The results of analogous experiments in photografting are shown in Table III, and compared with the corresponding gamma ray data. In

Table III. Comparison of UV and Ionizing Radiation for Acid Enhancement Effects in Grafting Styrene in Methanol to Cellulose

Styrene (% v/v)	Graft (%)									
	Gamma Ray[a]					UV[c]				
	N.A.	H_2SO_4 (1M)	HCl (2M)	HNO_3 (0.02M)	H_3PO_4 (0.67M)	N.A.	H_2SO_4	HCl	HNO_3	H_3PO_4
10	6	28	27	7	9	–	15	9	–	–
20	30	83	b	29	26	13	54	34	20	54
40	39	20	b	31	29	28	73	47	23	68
60	47	23	b	32	30	34	77	72	37	91
80	56	21	b	29	32	53	95	127	77	92
90	–	–	b	–	–	–	116	131	107	167

[a] Dose rate of 2.64×10^4 rad/hr to total dose of 2.0×10^5 rad in stoppered tubes

[b] Phase separation occurred

[c] Acid (1% v/v) with uranyl nitrate as sensitizer (1% w/v) and irradiated for 24 hr at 24 cm from 90 W high pressure UV lamp.

both radiation systems it is observed that all mineral acids studied increase the grafting yields at least at one particular monomer concentration. Sulfuric and hydrochloric acids are the most efficient enhancement acids with gamma ray grafting whereas all acids, particularly phosphoric, lead to very significant increases in copolymerization yields with UV. In the ionizing radiation system, with inclusion of acid, a peak in grafting occurs at low monomer concentrations (20% with sulfuric and 10% with hydrochloric acid). Addition of acid also results in grafting enhancement at monomer concentrations removed from the peak (10% with sulfuric, nitric and phosphoric acids). Analogous results are also observed in the UV system, except that the peak or Trommsdorff effect (6) in copolymerization shifts to high monomer concentrations (90%). These acid enhancement effects are also dependent on radiation dose as demonstrated by the data in Figure 1, the specific grafting yields at the various doses being shown in Table IV. With increasing dose, there is almost an exponential increase in enhancement with grafting at

Table IV. Effect of Acid as the Dose is Increased on the Radiation Grafting of Styrene in Methanol to Cellulose.[a]

Total Dose ($\times 10^5$ rad)	Graft (%)	
	Neutral	H_2SO_4 (1.0×10^{-1} M)
1.0	7	10
2.0	26	35
5.0	101	136
7.0	121	208
8.0	151	321
10.0	219	514

[a] Styrene (30% v/v) in methanol grafted at 3.0×10^4 rad/hr at 15°C

constant sulfuric acid addition. Grafting can also be performed in air with almost the same efficiency as in evacuated vessels, the gamma ray system being similar to UV grafting (Figure 2) in this respect.

Mechanistically the role of acid in enhancing gamma radiation copolymerization involves a number of competing processes. In previous work in this field with cellulose (6), it has been suggested that acid at the concentrations used does not markedly affect the precipitation of the grafted chains or the swelling of the backbone polymer, especially with cellulose. Instead it has been proposed that part of the acid enhancement can be attributed to a radiation chemistry phenomenon involving increased G(H) yields and thermalized electrons (Equations 7 to 9). These processes can lead

$$CH_3OH + H^+ \longrightarrow CH_3OH_2^+ \qquad (7)$$

$$CH_3OH_2^+ + e \longrightarrow CH_3OH + H \qquad (8)$$

$$\text{CelloH} + H^+ \longrightarrow \text{CelloH}_2^+ \qquad (9)$$

to increased numbers of grafting sites by abstraction reactions with the trunk polymer (Equation 10). More recent work on the acid effect

$$\text{CelloH} + H \longrightarrow \text{Cello}^\bullet + H_2 \qquad (10)$$

indicates that acid leads to an increase in styrene-solvent intermediates ($MS^\bullet$) in the grafting solution, such species resulting in more grafting sites (Equation 11). In these later studies (16), it

$$MS^\bullet + \text{CelloH} \longrightarrow MSH + \text{Cello}^\bullet \qquad (11)$$

has also been observed that in the presence of acid, the chain length of oligomer in the grafting solution is shortened and the numbers of these shorter chains are increased. Shorter oligomer chains could diffuse more readily into the swollen backbone polymer to achieve more efficient termination at a grafting site. The increase in concentration of oligomer chains in the bulk of the solution would also result in an increase in the viscosity of both the grafting solution and the solution that is absorbed within the backbone swollen polymer, thus leading to the enhanced Trommsdorff peak as observed experimentally in the presence of acid.

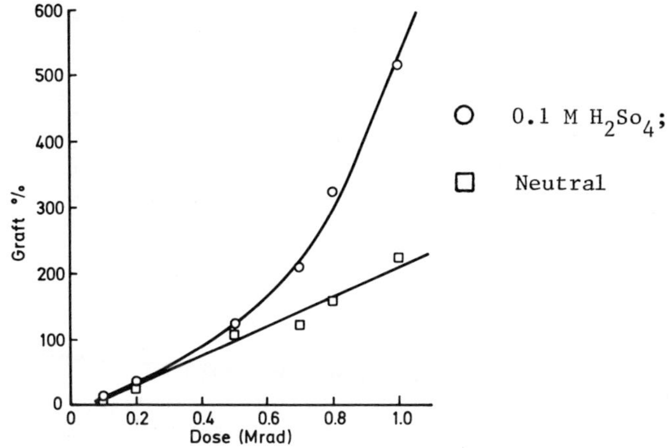

Figure 1. Acid enhancement in grafting of styrene to cellulose. Dose rate 0.03 Mrad hr^{-1}. Styrene 30% (v/v) in methanol.

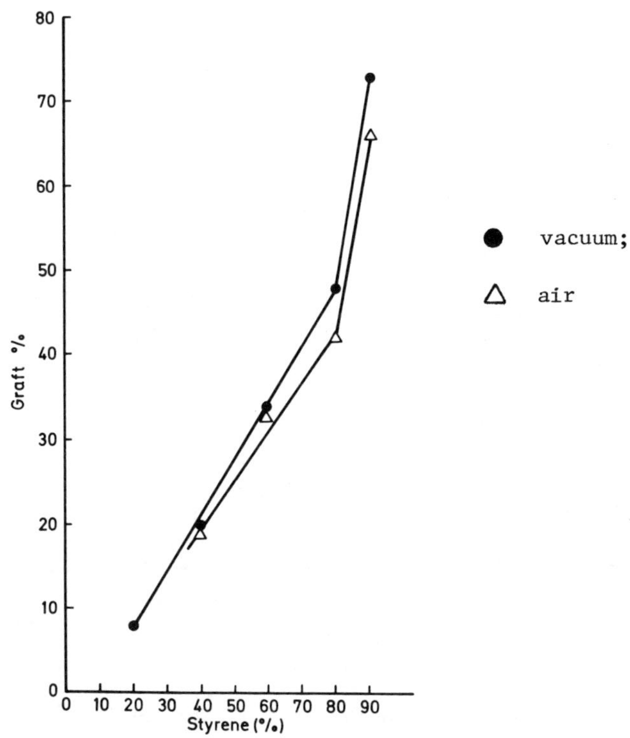

Figure 2. Grafting yield of styrene in methanol, in vacuum and in air for cellulose UV system. Sensitizer conditions (Table I).

The range of differences depicted in Figure 1 for the component affect grafting in a number of ways. Thus the presence of residual water in cellulose (∿ 7% at ambient conditions) can lead to acid enhancement in photosensitized grafting by processes similar to the sulfuric acid catalyzed photopolymerization of monomers in aqueous solution where, in the presence of acid (e.g. H_2SO_4), additional radicals are formed. These may subsequently react with trunk polymer by abstraction reactions to give increased numbers of grafting sites (Equations 12,13). Changes in the physical parameters of cellulose

$$SO_4^{2-} + H_2O \longrightarrow SO_4^{\overline{\cdot}} + H^{\cdot} + OH^- \quad (12)$$

$$SO_4^{\overline{\cdot}} + H_3O^+ \longrightarrow H_2SO_4 + OH^{\cdot} \quad (13)$$

by the inclusion of acid can also advantageously affect grafting. Thus acid can (i) produce both intercrystalline and intracrystalline swelling which loosens the "order-disorder" structure of the cellulose making it more accessible to reagents and (ii) act as a catalyst in the hydrolysis of the cellulose leading to uncoiling of the chains and improved monomer accessibility. Such hydrolysis reactions involve an intermediate complex between the glycosidic oxygen and the proton. In the presence of UV, breakage of the glycosidic bond of the complex is facilitated leading to additional grafting sites.

Comparison of Cellulose with Wool, the Polyolefins and PVC in Acid Enhanced Grafting. The results in Table V show that acid enhancement in grafting is a general phenomenon which is applicable to a range of

Table V. Comparison of Cellulose with Other Backbone Polymers (Wool, Polyethylene, Polypropylene and PVC) for Acid Enhanced Radiation Grafting of Styrene in Methanol

Styrene (% v/v)	Graft (%)									
	Cell.[a]		Wool[b]		Polyeth.[c]		Polyprop.[d]		PVC[e]	
	N.A.	H^+	N.A.	H^+	N.A.	H^+	N.A.	H^+	N.A.	H^+
20	8	24	8	38	57	63	29	218	0	10
30	17	33	-	-	75	130	94	150	13	34
40	22	31	27	58	70	100	50	85	66	70
50	21	27	-	-	68	75	37	55	6	25
60	28	25	28	73	60	83	36	45	3	10

[a] Dose rate of 4.0×10^4 rad/hr to 2.0×10^5 rad with 4.0×10^{-1} M H_2SO_4

[b] Dose rate of 2.5×10^4 rad/hr to 2.0×10^5 rad with 2.0×10^{-1} M H_2SO_4

[c] Dose rate of 4.0×10^4 rad/hr to 2.0×10^5 rad with 2.0×10^{-2} M H_2SO_4

[d] Dose rate of 4.0×10^4 rad/hr to 3.0×10^5 rad with 2.0×10^{-1} M H_2SO_4

[e] Dose rate of 1.0×10^5 rad/hr to 1.0×10^5 rad with 7.5×10^{-2} M H_2SO_4

backbone polymers of widely differing structures. One of the significant features of the data in the table is the differing radiation conditions required to observe the Trommsdorff peak for each of the polymer systems studied. Thus for PVC, the gel effect occurs at 40%

monomer concentration in both neutral and acid solutions whereas with polypropylene, using the same monomer/solvent system, the peak is observed at 30% styrene in neutral solution and at 20% in acidified methanol. With polyethylene, the Trommsdorff effect occurs at 40% monomer in neutral solution and 30% in acid, whilst with wool, no gel effect is seen in either neutral or acid solution. For all backbone polymers, the grafting yields, under the radiation conditions used, are of comparable orders of magnitude except for PVC where the higher yields may reflect the fact that during irradiation PVC generates HCl in situ which may act as acid additive and accelerate grafting. The acid results with wool indicate the importance of dose-rate effects in grafting. At much lower dose-rates, in the presence of acid, wool grafts to extremely high yields (17), indicating that the basic wool structure is susceptible to bond rupture at low radiation doses thus increasing the sites available for grafting in this trunk polymer.

The ease of grafting wool is exemplified in the UV system (Table VI) where yields of over 2400% are observed in the presence of acid. In the UV experiments, the ultraviolet light is delivered over 24 hours and this relatively slow radiation dose-rate accentuates the yield of graft with wool. The functional groups in wool also absorb UV strongly and assist the sensitized grafting, thus copolymerization yields in the presence of acid without sensitizer are an order of magnitude higher than when sensitizer alone is present. With both sensitizer and acid, yields are an order of magnitude higher again. Other authors have discussed UV grafting to wool (18-20) but have not reported the use of the sensitizers discussed here and have not examined acid effects. Their results confirm that UV grafting to wool is a facile reaction.

With respect to the other backbone polymers in UV grafting, PVC is as expected whereas the results for the polyolefins, especially polyethylene, are surprising when compared with cellulose. The relatively high copolymerization yields with polyethylene may be attributed to the presence of trace impurities in the backbone polymer which are present during synthesis. These impurities sensitize the grafting process.

Comparison of Acid with Polyfunctional Monomers as Additives in Polyethylene Grafting. New additives, as well as acid, have been found to accelerate grafting in the presence of ionizing radiation. Much of this recent work has been performed with polyethylene. Only very preliminary studies of these new additives have been carried out with cellulose, however these early results suggest that these latest additives will also be very valuable for copolymerization reactions with cellulose. The more comprehensive polyethylene data will therefore be presented here to act as a guide for what might be predicted when cellulose is used as backbone polymer.

The new additives are polyfunctional monomers, which are exemplified by divinylbenzene (DVB) and trimethylol propane triacrylate (TMPTA) and are used in concentrations of 1% (v/v). When these polyfunctional monomers are added to the monomer solution, the results show that the gamma ray grafting of styrene in methanol to polyethylene is enhanced to a comparable degree to that of inclusion of acid (Table VII). The polyfunctionality of DVB and TMPTA is obviously important in these reactions, since the number of grafting sites in the system is potentially enhanced by the increase in

Table VI. Comparison of Cellulose and Other Backbone Polymers (Wool, Polyethylene, Polypropylene and PVC) for Acid Enhanced UV Grafting of Styrene in Methanol

Styrene (% v/v)	Graft (%)															
	Cellulose[a]		Wool[b]				Polyethylene[c]				Polypropylene[c]				PVC[d]	
	N.A.	H^+	N.A.		H^+		N.A.		H^+		N.A.		H^+		N.A.	H^+
			N.S.	S.	N.S.	S.	N.S.	S.	N.S.	S.	N.S.	S.	N.S.	S.	S.	S.
20	13	54	2	23	8	160	0	26	18	21	0	5	17	66	18	26
40	28	73	29	22	170	-	2	149	48	132	1	163	38	166	67	44
60	34	77	55	17	246	1506	10	41	24	39	6	44	30	37	16	19
80	53	95	53	40	285	2487	8	18	15	22	6	26	13	15	15	11

a Conditions as in Table III N.A. = no acid

b Uranyl nitrate (1% w/v) as sensitizer. Irradiations at 30 cm for 24 hr from 90 W high pressure UV lamp. Sulfuric acid (1% w/v).

c Benzoin ethyl ether (1% w/v) sensitizer. Irradiated as in footnote b.

d Benzoin ethyl ether (1% w/v) sensitizer. Irradiated as in footnote b. At 30% monomer concentration, graft as follows (N.A. = 43, H^+ = 77).

Table VII. Comparison of Acid with Polyfunctional Monomers as Additives for Enhancing UV and Ionizing Radiation Grafting of Styrene in Methanol to Polyethylene Film

Styrene (% v/v)	Graft (%)							
	Gamma Ray[a]				UV[b]			
	Neutral	H^+	DVB	TMPTA	Neutral	H^+	DVB	TMPTA
20	14	19	15	-	28	14	58	28
30	37	51	41	39	101	126	74	52
40	76	81	74	73	189	193	203	321
50	109	134	136	137	124	107	385	412
60	89	119	121	-	-	-	264	322
70	89	73	-	105	37	31	191	133
80	68	62	74	59	32	39	77	-

[a] Dose rate of 4.1×10^4 rad/hr to 2.4×10^5 rad with sulfuric acid (2.0×10^{-1} M), divinylbenzene (DVB) and trimethylol propane triacrylate (TMPTA) at 1% (v/v)

[b] Irradiated at 24 cm for 24 hr from 90 W high pressure UV lamp. Other conditions as in footnote a. Benzoin ethyl ether (1% w/v).

unsaturation due to the presence of the type of additive. Thus with polyfunctional monomers such as DVB in additive amounts, branching of the grafted polystyrene can occur when one end of the DVB, immobilized during grafting, is bonded to the growing chain. The other end is unsaturated and free to initiate a new chain growth via scavenging reactions. The new branched polystyrene chain may eventually terminate, cross-linked by reacting with an adjacent polystyrene chain or immobilized DVB radical. Grafting is thus enhanced mainly through the branching of the grafted polystyrene chain. The results from the addition of TMPTA can be explained in an analogous manner.

In the UV initiated grafting reactions, the addition of DVB and TMPTA accelerate the process similarly to that observed in the gamma ray system. However the magnitude of the enhancement with the polyfunctional monomers with UV is much higher than with acid. The presence of these monomers also causes a shift in the Trommsdorff peak from 40 to 50% styrene concentration. The mechanism of the DVB and TMPTA enhancement in grafting with UV appears to be the same as for the gamma system.

Synergistic Effect of Acid and Polyfunctional Monomers as Additives in Polyethylene Grafting. As with the previous section, data in this section for grafting to polyethylene will be presented as a guide for what may be expected generally in copolymerization, especially for cellulose. Thus it is found that if acid and the polyfunctional monomers are included in the same grafting solution, a synergistic effect of the two additives is observed. For the gamma ray grafting of styrene in methanol to low density polyethylene, the effects of DVB and TMPTA are very much enhanced in the presence of acid (Table VIII). As previously discussed, the inclusion of acid increases the concentration of $H^{\bullet}$ and $MS^{\bullet}$ species (Equations 8 and 11) in the grafting solution. This increase in radical concentration when combined with increased unsaturation in the grafted chains due to the polyfunctional monomers leads to branched polystyrene chains of higher molecular weight and hence higher graft. An analogous explanation can be proposed for the synergistic effect of acid and polyfunctional monomer enhancement in grafting in the UV system (Table VIII).

Table VIII. Synergistic Effect of Acid and Polyfunctional Monomers as Additives for Enhancing UV and Ionizing Radiation Grafting of Styrene in Methanol to Polyethylene Film

Styrene (% v/v)	Graft (%)					
	Gamma Ray[a]			UV[b]		
	Neutral	DVB+H+	TMPTA+H+	Neutral	DVB+H+	TMPTA+H+
20	14	27	-	28	43	41
30	37	58	54	101	98	78
40	76	119	106	189	230	266
50	109	188	181	124	446	525
60	89	156	-	-	298	371
70	89	112	101	37	187	188
80	68	89	95	32	61	-

[a] Dose rate of 4.1×10^4 rad/hr to 2.4×10^5 rad with sulfuric acid (2.0×10^{-1} M), DVB and TMPTA at 1% (v/v).
[b] Irradiated at 24 cm for 24 hr from 90 W high pressure UV lamp. Other conditions as in footnote a. Benzoin ethyl ether (1% w/v).

Significance of Additive Enhancement Effects in Grafting. The use of additives such as acids and the polyfunctional monomers to enhance radiation and photografting of monomers to backbone polymers is important in a preparative context since any procedure for increasing yields at constant dose will be extremely valuable for those trunk polymers which are susceptible to degradation by ionizing radiation or UV. Thus, the lower the dose to achieve a particular graft, the more chemically stable the final copolymer should be. The fact that both acid and polyfunctional monomers can be used for this purpose is also advantageous, since each type of additive may give a copolymer possessing a different structure. With the polyfunctional monomers some cross-linking may be expected whereas, with acid alone, the level of cross-linking should be minimal. Finally when acid is used as an additive in these grafting reactions, particularly with cellulose, care should be exercised in the solvent extraction procedure for homopolymer removal. Thus when the crude graft copolymer directly from irradiation is being treated at elevated temperature in a soxhlet apparatus, any residual acid remaining in the cellulose will be concentrated and will subsequently attack the backbone polymer leading to severe acid degradation. Cellulose is particularly prone to this reaction especially when strong oxidizing acids such as nitric are present. It is therefore recommended that when acid additives are used, the backbone polymer be thoroughly washed with the appropriate cold solvent (methanol for cellulose) to remove all acid prior to extraction.

Value of these Copolymers in Art Restoration and Preservation. Currently, the present authors are involved in a unique application of radiation polymerization techniques where the present grafting work can be of value. Thus radiation polymerization is becoming increasingly useful in the restoration and preservation of art treasures (21, 22), particularly for materials of cellulosic origin. Methods now used include impregnation of cellulosics such as wood with acrylate monomers followed by irradiation. More recent developments involve radiation rapid cure (RRC) treatments where surface modification of cellulosics, wool and leather are important. In the RRC techniques both electron beam and UV are initiating sources and the polymer system includes a low molecular weight oligomer dissolved in an appropriate monomer.

The results of the present additive effect studies in radiation grafting are of significance in the above restoration work since, in both impregnation and RRC procedures especially with cellulose, wool and leather materials, there is the possibility that simultaneous grafting occurs with the polymer formed from the oligomer mix (23). Tritium labelling studies show that grafting can be appreciable in such situations (23). The fact that additives are available, especially acid type materials, which can protonate the OH groups of cellulosics to form potential sites for grafting is also important, since, by this mechanism, yields of graft can be increased. The occurrence of grafting, during RRC treatments particularly, is an advantage since properties such as adhesion, flexibility and toughness of the finished material can be significantly improved.

In a unique application of RRC, the present authors are preserving timber and leather artefacts from Henry VIII's warship, The Mary Rose, which sunk in the Solent and was recently recovered after 438 years on the seabed. The fact that, using the additives outlined in this paper, grafting may be increased during these RRC treatments is very important, since the enhancement effect may well improve the physical properties of the material and thus should assist the longevity of the final product of the preservation process.

Acknowledgments

The authors thank the Australian Institute of Nuclear Science and Engineering, the Australian Research Grants Committee and the Australian Atomic Energy Commission for financial assistance.

Literature Cited

1. Krassig, H.A.; Stannett, V.T. Advan. Polym. Sci. 1965, 4, 111.
2. Arthur, J.C. Jr. Advan. Chem. Ser. 1971, 99, 321.
3. Dilli, S.; Garnett, J.L.; Martin, E.C.; Phuoc, D.H. J. Polym. Sci. C 1972, 37, 57.
4. Nakamura, Y.; Schimada, M. In "Cellulose Chemistry and Technology; Arthur, J.C. Jr. Ed.; ACS SYMPOSIUM SERIES No. 48, American Chemical Society: Washington, D.C., 1977; p.298.
5. Hebeish, A.; Guthrie, H.T. "The Chemistry and Technology of Cellulosic Copolymers", Springer-Verlag, Berlin Heidelberg, 1981.
6. Garnett, H.L. In "Cellulose Chemistry and Technology"; Arthur, J.C., Jr., Ed.; ACS SYMPOSIUM SERIES No. 48, American Chemical Society: Washington, D.C., 1977; p.334.
7. Geacintov, N.; Stannett, V.T.; Abrahamson, E.W.; Hermans, J.J. J. Appl. Polym. Sci. 1960, 3, 54.
8. Arthur, J.C. Jr. Polym. Preprints 1975, 16, 419.
9. Kubota, H.; Murata, Y.; Ogiwara, Y. J. Polym. Sci. 1973, 11, 485.
10. Tazuke, S.; Kimura, H. Polym. Lett. 1978, 16, 497.
11. Ang, C.H.; Garnett, J.L.; Levot, R.; Long, M.A. J. Polym. Sci. Polym. Lett. 1983, 21, 257.
12. Chappas, W.J.; Silverman, J. Rad. Phys. Chem. 1979, 14, 847.

13. Barker, H.; Garnett, J.L.; Levot, R.; Long, M.A. J. Macromol. Sci-Chem. 1978, A12(2), 261.
14. Garnett, J.L. Rad. Phys. Chem. 1979, 14, 79.
15. Dilli, S.; Garnett, J.L. J. Polym. Sci. 1966, A-1, 2323.
16. Ang, C.H.; Garnett, J.L.; Jankiewicz, S.J.; Sangster, C. In "Graft Polymerization of Lignocellulosic Fibers;" Hon, D.N.S., Ed.; ACS SYMPOSIUM SERIES No. 187, American Chemical Society: Washington, D.C., 1982, p. 141.
17. Garnett, J.L.; Leeder, J.D. In "Textile and Paper Chemistry and Technology;" Arthur, J.C. Jr., Ed.; ACS SYMPOSIUM SERIES No. 49, American Chemical Society: Washington, D.C., 1977, p. 197.
18. Needles, H.L.; Wasley, W.L. Text. Res. J. 1969, 39, 97.
19. Bamford, C.H.; Crowe, P.A.; Wayne, R.P. Proc. Roy. Soc., Ser. A. 1965, 284, 455.
20. Ishibashi, H.; Oku, M. Cirtel III 1965, 385.
21. Garnett, J.L.; Major, G. Inst. Conserv. Cultural Material Bull. 1979, S(2), 49.
22. Davis, N.P.; Garnett, J.L.; Long, M.A.; Major, G.; Nicol, K.J. In "Preservation of Paper and Textiles of Historic and Artistic Value II;" Williams, J.C., Ed.; ADVANCES IN CHEMISTRY SERIES No. 193, American Chemical Society: Washington, D.C., 1981; p.223.
23. Ang, C.H.; Davis, N.P.; Garnett, J.L.; Yen, N.T. Rad. Phys. Chem. 1977 9(4-6), 831.

RECEIVED May 15, 1984

19

The Mechanism of Surface Cavitation in Polyethylene Terephthalate Fibers

C. J. DURNING[1], H. D. WEIGMANN, L. REBENFELD, and W. B. RUSSEL

Textile Research Institute and Department of Chemical Engineering, Princeton University, Princeton, NJ 08544

Treating as-spun polyethylene terephthalate (PET) filaments with interactive solvents can cause surface cavitation, depending on the fiber/solvent contacting conditions. In this report, we discuss the preliminary results of an experimental investigation of cavitation and outline a theory for the process. The mathematical representations of four fiber/solvent treatment conditions are discussed: strongly interacting liquids and vapors, weakly interacting liquids, and mixtures of strongly and weakly interacting liquids. The preliminary experimental and theoretical results suggest that surface modification can be controlled by using liquid mixtures.

Predominantly chemical methods are used to modify polyethylene terephthalate (PET) surfaces. Finishing, coating and grafting processes (1) impart antistatic properties, soiling resistance and flame resistance to polyester textiles. Plasma discharge treatments (2,3) can improve the adhesion properties of polyester films and fibers. Recently, Weigmann et al. (4,5) proposed a simple process for physically modifying polyester fiber surfaces via solvent induced crystallization (SINC). The treatment can increase the fiber specific surface area several hundredfold (6) without compromising the mechanical properties. The resulting fibers may find useful applications in composites or in filtration.

The current version (6) of the fiber treatment process employs a drawing rig equipped with a solvent treatment tube where an as-spun PET monofilament briefly contacts a swelling agent, causing rapid crystallization of surface layers. Solvent exchange and drying steps quench the crystallization before the fiber is drawn between feed and take-up rolls. A skin/core morphology results

[1] Current address: Department of Chemical Engineering and Applied Chemistry, Columbia University, New York, NY 10027.

0097-6156/84/0260-0309$06.00/0

(Figure 1a); the solvent crystallized sheath, B, may possess extreme surface porosity, A, while the unpenetrated core, C, apparently develops a fibrillar morphology during the drawing step, imparting high tenacity to the modified filaments. Fiber-solvent contact times as short as 0.2 seconds are possible, allowing surface modified filaments in the 20-30 denier range to be produced.

Earlier workers (7-9) established phenomenologically the mechanisms of mass transport and crystallization during SINC of amorphous PET in "interactive" solvents. (Interactiveness indicates the ability of a solvent to induce structural changes in PET, which correlates with the liquid's solubility parameter (10) and with the glass transition temperature of the fully swollen polymer (5).) During the sorption of interactive solvents, discontinuous swelling (11) occurs (Figure 1b), with solvent diffusion through the swollen surface layers (region B in Figure 1a) controlling the transport process. A semicrystalline, spherulitic morphology develops rapidly in the swollen portion of the fiber since the solvent enhances greatly the polymer crystallization rate.

One must carefully control the solvent contacting conditions to ensure cavitation of the surface layer (6). This article discusses the effects of several treatment conditions on the surface morphology.

Experimental Investigations

Because of the intimate coupling among sorption, crystallization and cavitation during SINC, one must examine all three to fully characterize the fiber treatment process. The preliminary results of our experimental investigations of PET exposed to methylene chloride at unit activity reveal several features of SINC not previously discussed in the literature.

Sorption. We recorded sigmoidal (i.e., "S"-shaped) plots of relative weight gain versus the square root of the exposure time ($\sqrt{t}$) for $MeCl_2$ in PET films at 0°C. These indicate non-Fickian diffusion (12). Although generally sigmoidal, the curve shapes depend somewhat on the film thickness; thick films ($\cong$ 0.09 cm) show a significant portion increasing linearly with $\sqrt{t}$ while thinner films ($\cong$ 0.003 cm) do not. Plotting the relative weight gain against the parameter, $\sqrt{t}/2\Lambda_p$, where $2\Lambda_p$ is the film thickness, indicates that the diffusion is "history dependent" (13,14) since the curves for thin films fall below those for thick films (Figure 2). The effect is consistent with several theories for non-Fickian transport based on diffusion with a simultaneous time-dependent transformation from glass to rubber-like states (15-17).

Crystallization. We measured the density of initially amorphous PET films following exposures to methylene chloride at unit activity, thereby obtaining crystallization kinetics. The density gradient technique was used (18) with carbon tetrachloride/n-heptane mixtures as the immersion media. To avoid spurious readings, efforts were made to remove residual solvent from the crystallized samples and to prevent the entrapment of air in surface cavities during the density

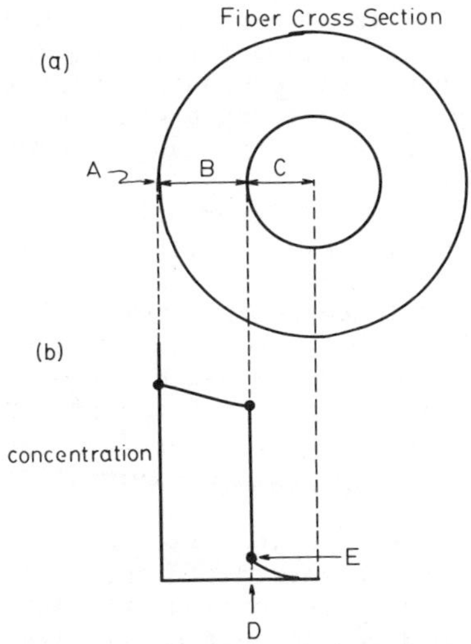

Figure 1a. Schematic representation of skin/core morphology resulting from solvent contact. (A) surface layer, possibly cavitated; (B) penetrated region, solvent crystallized; (C) dry core, subsequently drawn.

Figure 1b. Schematic representation of discontinuous swelling accompanying sorption of an interactive solvent during the fiber treatment process. (D) moving boundary separating swollen from unswollen polymer; (E) threshold concentration for polymer swelling.

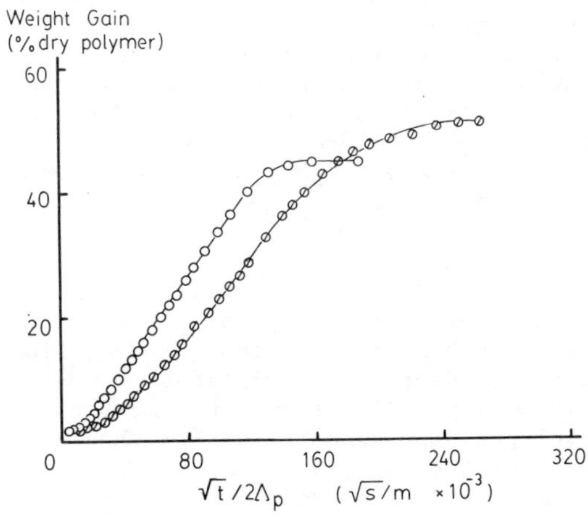

Figure 2.

Relative weight gain versus $\sqrt{t}/2\Lambda_p$ for initially amorphous PET films exposed to methylene chloride at 0°C and unit activity. $2\Lambda_p$ = film thicknesses: 0.03 cm (⊘); 0.09 cm (○).

measurements. After crystallization, samples were dried under high vacuum and then prewetted with carbon tetrachloride before placement in the density column. Using the constants developed by Kilian (19), the densities measured by our procedure provide quite reasonable values (20) for the ultimate degree of crystallinity induced (25-35 percent).

The shape of the kinetic curves for crystallization depend on the film thickness (Figure 3); thicker films show a linear increase in density with $\sqrt{t}$, while thin films give sigmoidal, "S"-shaped curves resembling those predicted for bulk crystallization by Avrami's theory (21). The former result corroborates earlier work by Desai and Wilkes (7). Sigmoidal kinetic curves have been observed during SINC in thin polycarbonate sheets (22); these suggest decoupling between solvent transport and local crystallization.

Cavitation. We also obtained scanning electron micrographs of PET films or fibers crystallized in four different environments: strongly interacting liquids and vapors, weakly interacting liquids, and mixtures of strongly and weakly interacting liquids. The strongly interactive and weakly interactive solvents were methylene chloride and methanol, respectively.

As found by Wilkes et al. (7,8), brief contact with liquid methylene chloride produced severe surface cavitation (Figure 4). Fracture cross-sections showed that the cavitation is confined to the immediate vicinity of the surface (Figure 5). Despite inducing discontinuous swelling and significant crystallinity, saturated methylene chloride vapor did not induce appreciable surface cavitation, confirming our earlier findings (5). (Very early work (7) suggested that swelling vapors could induce surface porosity comparable to that resulting from liquid contact; however, our investigations do not support this conclusion.) Similarly, contact with weakly interactive, liquid methanol also causes swelling and crystallization (23), but only minor surface roughening results.

Liquid mixtures of methylene chloride and methanol apparently cause intermediate levels of surface modification, depending on the composition of the mixture (Figure 6). The micrographs indicate a continuous decline in surface roughness as the concentration of the strong swelling agent decreases.

The Mechanism of Cavitation

A theory for SINC must predict simultaneously the sorption, crystallization and cavitation behavior mentioned above. To accomplish this, one must combine adequate descriptions of non-Fickian diffusion, polymer/diluent crystallization, and local cavitation. Although well developed approaches exist for the former two (9,11-17) none are available for the latter.

We postulate the cavitation results from the accumulation of the pure liquid solvent during the development of polymer crystallites. Since the solvent cannot remain in crystalline regions, crystallite development expels the diluent to neighboring amorphous polymer, increasing the local concentration of the penetrant.

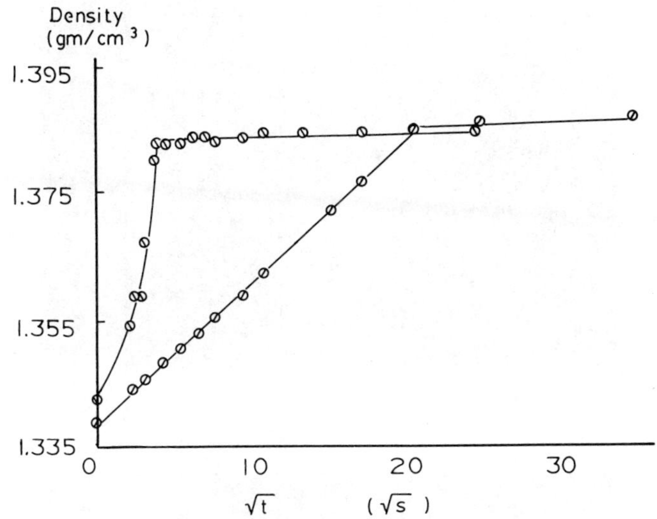

Figure 3.

Apparent density versus $\sqrt{t}$ for initially amorphous PET films exposed to methylene chloride at 38°C. Film thicknesses: 0.0025 cm (⦸), 0.03 cm (○).

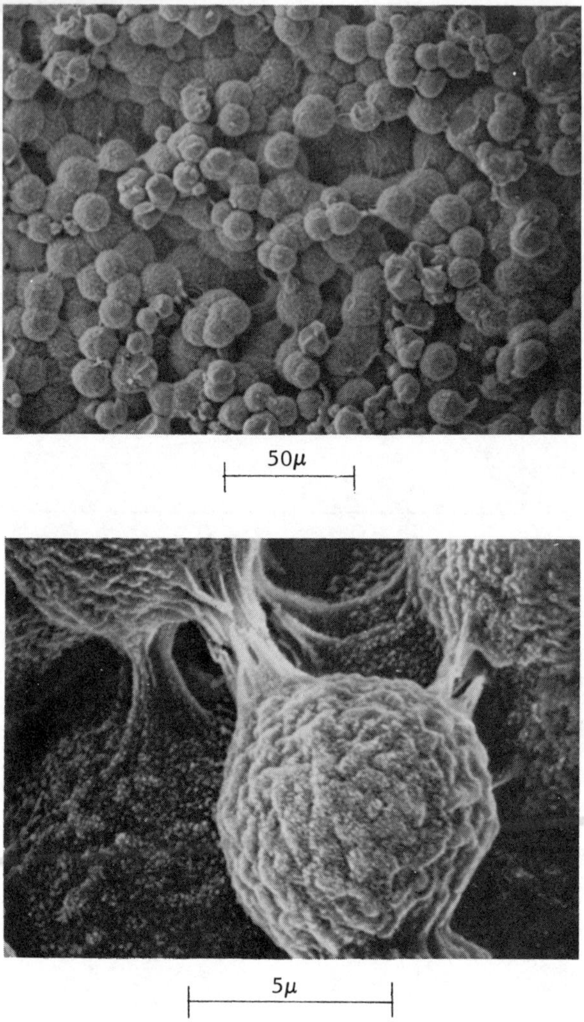

Figure 4.

Surface cavitation of 0.09 cm PET film resulting from brief contact (3 min) with methylene chloride liquid at room temperature.

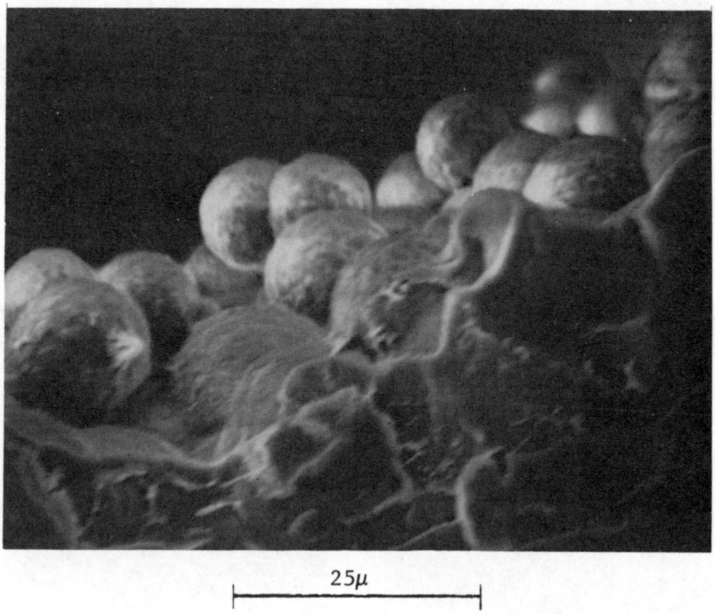

Figure 5.

Fracture cross-section showing surface cavitation of 0.09 cm PET film resulting from brief contact (3 min) with methylene chloride at room temperature.

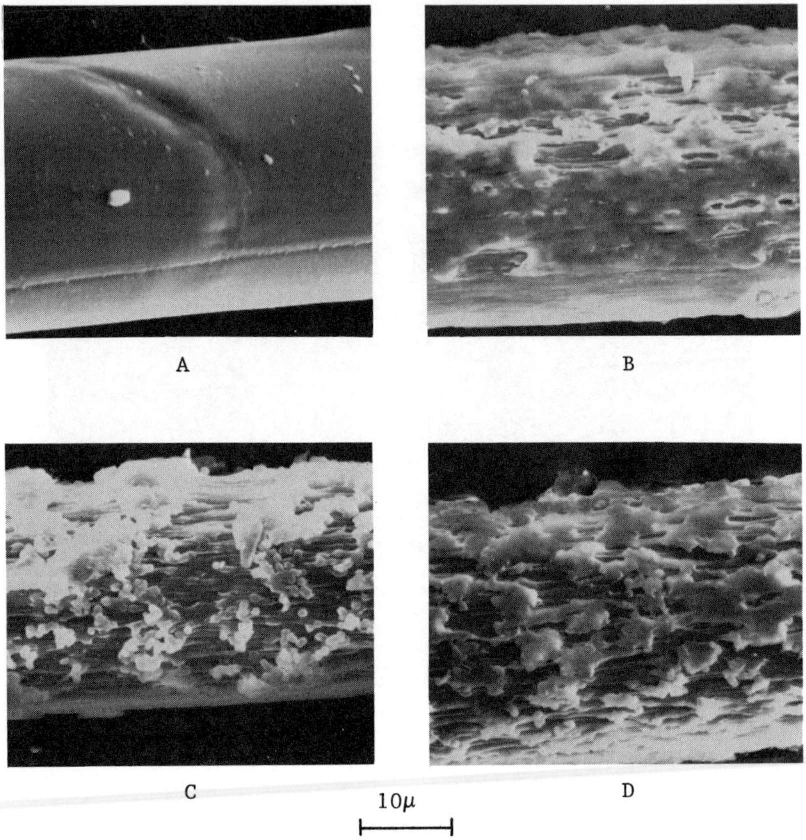

Figure 6.
Surface modification of 360 denier as-spun PET monofilament resulting from brief contact (4.8s) with methylene chloride/methanol mixtures at 13°C. Filaments were drawn to a draw ratio of four after contacting the liquid mixture. Mixture compositions (volume percent methanol): (A) 100; (B) 30; (C) 20; (D) 0.

(Physical evidence supporting the idea of solvent accumulation in amorphous regions due to crystallization is discussed by Zachmann (23).) If this causes local saturation of the amorphous polymer, then continued crystallization would produce local phase separation, creating pockets of pure liquid solvent and a network of holes and cavities in the dried specimen. Accepting this, an immediate conclusion is that saturation of the amorphous component must precede complete crystallization for cavitation to develop.

We now relate the rate of liquid accumulation in cavities to the crystallization rate. Within a crystallizing volume element having a saturated amorphous component (Figure 7), the masses of the solvent, M_s, and polymer, M_p, are

$$M_s = \rho_s V_s + c_o V_a \tag{1a}$$

$$M_p = \rho_c V_c + \rho_a V_a \tag{1b}$$

where V_s, V_a and V_c are the volumes of the pure liquid, mixed and crystalline phases, respectively; ρ_s, ρ_a and ρ_c are their overall densities; and c_o is the solvent's concentration in the mixed phase at saturation (i.e., the solvent's solubility in the amorphous component of the polymer). The total volume of the element, V_e, is

$$V_e = V_s + V_a + V_c$$

Next, we assume that crystallization does not alter the solubility c_o, and does not cause substantial hydrodynamic flow of the solvent. Flory's theory (24) for semicrystalline polymer/solvent solutions shows that the former is valid for small changes in the crystalline volume fraction. Hydrodynamic flow, resulting from a hydrostatic pressure developed during densification, can also be neglected for small changes in crystallinity. Since the total change in the crystalline volume fraction for solvent induced crystallization is only 25-35 percent (20), both assumptions represent reasonable first approximations in describing the cavitation process. In this case, a change in the phase volumes within the crystallizing element must satisfy the restrictions

$$\frac{\partial M_s}{\partial t} = 0 = \rho_s \frac{\partial V_s}{\partial t} + c_o \frac{\partial V_a}{\partial t} \tag{2a}$$

$$\frac{\partial M_p}{\partial t} = 0 = \rho_c \frac{\partial V_c}{\partial t} + \rho_a \frac{\partial V_a}{\partial t} \tag{2b}$$

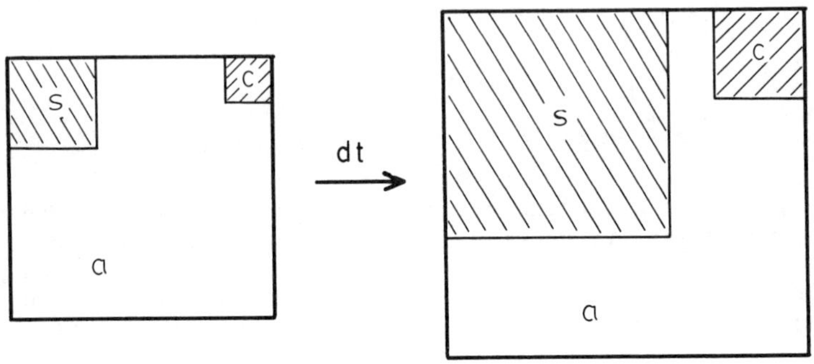

Figure 7.

Schematic of a small, crystallizing volume element within a sample undergoing SINC, having a saturated amorphous component.

(a) saturated, amorphous phase; (c) pure crystalline phase; (s) pure liquid phase.

$$\frac{\partial V_e}{\partial t} = \frac{\partial V_s}{\partial t} + \frac{\partial V_a}{\partial t} + \frac{\partial V_c}{\partial t} \tag{2c}$$

Partial differentials appear because the volume element has a specific location within the sample. Combining and rearranging Equations 2 gives

$$\frac{\partial V_e}{\partial t} = \frac{\partial V_s}{\partial t}\left[1 - \frac{\rho_s}{c_o} + \frac{\rho_s \rho_a}{\rho_c c_o}\right] \tag{3a}$$

$$\frac{\partial V_e}{\partial t} = \frac{\partial V_c}{\partial t}\left[1 - \frac{\rho_c}{\rho_a} + \frac{\rho_c c_o}{\rho_s \rho_a}\right] \tag{3b}$$

$$\frac{\partial V_s}{\partial t} = \frac{\rho_c c_o}{\rho_s \rho_a}\frac{\partial V_c}{\partial t} \tag{3c}$$

One can observe experimentally the crystalline volume fraction, f, and the void volume fraction, ε, defined by

$$f \equiv \frac{V_c}{V_e} \qquad \varepsilon \equiv \frac{V_s}{V_e}$$

so that changes in the volume fractions are defined by

$$\frac{\partial f}{\partial t} = \frac{1}{V_e}\left(\frac{\partial V_c}{\partial t} - f\,\frac{\partial V_e}{\partial t}\right) \tag{4a}$$

$$\frac{\partial \varepsilon}{\partial t} = \frac{1}{V_e}\left(\frac{\partial V_s}{\partial t} - \varepsilon\,\frac{\partial V_e}{\partial t}\right) \tag{4b}$$

Substituting for $\partial V_e/\partial t$ in Equations 4a and 4b using Equations 3a and 3b, respectively, then employing Equation 3c, gives

$$\frac{\partial \varepsilon}{\partial t} = \frac{\partial f}{\partial t}\left(\frac{c_o\rho_c - \varepsilon(c_o\rho_c - \rho_s\rho_c + \rho_s\rho_a)}{\rho_s\rho_a - f(\rho_s\rho_a - \rho_s\rho_c + \rho_c c_o)}\right) \tag{5}$$

relating the rate of cavitation, $\partial\varepsilon/\partial t$, to the crystallization rate, $\partial f/\partial t$, when the amorphous component is saturated. By assuming that the total volume of the element remains constant (an approximation) and that the polymer and solvent volumes in the mixed phase are additive, one can simplify Equation 5 to

$$\frac{\partial\varepsilon}{\partial t} = \frac{\partial f}{\partial t}\left(\frac{v_1^0}{(1 - v_1^0)}\right) \tag{6}$$

where v_1^0 is the volume fraction of the solvent in the mixed phase at saturation.

To calculate cavitation in a macroscopic specimen, one must determine the concentration and crystallization history of each volume element in the sample during the sorption process and apply Equation 6 to each from the time saturation occurs. This has been done for films by solving a non-Fickian transport equation, an equation for local crystallization and Equation 6 simultaneously. The transport and crystallization equations (without their initial and boundary conditions) are

$$D \frac{\partial}{\partial x} (1 - f) \frac{\partial c}{\partial x} = \frac{\partial}{\partial t} (1 - f)\ c \tag{7}$$

$$\frac{\partial f}{\partial t} = A\ f^{2/3}\ (f_0 - f) \tag{8}$$

respectively (25). The symbols are those usually associated with diffusion problems: D is the diffusivity, c the concentration, x the distance, etc. A is the polymer crystallization rate, while f_0 is the ultimate crystalline volume fraction.

Equations 6-8 apparently represent the sorption and crystallization behavior in films reasonably well (26), and should adequately represent fiber treatments for short contact times. In the remaining section, we discuss some preliminary calculations with Equations 6-8, employing physical constants characteristic of SINC of PET at room temperature.

Surface Conditions during Fiber Modification

One can simulate fiber treatment conditions by specifying appropriate surface boundary conditions. We consider the situations encountered experimentally.

Strongly interactive liquids. Here we assume that the fiber surface layers achieve equilibrium instantaneously, so the surface boundary condition is constant surface concentration (27), $c_s = c_o$. With this, Equations 6-8 predict exactly the results depicted in Figures 4 and 5: severe surface porosity confined to the surface layer.

Strongly interactive vapors. For saturated vapors, a large density difference exists across the surface interface, and an external mass transfer resistance is present since solvent molecules must migrate through the background air before reaching the fiber surface. The mathematical representation of these conditions is (27)

$$k\,\kappa(c_o - c_s) = -D\left.\frac{\partial c}{\partial x}\right|_{surface} \tag{9}$$

where k is the external mass transfer coefficient, κ is the ratio of the external to internal solvent density at equilibrium, and c_s is the surface concentration. Using a stagnant film model, we estimated k for a fiber in quiescent solvent vapor 1 cm above the liquid solvent, which corresponds to the conditions for the vapor treatment procedure used in our earlier work (5). We estimated κ from the ideal gas law and typical equilibrium weight gains for solvents in PET. The surface concentration predicted by Equations 6-9 remains below the saturation value throughout the sorption process (Figure 8), thereby preventing significant cavitation.

Weakly interactive liquids. Such solvents do not provide sufficient free volume for polymer chain mobility characteristic of the rubbery state (5). The PET/methanol system illustrates this; amorphous PET imbibed with methanol at ≅ 25°C remains noncrystalline indefinitely (20), indicating severely restricted chain mobility even in the swollen state. For such systems, the kinetic restrictions to chain rearrangement prevent the rapid achievement of equilibrium in surface layers (5). As a result, the solvent surface concentration increases slowly during the sorption process. Several authors (12, 17,28-30) have suggested specific relationships governing the time dependence of the surface concentration in such cases; we have employed a simple exponential increase, viz.:

$$\frac{c_s - c_1}{c_o - c_1} = 1 - \exp\left(-\frac{t}{\beta t^*}\right) \quad \text{for } t < 4\beta t^*$$

$$c_s = c_o \quad \text{for } t \geq 4\beta t^* \tag{10}$$

where c_s is the surface concentration at time t and c_1 is the value immediately after contacting the solvent. The characteristic time t^* is the time scale for the non-Fickian transport (defined in references 16 and 25), so that β is the time constant for attaining equilibrium at the surface, scaled by t^*. The constant 4β therefore represents the number of characteristic times required for surface saturation.

Figure 9 shows the surface cavitation predicted by Equations 6-8 and 10, setting c_1 equal to the "threshold" concentration (5) for PET swelling. As the time scale for surface saturation

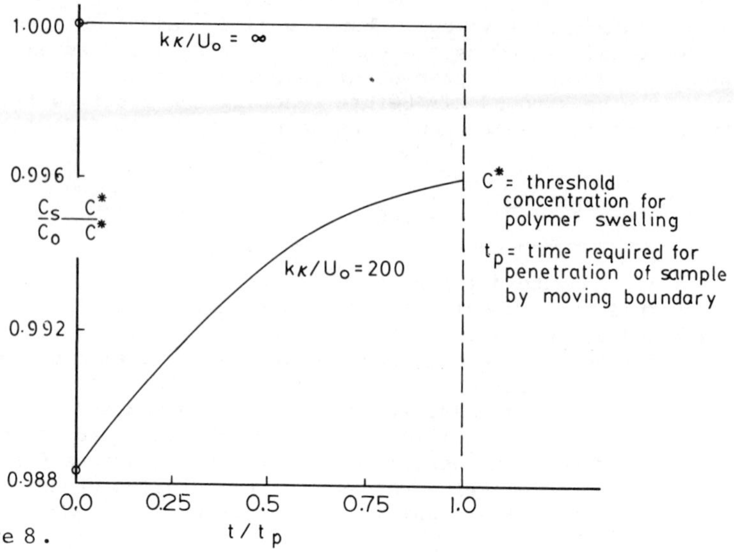

Figure 8.

Prediction of the fiber surface concentration during brief exposure to the saturated vapor of a strong swelling agent. The model parameters are estimated for the system PET/methylene chloride vapor at room temperature. The combination $k\ \kappa/U_0$ is the dimensionless external mass transfer coefficient; $k\ \kappa/U_0 = \infty$ corresponds to a constant surface concentration.

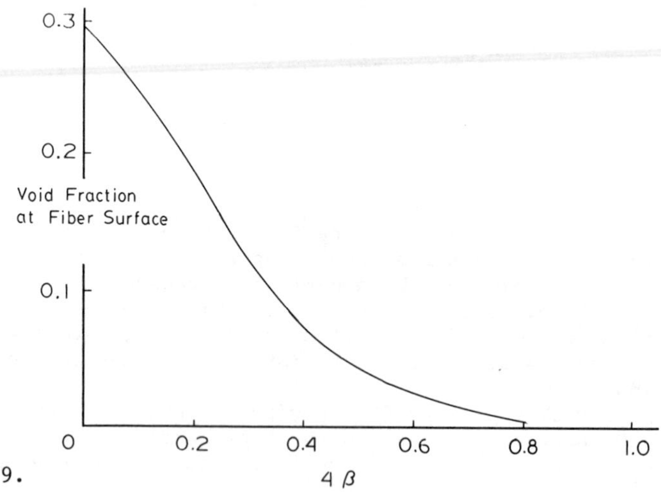

Figure 9.

The effect of a time dependent surface concentration on fiber surface morphology.

approaches that for transport (i.e., $4\beta \rightarrow 1$), the theory predicts significantly less cavitation. Assuming $4\beta \cong 1$ represents the surface history for weakly interactive solvents, Equation 10 offers a tentative explanation of the experimental results.

Liquid mixtures of strongly and weakly interactive solvents. Because of the external mass transfer resistance imposed by the inert component, the strongly interactive solvent cannot readily access the fiber surface. The extent of this effect depends on the Schmidt and Reynolds number (31) in the treatment zone. Also, because of its reduced activity, the strongly interactive component may not induce the chain mobility required for rapid achievement of surface equilibrium. By analogy with the preceding cases, fiber treatment with such mixtures should induce less cavitation than the strongly swelling agent alone. By adjusting the composition and flow rate of the mixture in the treatment tube (6), it should be possible to control the surface conditions and thereby the surface cavitation. Our preliminary experimental results (Figure 6) support this conclusion.

Summary

This report summarizes recent work on a process for the physical modification of polyester fiber surfaces via solvent induced crystallization (SINC). New experimental results show:

- Sorption during SINC is non-Fickian;
- Local crystallization and transport decouple in thin specimens;
- Liquid mixtures of strong and weak swelling agents induce intermediate levels of surface cavitation.

A theory for cavitation was outlined, and preliminary calculations for four fiber treatment conditions were discussed. The theory corroborates the experimental results and reinforces the idea that liquid mixtures of strong and weak swelling agents can be used to control surface modification.

Acknowledgments

The authors thank Dr. A. Goźdź for helpful discussions, and Ms M.G. Scott and Ms S. Reutsch for experimental assistance. We gratefully acknowledge the National Science Foundation (Grant #DMR-7905980) for financial support.

Literature Cited

1. Szegö, L. Adv. Polym. Sci. 1979, 31, 89.
2. Briggs, D.; Rance, D.G.; Kendall, C.R.; Blythe, A.R. Polymer 1980, 21, 895.
3. Amouroux, J.; Goldman, M.; Revoil, M.F. J. Polym. Sci. Chem. Ed. 1982, 19, 1373.

4. Gerold, E.A.; Rebenfeld, L.; Scott, M.G.; Weigmann, H.-D. Text. Res. J. 1979, 49, 652.
5. Durning, C.J.; Scott, M.G.; Weigmann, H.-D. J. Appl. Polym. Sci. 1982, 27, 3597.
6. Goździ, A.; Weigmann, H.-D. Text. Res. J., in press.
7. Desai, A.B.; Wilkes, G.L. J. Polym. Sci. Symp. 1974, 46, 291.
8. Makarewicz, P.J.; Wilkes, G.L. J. Polym. Sci. Phys. Ed. 1978, 16, 1529.
9. Makarewicz, P.J.; Wilkes, G.L. J. Polym. Sci. Phys. Ed. 1978, 16, 1559.
10. Knox, B.H.; Weigmann, H.-D.; Scott, M.G. Text. Res. J. 1975, 45, 203.
11. Peterlin, A. J. Res. Nat. Bur. Std. 1977, 81A, 243.
12. Fujita, H. Fortschr. Hoch. Polym. Forsch. 1964, 3, 1.
13. Crank, J. J. Polym. Sci. 1953, 11, 151.
14. Park, G.S. In "Diffusion in Polymers"; Crank, J.; Park, G.S., Eds.; Academic Press: New York, 1968.
15. Frisch, H.L. J. Chem. Phys. 1964, 41, 3679.
16. Astarita, G.; Sarti, G.C. Polym. Eng. and Sci. 1978, 18, 388.
17. Thomas, N.; Windle, A.H. Polymer 1982, 23, 529.
18. ASTM Standard #D1505.
19. Kilian, H.G. Kolloid Zeit. 1961, 176, 49.
20. Zachmann, H.G. Faserforschung und Textiltechnik 1967, 18, 95.
21. Avrami, M. J. Chem. Phys. 1939, 7, 1103.
22. Turska, E.; Pryzgocki, W. Faserforschung und Textiltechnik 1967, 18, 91.
23. Zachmann, H.G.; Konrad, G. Die Makromol. Chemie 1968, 118, 189.
24. Flory, P.J. J. Chem. Phys. 1949, 17, 223.
25. Durning, C.J.; Russel, W.B., Polymer, submitted for publication.
26. Durning, C.J.; Russel, W.B. Polymer, submitted for publication.
27. Crank, J. "The Mathematics of Diffusion", 2nd Ed.; Oxford Press: London, 1975; Chap. 1.
28. Joshi, S.; Astarita, G. Polymer 1979, 20, 455.
29. Long, F.A.; Richman, D. J. Am. Chem. Soc. 1960, 82, 513.
30. Petropoulos, J.H.; Roussis, P.P. In "Permeability of Plastic Films and Coatings"; Hopfenberg, H.B., Ed.; Plenum Press: New York, 1974.
31. McGregor, R. "Diffusion and Sorption in Fibers and Films"; Academic Press: New York, 1974; Chap. 14.

RECEIVED January 11, 1984

20

Fluorenone Polymers and Fibers

A. M. USMANI

The Research Institute, University of Petroleum and Minerals, Dhahran, Saudi Arabia

Fluorenone polymers as typified by bisphenol fluorenone copolyester of terephthalic and isophthalic acids have many outstanding properties, e.g., heat resistance and high glass transition temperature. They do not generate HCN during burning since the polymer structure does not contain nitrogen and this is a potential advantage in fire-resistant fiber applications. Fluorenone polyesters can be toughened with minor percentages of ethylene/vinyl acetate/vinyl alcohol terpolymer. Potentially they also can be toughened with small amounts of an acetylene terminated bifunctional monomer of fluorenone which can react at high temperatures to form an improved thermally stable product.

The synthetic fiber industry is only about fifty years old yet the annual production is in billions of lbs. The development of fibers resulted due to advances in polymer synthesis and new spinning methods. At the present time nylons, polyesters, acrylics and polyolefins are major classes of synthetic fibers. Fibers have also been made from polymers, e.g., polyvinylidene chloride and polyvinyl alcohol but their commercialization has not materialized. Recently, we have made fibers from fluorenone polyesters which have good potentials and should be further developed.

Fluorenone polyesters are made from isophthalic acid or terephthalic acid/isophthalic acid and bisphenol fluorenone and a typical formula is as follows:

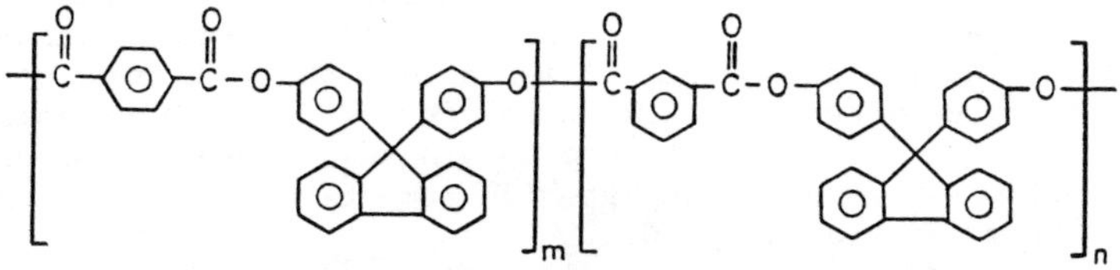

0097-6156/84/0260-0325$06.00/0

Linear aromatic polyesters prepared from diphenols and dibasic acids have been reported in some early patents (1-3). More recently, Morgan (4) has synthesized aromatic polyesters by interfacial and solution methods from aromatic diacid chlorides and bisphenols that have essentially planar, doubly attached groups on the methylene units between the phenylene rings. Bier (5) has recently reviewed polyesters made from aromatic dicarboxylic acids and bisphenols and describes the preparation of some bisphenol monomers and polyesters made therefrom by interfacial condensation, solution condensation, melt condensation, and re-esterification reaction methods.

In this paper we will describe the chemistry and characterization of bisphenol fluorenone polyesters and copolyesters that were found to be suitable for conversion into fiber. We will also describe the synthesis and utility of acetylene terminated fluorenone to improve deficient properties of fluorenone polyesters.

Fluorenone Polyesters

Flourenone polyester synthesis has much latitude. The ratio of dibasic acids (terephthalic and isophthalic) and bisphenols (bisphenol fluorenone and bisphenol A) can be varied depending upon desired properties. Molecular weight can also be regulated. We can expect good processability with fluorenone polyesters rich in bisphenol A whereas polyesters containing large amounts of bisphenol fluorenone will have marginal processability. Marginal processability can be improved by using a reactive plasticizer. We have characterized fluorenone polyesters, e.g., bisphenol A/bisphenol fluorenone (90/10) polyterephthalate (FPE-1), bisphenol fluorenone polyisophthalate (FPE-2), bisphenol A/bisphenol fluorenone (20/80) polyisophthalate (FPE-3) and bisphenol fluorenone copolyester of terephthalic and isophthalic acid (50/50) (FPE-4) and evaluated their potential as fibers. We also synthesized an acetylene terminated bifunctional monomer of fluorenone which can plasticize fluorenone polyesters at moderately elevated temperatures and can react at high temperatures to form an improved thermally crosslinked network.

Characterization of Polyesters

Fire reistance, chemical composition by infrared (IR) and nuclear magnetic resonance (NMR) spectroscopy, thermal analyses, Clash-Berg moduli determination and dynamical mechanical analyses were determined. The fluorenone polyesters were spun as fibers from solution. They were blended with an acetylene terminated fluorenone monomer for plasticization and crosslinking at high temperatures to form an improved thermally stable product.

Thermomechanical Analyses (TMA) were recorded on a DuPont thermal analyzer, Model 943 TMA fitted with a penetrating tip probe, at 2g load and 10°C/min temperature rise. Vertical displacement and the first derivative of that displacement with respect to time were recorded as a function of temperature. Dynamic Mechanical Analyses (DMA) were obtained on a DuPont 981 DMA instrument.

Monomer Synthesis

Acetylene terminated monomers, e.g., acetylene terminated sulfone are known to exhibit high thermal stability, good adhesion characteristics and high moisture resistance (6,7). Starting with bisphenol fluorenone we synthesized acetylene terminated fluorenone (ATF).

ATF was prepared by the reaction route shown below. p-Fluorophenyl acetylene (6.00 g) was added to 9,9-bis(4-hydroxyphenyl) fluorenone (8.50 g) in dimethylsulfoxide (230 ml) containing 25.8 g of potassium carbonate. The reaction was carried out under an inert atmosphere at 130°C for 4 days. The progress of the reaction was followed by monitoring disappearance of hydroxyl band at 3400 cm^{-1} and appearance of acetylene band at 3300 cm^{-1}. The reaction mass was then mixed with aqueous potassium hydroxide to remove

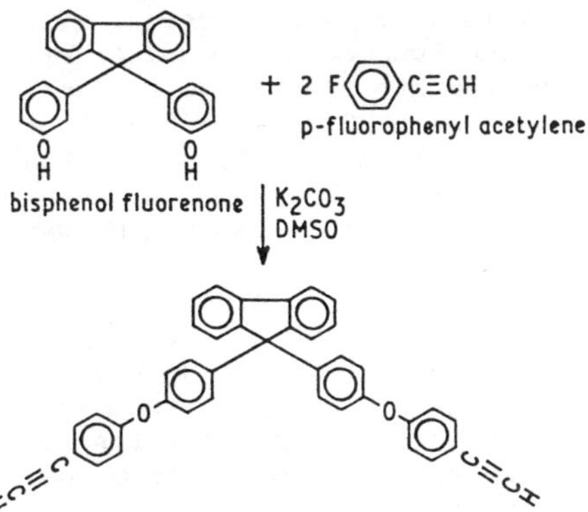

any unreacted bisphenol and added to 900 ml diethyl ether. The separated ether layer was washed several times with water to remove DMSO and inorganic materials. Ether was flashed off to recover crude ATF (7.1 g). ATF was crystallized from ethanol/water (90/10) to give a brownish yellow powder. The recrystallized product showed an acetylenic hydrogen peak at 3300 cm^{-1} (IR) and 3.0 ppm (NMR).

p-Fluorephenyl acetylene, used in ATF synthesis, was prepared as follows:

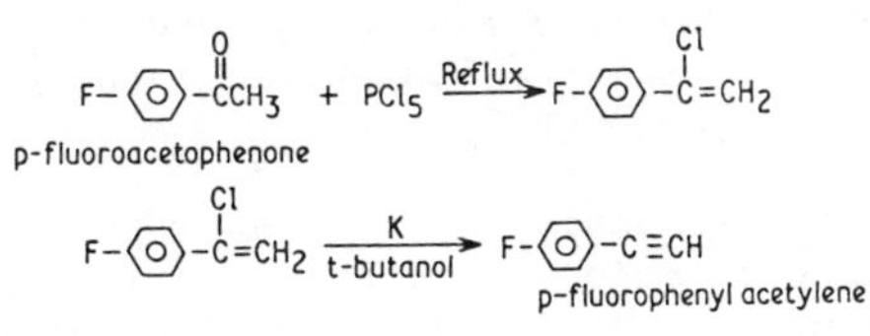

A reversed phase high performance liquid chromatography (RP-HPLC) was run on ATF. It revealed that besides ATF other low molecular weight bifunctional oligomers were also formed in small amounts. Formation of reactive oligomer does not reduce utility of ATF as a reactive plasticizer for fluorenone polyesters.

Results and Discussion

Fire Resistance. Fluorenone polyester fibers were tested for fire resistance and compared against a polycarbonate control. FPE-2, -3, and -4 had good flame resistance. They did not support combustion, did not drip, and were nearly self-extinguishing. When they burned, black soot was given off. FPE-1 was slightly inferior to the others in flame resistance. The control polycarbonate burned with excessive dripping and considerable black soot. Thus the fluorenone polyesters have potentially outstanding fire resistance properties.

IR and NMR Spectroscopy. Infrared spectrum of FPE-1 was recorded on a Perkin-Elmer Model 137 spectrophotometer. FPE-1 was dissolved in methylene chloride, cast on sodium chloride, dried in a vacuum oven at 40°C for 2 hours, and its spectrum was then recorded (Figure 1). A weak absorption band at 3.38 μ, due to the presence of an aliphatic CH from bisphenol A, is seen. Also a weak absorption band at 7.35 μ is due to the $C-CH_3$ of bisphenol A in the polyester. The ratio of bisphenol A to bisphenol fluorenone was established by NMR measurement.

NMR spectrum of FPE-1 was recorded on a Varian Associates Model CFT-20 spectrometer equipped with a proton probe operating at a frequency of 80 MHZ. The polymer was dissolved in $CD_2C\ell_2$ and the NMR spectrum is shown in Figure 2. Methyl resonance (from bisphenol A) occurs at 1.7 ppm. At ∿1.6 ppm, a minor impurity in $CD_2C\ell_2$ seems to appear. The aromatic protons stemming from the bisphenol A and bisphenol fluorenone portions of the polyester appear at 7.2 and 8.3 ppm. The ratio of the methyl-to-aromatic protons suggests that the polyhydroxy residue of FPE-1 contains 90 wt % bisphenol A and 10 wt % bisphenol fluorenone.

Thermogravimetric Analysis. Thermogravimetric analysis (TGA) on the four polyesters were performed in air and helium. The TGA curves are shown in Figures 3 (FPE-1 and -4) and 4 (FPE-2 and -3); all scans were taken at a heating rate of 10°C/min. The data for FPE-4 are rather typical, showing weight loss starting at 450°C with pyrolysis continuing up to 600°C. The advanced stages of weight loss in air involved char 'burnoff' as indicated by the differences between the helium and air traces in Figure 3 and the second stage decomposition inflections in Figure 4. From the TGA thermograms, char formation levels are seen to be in the 30-60% range.

Solubility and Intrinsic Viscosity. Fluorenone polyesters are readily soluble in methylene chloride, dimethylformamide, dimethylacetamide, and tetrahydrofuran. Intrinsic velocities [η] in methylene chloride were determined at 25°C (see Table I).

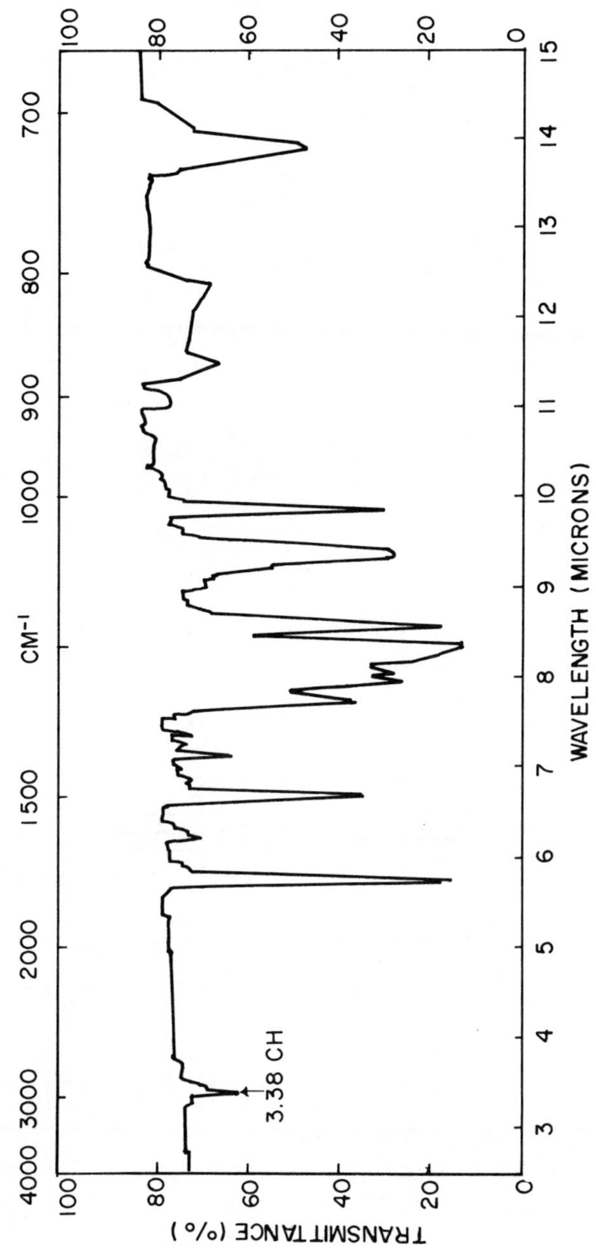

Figure 1. Infrared Spectrum of FPE-1.

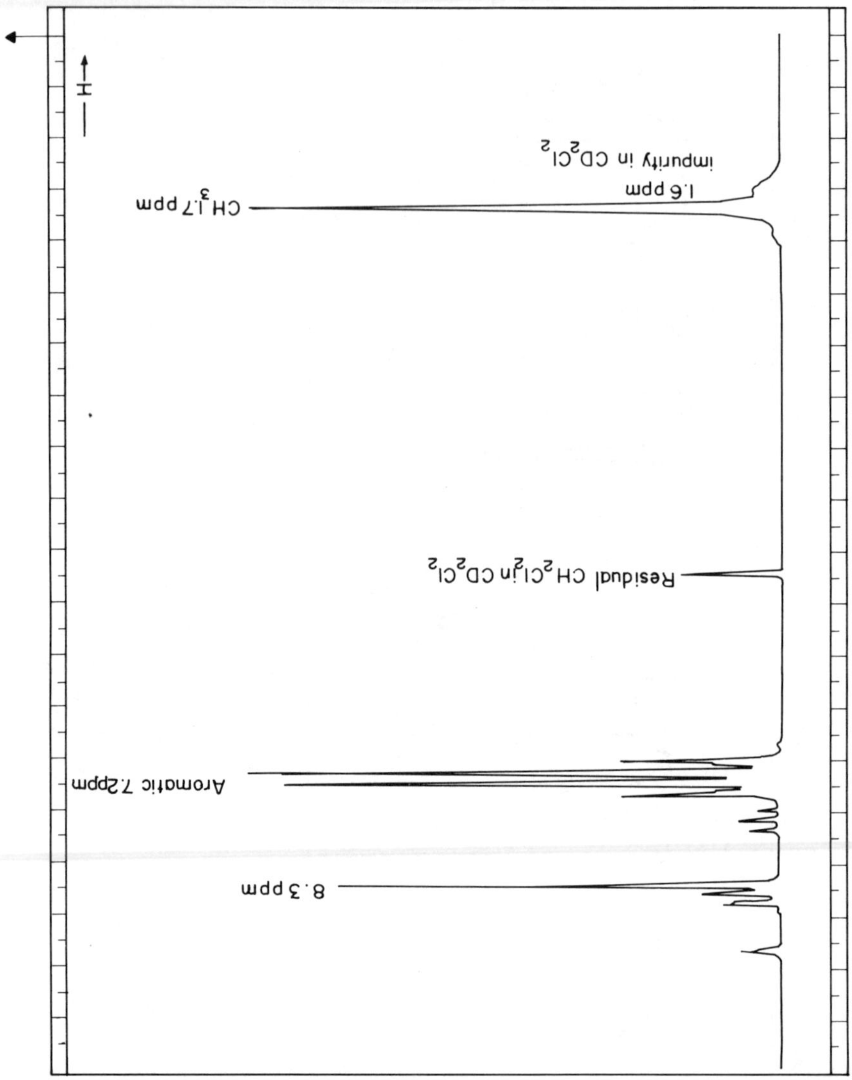

Figure 2a. NMR Spectrum of FPE-1.

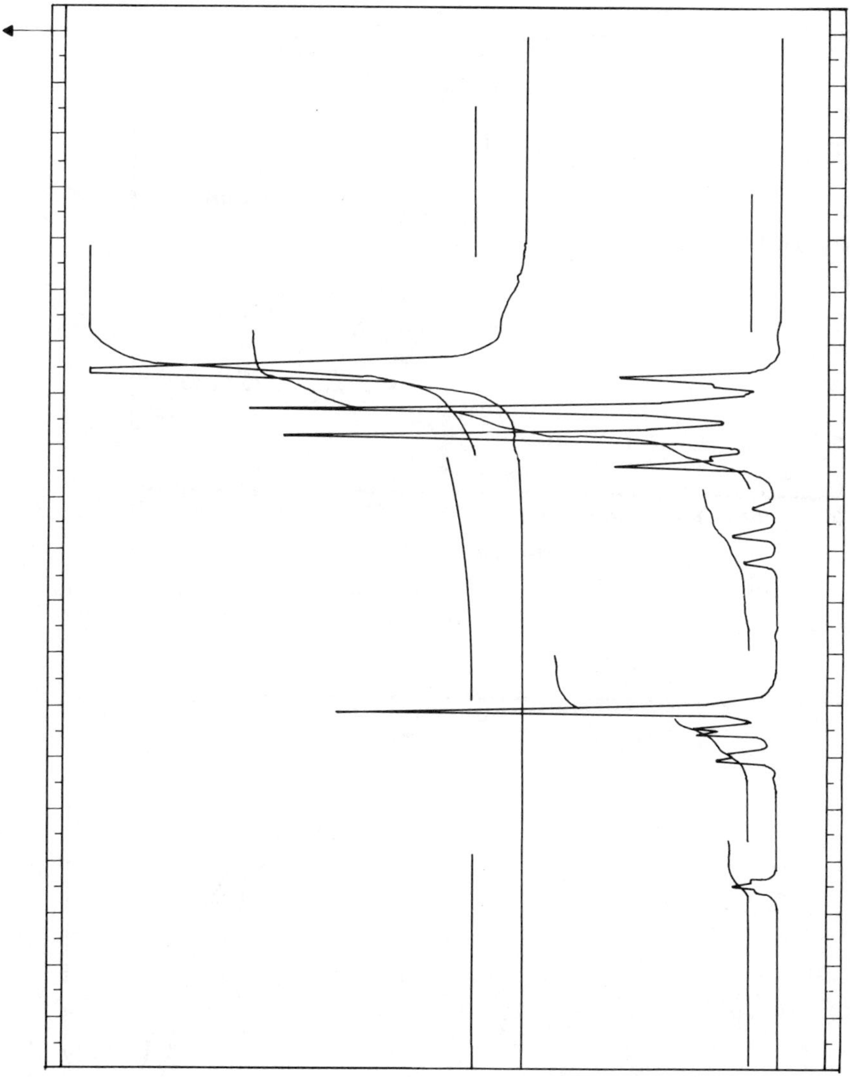

Figure 2b. NMR Spectrum of FPE-1. Expanded Plot Integral.

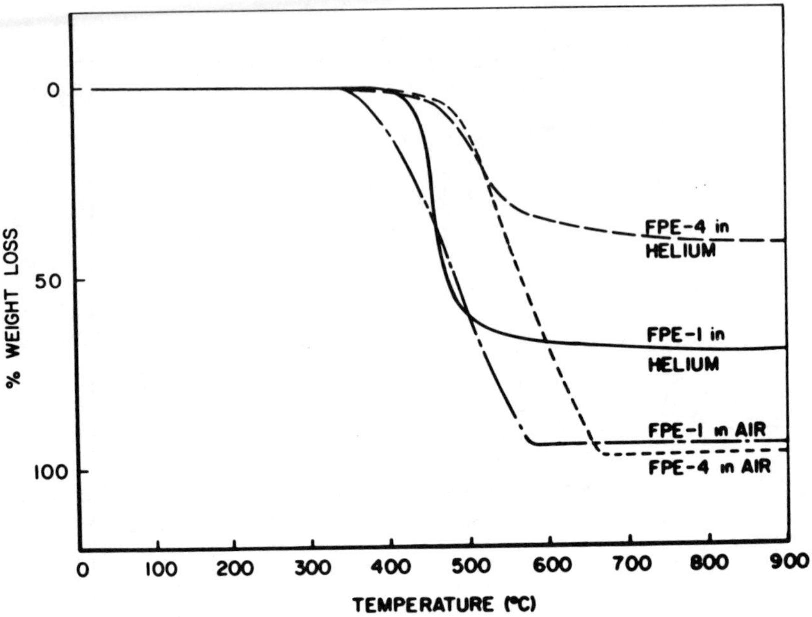

Figure 3. TGA of Samples FPE-1 and -4.
Reproduced from Ref. 9. Copyright 1980 American Chemical Society.

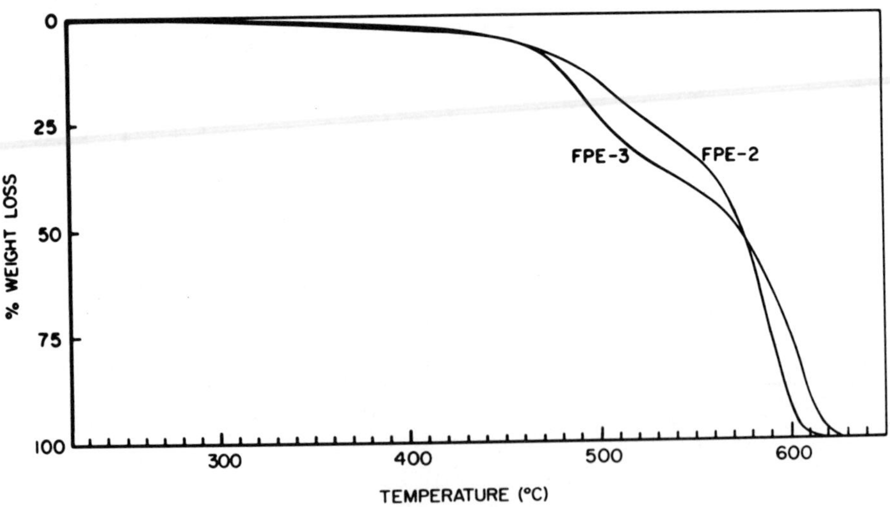

Figure 4. TGA of Samples FPE-2 and -3.
Reproduced from Ref. 9. Copyright 1980 American Chemical Society.

Table I. Intrinsic Viscosities of Fluorenone Polyesters in Methylene Chloride

Sample	[η]
FPE-1	0.720
FPE-2	0.501
FPE-3	0.537
FPE-4	0.351

Preparation of Fibers from Solution. Fibers can be spun (wet or dry) from 10% polymer solution in methylene chloride or tetrahydrofuran (THF). An advantage of THF is that it is a good solvent for the ethylene terpolymer (E/VA/VOH) as well, which we used to modify the fluorenone polyesters. The fluorenone polyester fibers formed were transparent (Figure 5). The polyesters can be blended with 10% E/VA/VOH and they formed slightly cloudy films. Blending in this case was carried out with FPE-4 in an attempt to toughen the polymer. However, the cloudy appearance of the films indicates that some phase separation occurs and toughening is not likely to occur. Tensile data determination confirmed our speculation.

Thermomechanical Analysis. The TMA scan on a molded FPE-1 film is shown in Figure 6. A glass transition temperature (T_g) of 180°C was observed. FPE-4 did not mold well so TMA was run on a sintered sample of FPE-4. A T_g of 270°C was observed (Figure 6). The TMA of FPE-4 blended with 10% E/VA/VOH is also included in Figure 6. Reduction in T_g to 238°C was noted for the polyblend. The polyblend should exhibit two distinct glass transition temperatures. The ethylene terpolymer can engage in transesterification and crosslinking reactions with fluorenone polyesters under elevated temperature during TMA runs. The crosslinking reactions are responsible for a single T_g.

Dynamic Mechanical Analysis. The modulus loss tangent data for FPE-2 and -3 are shown in Figures 7 and 8. The data for FPE-2 were obtained from a compression molded film and those for FPE-3 were obtained from a film cast from methylene chloride.

The glass transitions of FPE-2 and -3 (as indicated by loss modulus maxima, not shown) are 270 and 280°C, respectively. There are also various sub-T_g transitions apparent for each polymer. The peak damping value ($\tan \delta > 1.0$) observed for FPE-3 is comparable to that for polymers which exhibit a sharp transition from glassy to rubbery behavior.

Clash-Berg Moduli. The Clash-Berg moduli obtained for FPE-1 and -4 over a temperature range of 0 to 280°C are shown in Figure 9. The Clash-Berg T_f, the temperature where the rigidity modulus equals 45,000 psi, gives a measure of the low temperature limit of application. Below this temperature, the polymer will be glassy. T_{675} is the temperature where the rigidity modulus falls to 675 psi.

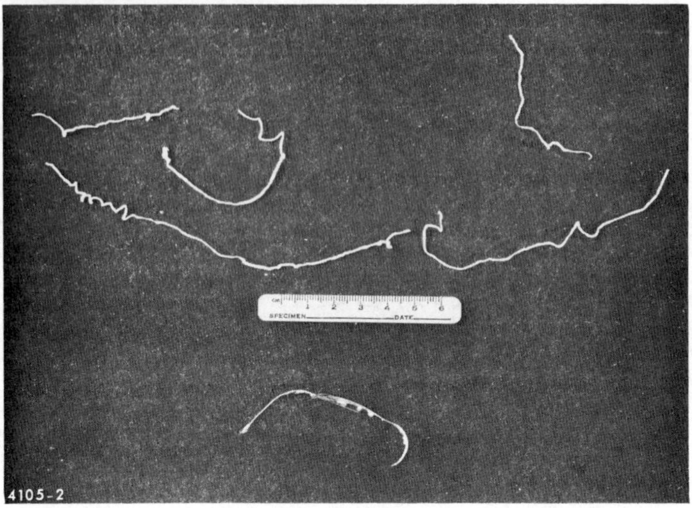

Figure 5. Photographs of FPE fibers.

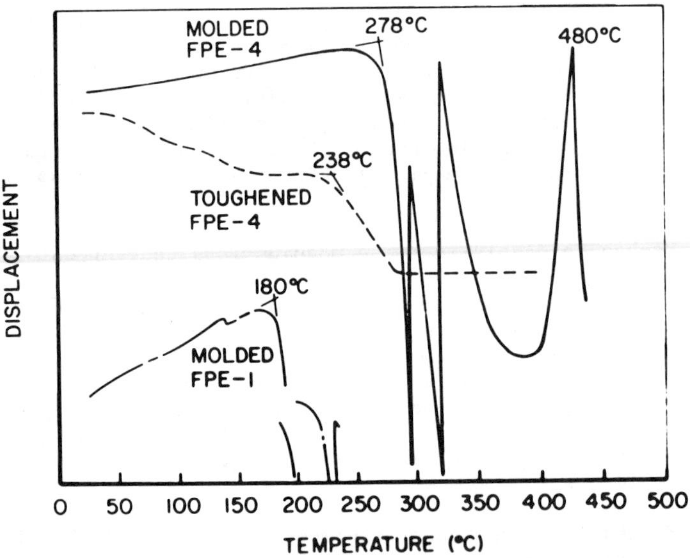

Figure 6. TMA of FPE-1, -4 and FPE-4 Plus Ethylene Terpolymer Blend (90/10).

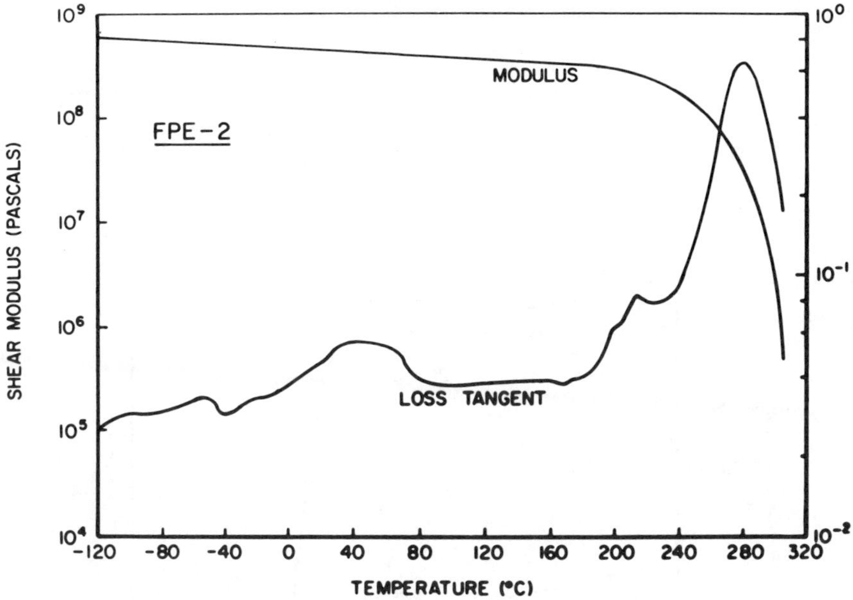

Figure 7. DMA of Sample FPE-2.
Reproduced from Ref. 9. Copyright 1980 American Chemical Society.

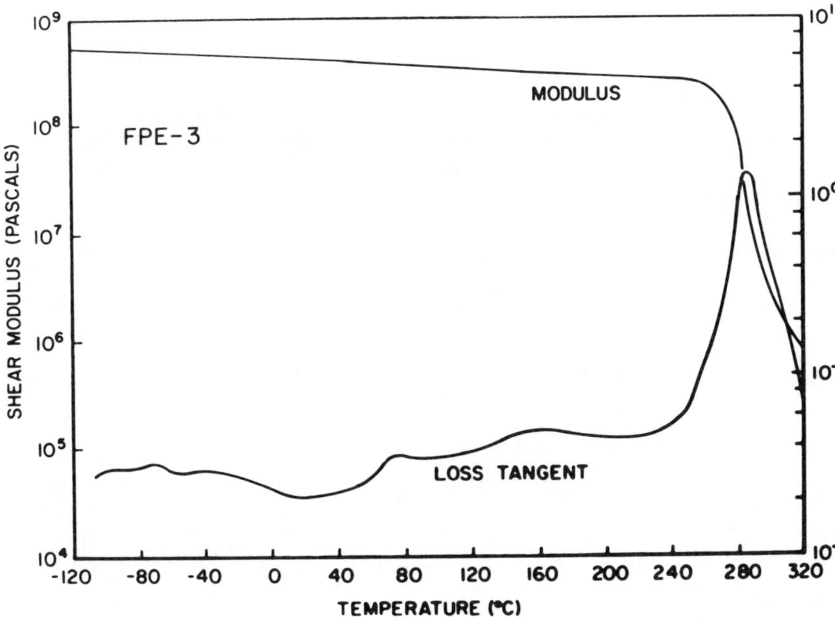

Figure 8. DMA of FPE-3.
Reproduced from Ref. 9. Copyright 1980 American Chemical Society.

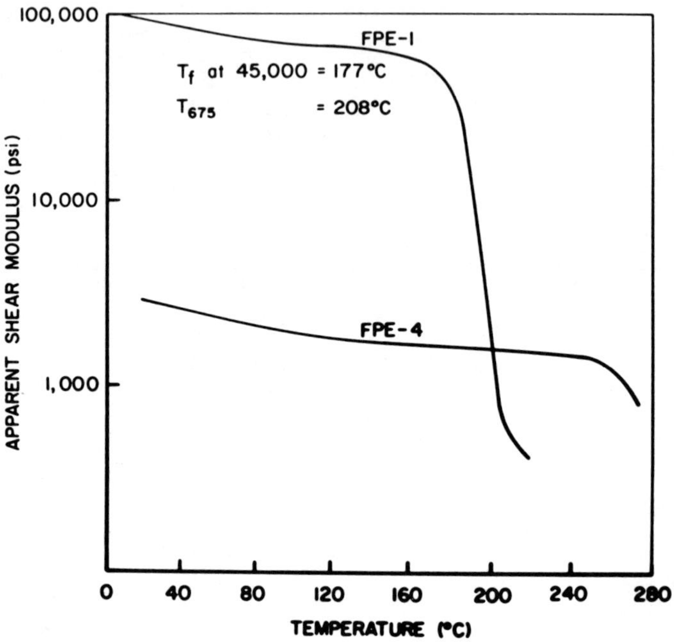

Figure 9. Clash-Berg Modulus Curves of FPE-1 and -4.

FPE-1, the bisphenol A/bisphenol fluorenone copolyester had a T_f (indicative of glass transition) at 177°C. Polymer FPE-4 could not be molded into a Clash-Berg specimen. Therefore, a filter paper was coated by THF solution of sample FPE-4 and dried. Eight strips were compiled and made into one test specimen and the Clash-Berg test was run. A T_g of 270°C is indicated (Figure 9). This is the same value as that determined by thermal mechanical analysis.

ATF. ATF has ether linkages and will plasticize fluorenone polyesters at moderately elevated temperatures. At high temperature (300°C), ATF will form a crosslinked network and thereby improve thermal stability. In Figure 10 thermogravimetric analysis (TGA) of ATF is shown and compared with a fluorenone polyester. The superior thermal stability of ATF is evident.

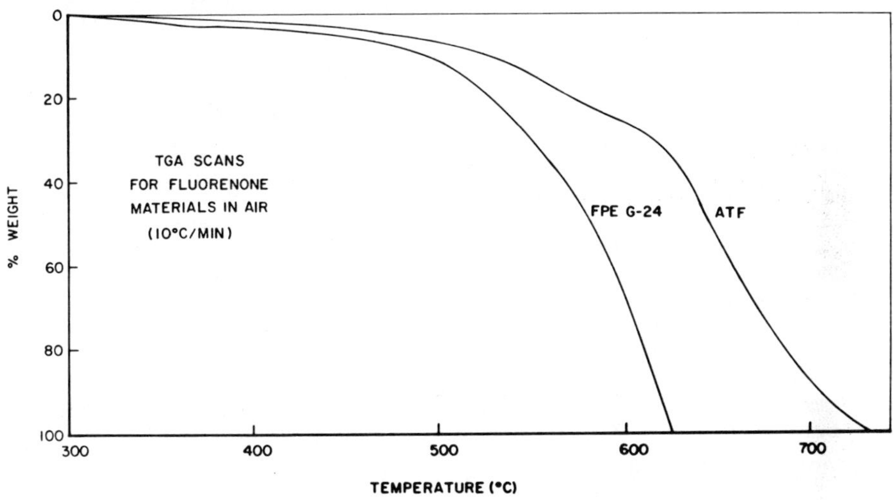

Figure 10. TGA of ATF and FPE.

Reproduced with permission from Ref. 8. Copyright 1983.

Conclusions

The fluorenone polyesters have very good thermal and flame resistance and produce medium char yields in the 30-60% range. They do not drip at high temperatures and will not ignite under sustained external flame exposure. Fibers can be spun from solutions of these polymers (wet or dry). They can be blended with 10% ethylene terpolymer without much loss in heat resistance properties, although not much toughening is achieved.

We have successfully synthesized ATF which can be used with FPE up to 80% but no less than 20%. ATF improves processability. Cross-linked ATF provides flexibility and thermal stability to the fluorenone polyester. The fact that the polyester structure does not contain nitrogen is a potential advantage in fire-resistant fiber applications, since there is no likelihood of HCN generation during burning. Because of high glass transition temperature of fluorenone polyesters, these polymers can be used as heat-resistant fibers.

Acknowledgment

The author expresses his appreciation to Nihal Ahmad for organizing and typing this paper.

References

1. Drewitt, J.G.; Lincoln, J. British Patent 621 102, 1949.
2. Wagner, F. C. French Patent 2 035 578, 1933.
3. Wallsgoove, E. R.; Reeder, F. British Patent 639 429, 1950.
4. Morgan, P. W. Macromolecules 1970, 3(5), 536.
5. Bier, G. Polymer 1974, 15, 527.
6. Kovar, R. F.; Ehlers, G. F.; Arnold, F. E. J. Polym. Sci. Polym. Chem. Ed. 1977, 15, 108.
7. Arnold, F. E.; Hedberg, F. L. in "Polymer Science"; Pearce, E. M. Ed.; Plenum, 1978, Vol. II.
8. Usmani, A. M. J. Mater. Sci. Lett. 1983, 2, 37.
9. Usmani, A. M.; Salyer, I. O.; Cartoff, R. P. Preprints of the ACS Division of Organic Coatings and Plastics, 1980, 43, 87.

RECEIVED February 28, 1984

21

Polyester Processability in Texturing as a Function of Spun Yarn Morphology

J. H. SOUTHERN, R. W. MILLER, and R. L. BALLMAN

Monsanto Company, Pensacola, FL 32575

The effects of polyester (polyethylene terephthalate) spinning variations on continuous filament spun yarn (POY) morphology and the resulting tensile properties and shrinkage have been previously reported, together with the textured yarn (PTY) dyeability dependency on POY and PTY structure. (1) The research reported herein describes representative results concerning textured yarn processability, as defined by the number of broken filaments per unit length. The PTY broken filament number is examined as a function of POY spinning conditions, tensile properties, and structure.

Single position spinning machine and texturing units have been previously described in detail. (1) For this study, spinning variables are defined in Table I. The texturing unit is operated at 600 meters per minute and 29 ± 2 gram T_2 tension (measured between the cooling plate and Barmag Type 5 9-disk twist aggregate; see Figure 1 schematic). All textured yarns have 70 denier, 34 filaments, 3.3 (± 1) gm. tenacity, and 18 (± 2) % elongation. Instron tensile breaks are measured at 30.5 cm/min. extension rate with 25.4 cm initial gauge length. PTY broken filaments are defined with a Toray Fray Counter (model DT-104, Type F detector using 60 meters-per-minute yarn transport speed at 15 grams tension, with broken filaments averaged over a 10 minute interval). POY yarns were also examined for broken filaments, with negligible counts observed (0-0.8/100 meters).

X-Ray, birefringence, and density techniques used to obtain crystalline and amorphous contents and Hermans-Stein orientation functions have also been defined in detail previously. (1-2) The orientation function is directly related to the angle (ϕ) that the average chain segment makes relative to the fiber axis and is defined as follows. (3)

$$f = \frac{3 \langle \cos^2\phi \rangle - 1}{2} \qquad (1)$$

0097-6156/84/0260-0339$06.00/0

TABLE I. PTY BROKEN FILAMENTS VS. POY PROCESS AND MORPHOLOGY

Flake IV	Melt Temperature °C	Spin Speed (Yds./Min.)	Capillary Diameter (mils)	Morphology X*	Morphology fa*	Broken Filaments (#/100 meters)	Tenacity (grams/denier)	% Elongation
0.68	290	3400	15	4.3	0.31	6.2	2.60	--
0.68	305			3.8	0.28	7.8	2.57	--
0.62	290			3.3	0.26	10.0	2.38	--
0.62	305	3400	15	3.4	0.23	11.5	2.29	--
0.68	295	3400	15	5.3	0.30	4.2	2.61	111
		4000		11.2	0.44	3.8	2.87	84
		4500		21.1	0.53	--	2.97	73
		5000		26.0	0.58	2.2	3.14	61
		5500	15	29.3	0.62	2.2	3.36	55
		3400	7	3.2	0.24	10.7	2.59	134
		4500		7.3	0.50	1.7	3.14	77
0.68	295	5500	7	25.8	0.62	1.0	3.51	55

*X = crystalline content (%)
fa = amorphous orientation

where f = -0.5; all chain segments perpendicular to fiber axis.

= 0; average chain segment orientation is random.

= +1; all chain segments parallel to the fiber axis.

The percent crystallinity data herein are derived from density measurements (1), while the amorphous orientation data are derived from a combination of sonic modulus, density and wide angle x-ray measurements (2).

Discussion

As evidenced by the Table I data, POY tensile strength increases and % elongation decreases with increasing spinning speed (consistent with earlier studies) (4-5), polymer molecular weight (proportional to IV) and reduced melt temperatures; all spinning changes yielding increased threadline tension. POY structure, as measured by crystalline content and amorphous orientation, generally increases with the above spinning variable changes (e.g., Fig. 2).

For the various POY yarns textured at constant tension to a fixed tenacity and elongation target (3.3 grams per denier tenacity; 18% elongation), the number of broken filaments generated by the texturing operation decreases with increased yarn structure (Table 1). Interestingly, the best POY structure versus PTY broken filament number correlation is achieved with the reciprocal product of amorphous orientation (fa) and crystalline content (χ):

$$\text{Frays} = -4.47 + 3.485\ (1/fa) \qquad r^2 = 0.910 \tag{2}$$

$$\text{Frays} = -.048 + .320\ (1/\chi) \qquad r^2 = 0.863 \tag{3}$$

$$\text{Frays} = 1.024 + 0.0745\ (1/\chi fa) \qquad r^2 = 0.945 \tag{4}$$

where r^2 is the amount of variance accounted for by each of the above equations. The Figure 3 ($1/\chi fa$) correlation is best understood from the following observations and thesis. Based on previous observations (2-6), increased amorphous orientation yields increased yarn tenacity at room temperature test conditions; however, for conventional polyester texturing conditions in the 200-240°C range, yarn tenacity should also be governed by crystallites, which act as the physical crosslinks preventing chain slippage for temperature between the polyester glass (110°C) and melt (255°C) transitions, as well as amorphous orientation; hence, the excellent broken filament correlation with the product of crystallinity and amorphous orientation. Essentially, polyester can be treated to a first approximation at conventional texturing temperatures as a rubbery network (7), with the crosslink density and molecular extension increasing with crystallinity content and amorphous orientation, respectively. Thus the mechanism for reducing PTY broken filaments is straightforward: increased POY structure implies increased POY tenacity at texturing temperatures which, in turn, yields fewer breaks under the constant stress condition used to texture the Table I samples.

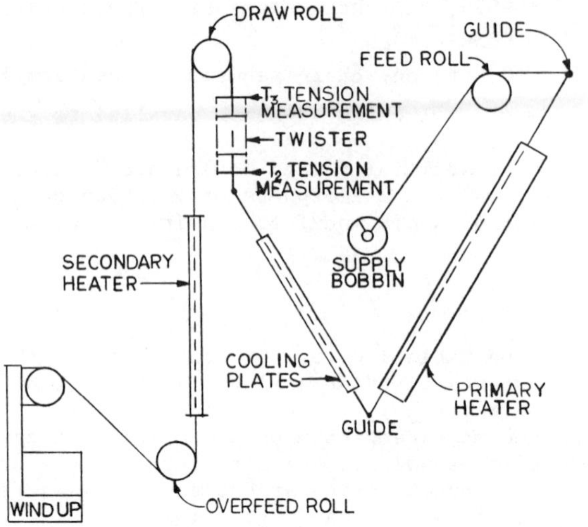

Figure 1. Single position drawtexturing unit schematic.

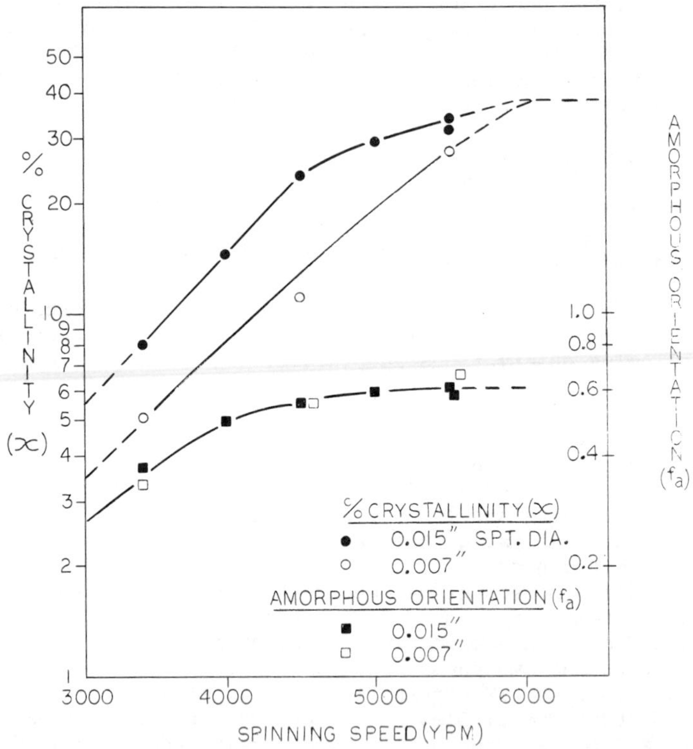

Figure 2. POY crystallinity and amorphous orientation versus spinning speed.

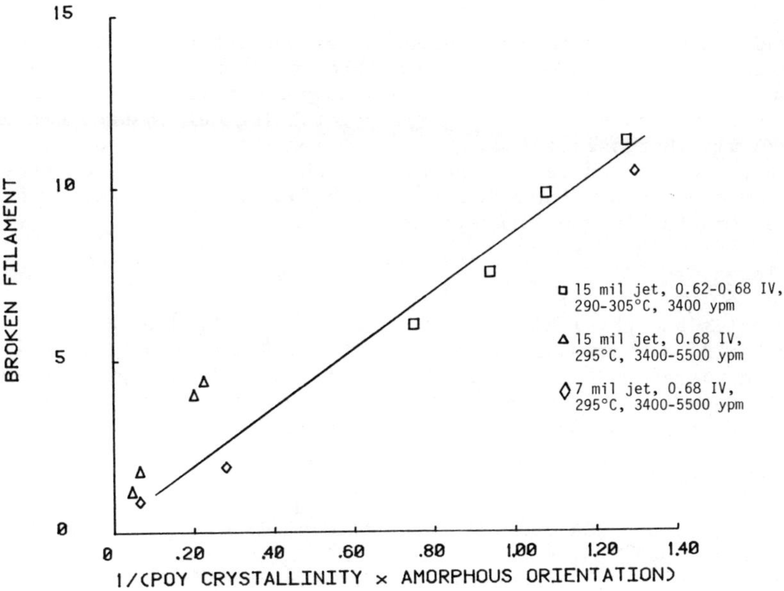

Figure 3. PTY broken filaments versus POY morphology.

It is of interest to point out the Figure 2 spin line capillary diameter effect on the POY yarn structure, where the 7 mil (0.018 cm) diameter capillary yielded no amorphous orientation change, but a significant crystallinity reduction relative to the 15 mil (0.38 cm) capillary. From the above discussion, the generally equal POY tensile properties for yarns spun from the two capillaries are predicted from the equal amorphous orientation values at equivalent spinning speeds (Table I). The greatly reduced PTY broken filament count for the yarn textured from the higher crystallinity POY spun from the 15 mil capillary is also consistent with the above discussion.

Thus, the results herein support the conclusion that improved POY texturing, as measured by reduced broken filament counts, can be attained via increases in either the spun yarn amorphous orientation or crystalline content. These desired morphological changes can be achieved via increased spinning speed, molecular weight, capillary diameter, and reduced melt temperature. The applicability of these polyester conclusions for other synthetic fibers is under study and will be reported in the future.

Acknowledgments

Yarn samples and physical property data were provided by J. A. Burroughs, D. R. Gates, J. W. Secrist, G. T. Reeves and L. J. Hill; morphology data by B. J. Senn. We wish to express our appreciation to J. H. Saunders for his support of this work and to the Monsanto Textiles Company for permission to publish it.

Literature Cited

1. Miller, R. W.; Southern, J. H.; Ballman, R. L.; Textile Research Journal 1983, 53, 670.
2. Samuels, R. J., "Structured Polymer Properties"; Wiley-Interscience, New York, 1974.
3. Stein, R. S.; J. Polymer Sci. 1958, 31, 327.
4. Huisman, R.; Heuvel, H. M.; J. Appl. Polymer Sci. 1978, 22, 943.
5. Heuvel, H. M.; Huisman, R.; J. Appl. Polymer Sci. 1978, 22, 2229.
6. Simpson, P. G.; Southern, J. H.; Ballman, R. L.; Textile Research Journal 1981, 50, 7.
7. Bueche, F.; "Physical Properties of Polymers"; John Wiley, New York, 1962, p. 37.

RECEIVED January 10, 1984

22

Effects of Annealing Below T_g on the Properties of Poly(vinyl chloride)

DEREK A. OTT and ECKHARD W. HELLMUTH

Polymer Section, Department of Chemistry, University of Missouri–Kansas City, Kansas City, MO 64110

The relationship between observed enthalpy-volume relaxations and thermal treatment of slightly oriented industrial PVC films was investigated. Differential scanning calorimetry at 20°C per minute and specific volume analysis (density gradient column) were used to study the effects of annealing near and below Tg. Nonlinear effects in the volume relaxation at relatively long times and temperatures close to the glass transition produce deviations in the specific heat curves at temperatures far above Tg in addition to the normal overshoot effects.

For many years, nonlinear relaxation effects have been observed in thermally treated poly (vinyl chloride) (PVC), a common commercial plastic. The influence of annealing in and below the glass transition temperature (Tg) on the physical and structural properties of PVC are well established and documented (1-8). Numerous investigations using differential scanning calorimetry (DSC) of PVC annealed beneath Tg have revealed a variety of enthalpy relaxations occurring near or below Tg (1,3-6). Workers have also discovered density alterations which were induced by suitable heat treatment below Tg (1-2,6). Since the polymer is generally considered amorphous, theories which offer explanation of this phenomenon usually involve free volume changes and molecular clusters. However, little agreement exists concerning the exact mechanisms behind these effects and the rate of their influence.

This project was carried out as an investigation of the relationship between thermal prehistory or treatment of the sample and properties of poly vinyl chloride. The rate of annealing on effects such as enthalpy relaxations and density reduction was investigated.

0097–6156/84/0260–0345$06.00/0

Experiments and Results

Materials. Clear, unplasticized PVC was obtained from B.F. Goodrich Vynaloy V 453 as .35mm sheets. Density (determined by a density gradient column prior to annealing) was 1.3342 g/ml (± .0006 g/ml). C^{13} NMR was used to determine the tacticity of the material. The PVC used is 46.4% heterotactic, 38.0% syndiotactic, and 15.6% isotactic. Tg was approximately 85°C.

Annealing. Samples of PVC were annealed at 65, 70, 75, 80, and 85°C (±.3°C). Annealing times were based on a logarithmic progression (10^0, $10^{0.5}$, 10^1, $10^{1.5}$, 10^2, $10^{2.5}$, and 10^3 hr.). Thus, the given storage times were 1.00, 3.16, 10.0, 31.6, 100, 316, and 1000 hours.

Differential Scanning Calorimetry. Thermograms were recorded with a Perkin Elmer Differential Scanning Calorimeter 1-b. All scanning rates were 20°C/min. The DSC results are illustrated in Figures 1-8. The reference line which appears in each figure is used as the baseline for comparison, because its prehistory is known (maintained at 150°C for 7 minutes and then cooled at the rate of 40°C/min through the glass transition temperature).

Annealing at 65°C produced the characteristic endothermic peaks ("overshoot"). Annealing times ranging from 10° to 10^3 hours are illustrated in Figures 1a-b. This figure shows the increase in peak height at a constant temperature. At longer annealing rates, the enthalpy peaks shifted to higher temperatures and the magnitude continued to increase as at shorter times.

This behavior of the DSC curves at 70°C (Figure 2) is similar to those at 65°C. Initial increase in the maxima is followed by a shift of the peak to a higher temperature which remained stationary between 31.6 and 1000 hours.

At 75°C (Figure 3) the trend continued as at the lower temperatures, but a significant change occurred at longer times. The minima following the peaks disappeared and the peaks split into two portions with a broad component at a higher temperature.

Figures 4a-b shows similar behavior for 80°C. At this temperature the splitting of the peak occurred at a shorter annealing time. The minimum disappeared as the second relaxation process began.

The effects of annealing at 85°C (Figure 5) seems much more complicated. The peak at the glass transition temperature decreased rapidly and the broad peak dominated behavior above Tg.

This trend is more easily seen in Figures 6-8. Specific heat curves of PVC annealed at 1, 31.6, and 1000 hours at the various annealing temperatures are illustrated. In the 1 hour figure, the specific heat peak increased from 65-80°C. However, at 85°C the magnitude of the peak suddenly decreased and became much broader. For those samples annealed for 31.6 hours, the process appeared at a lower temperature (75°C). Once again, the peak increased then suddenly decreased and broadened. The 1000 hour graph is dominated by the overall change. A slight decrease in temperature for the first peak and the Tg step is followed by a pronounced maxima at higher temperatures.

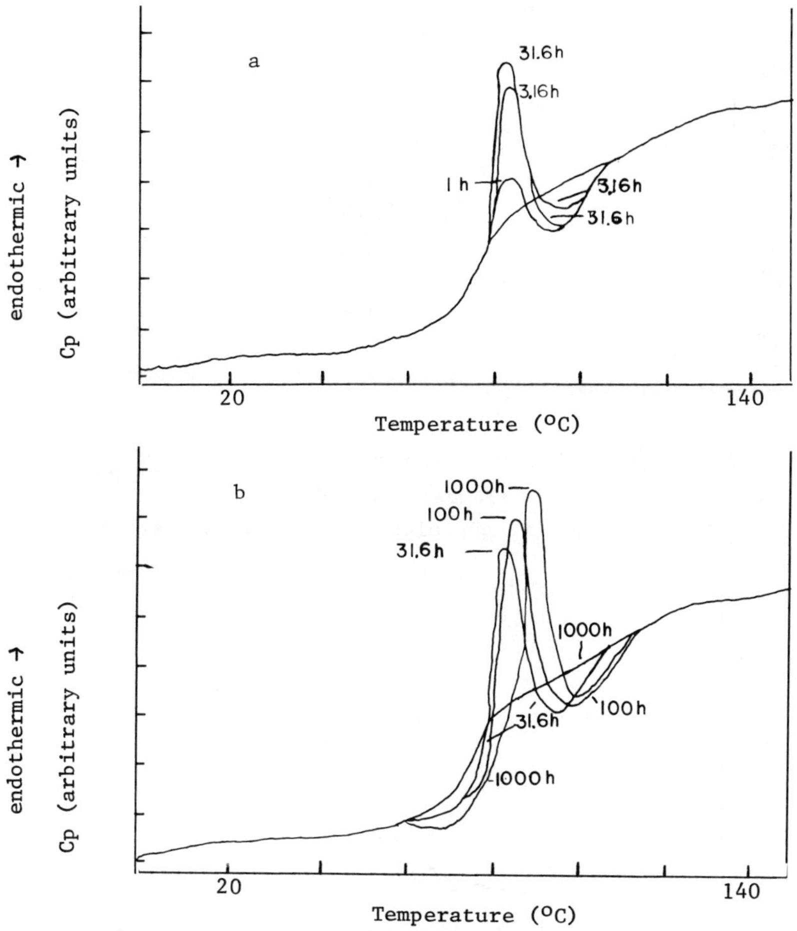

Fig. 1a and 1b DSC curves of PVC annealed at 65°C for different times

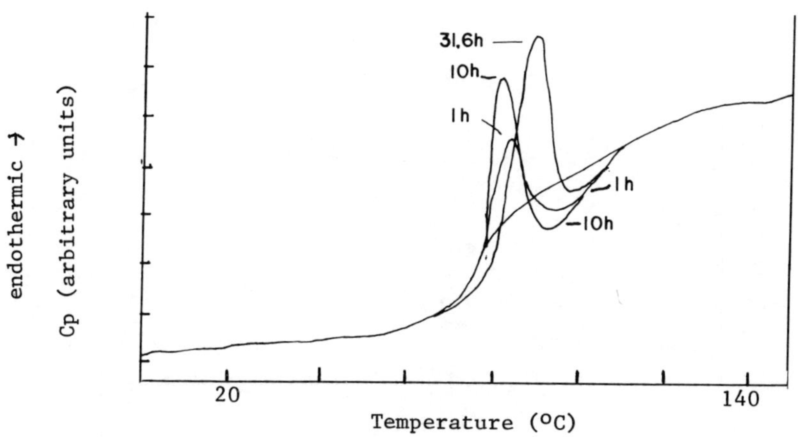

Fig. 2 DSC curves of PVC annealed at 70°C for different times

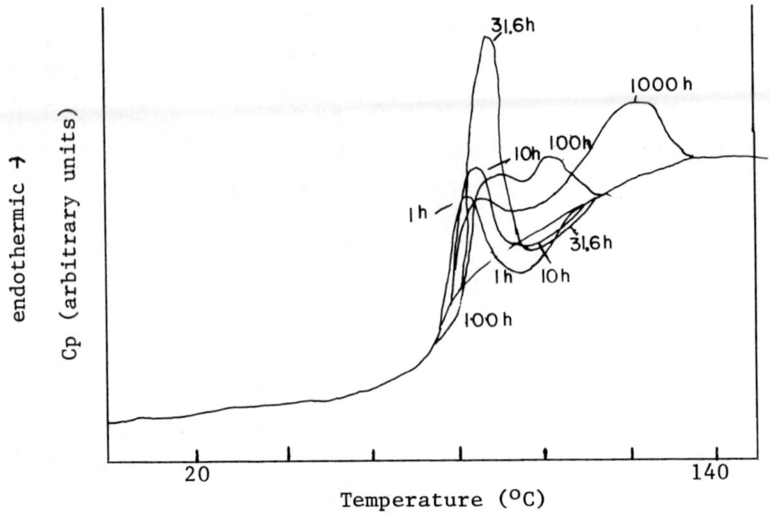

Fig. 3 DSC curves of PVC annealed at 75°C for different times

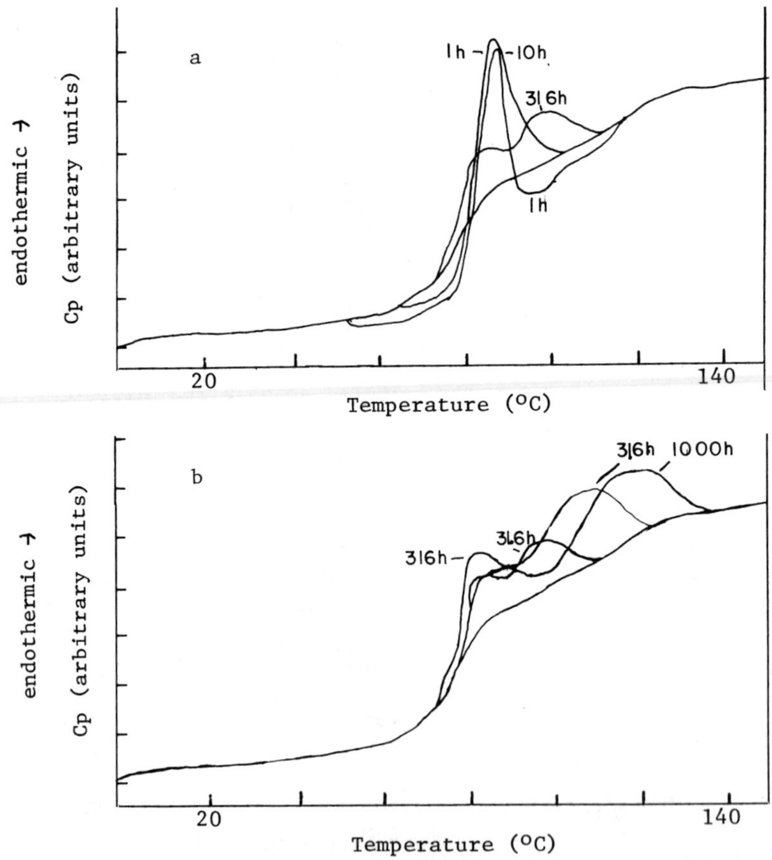

Fig. 4a and 4b DSC curves annealed at 80°C for different times

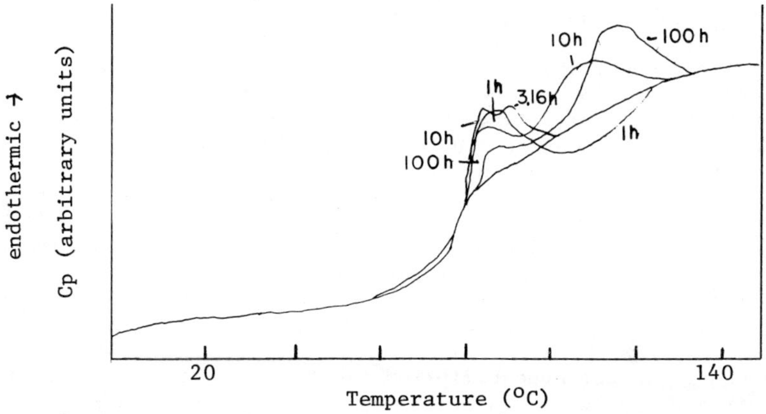

Fig. 5 DSC curves of PVC annealed at 85°C for different times

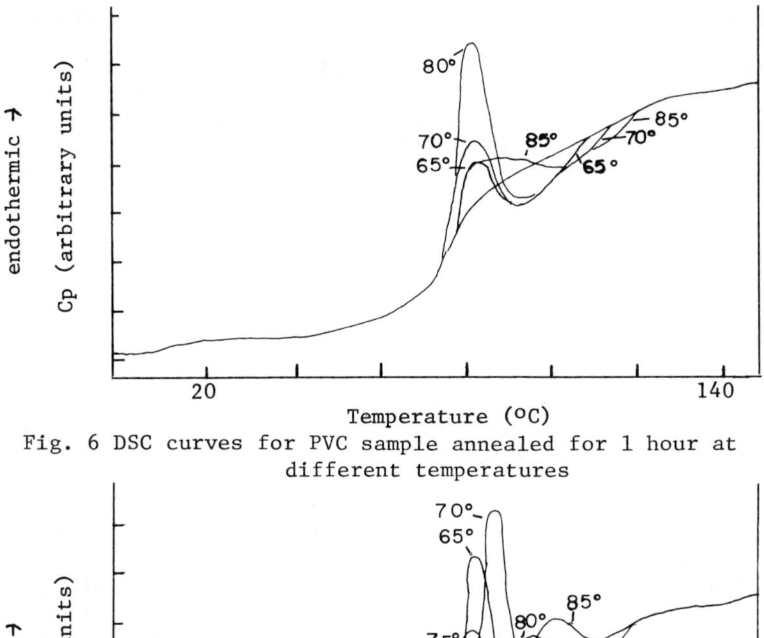

Fig. 6 DSC curves for PVC sample annealed for 1 hour at different temperatures

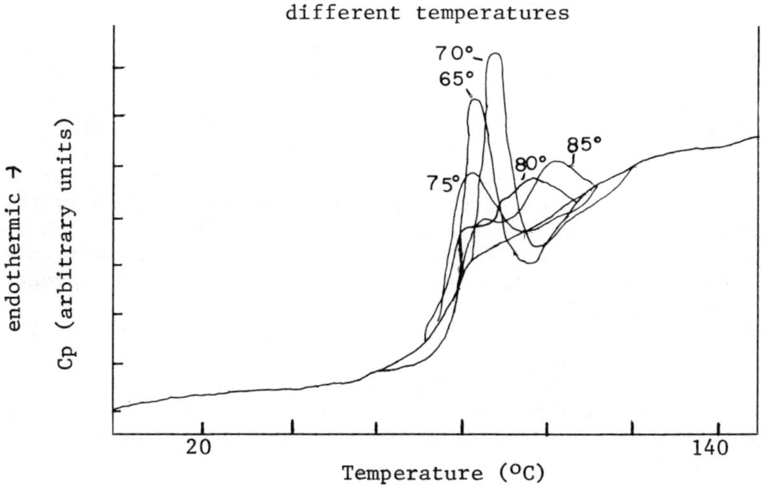

Fig. 7 DSC curves for PVC samples annealed for 31.6 hours at different temperatures

Specific Volume. Sample densities were obtained at 25.0°C using a density gradient column of an aqueous solution of calcium chloride. The measured specific volumes are shown in Figure 9. Annealing decreased the volume of the samples. For each curve or annealing temperature, a crucial time period where the volume decreased at a different rate process seemed to exist. The volume of untreated samples is also included in the figure.

During annealing, molecules moved more often and more freely than at room temperature. After initial expansion, the density increased, which has also been seen in polystyrene by Ibar (9) and Kovacs (10). At higher temperatures (75-85°C) relaxations occurred in at least two stages. The material seemed to relax to apparent constant values with additional density changes at longer annealing times.

A change in the nonequilibrium substructures of the material with different packing densities may influence the complex rate of these volume changes. A simplified model might involve an alteration of the hole structure and the clustering of the molecules with a smaller overall free volume distributed into larger microscopic voids.

The results of wide angle X-ray diffraction confirmed that crystallinity did not change during annealing below Tg. Structural changes in the microsystem of the glassy structure could not be observed by wide or small angle X-ray diffraction analysis.

However, some alterations (not necessarily crystalline changes) are suggested by preliminary positronium lifetime measurements. Results indicate an increase in the size of individual microscopic voids between molecular clusters, but an accompanying overall decrease in void volume.

Discussion

The normal explanation for specific heat deviation above Tg as effect of crystallization or melting can not be used here. A possible explanation of the observed data could involve cluster formation. The reported specific heat and volume data indicate structural changes in the glassy state with different mobilities nonequilibrium vitrification process and altered by the following thermal treatment of the samples. This process, an alteration of cooperative motions due to cluster formation, leads to a decrease in mobility of the chains and prevents equilibrium enthalpies above Tg to be directly attained at the selected scanning rate.

A comparison of the volume and enthalpy data substantiates this hypothesis. The second step of volume reduction is clearly related to the development of the specific heat peaks above Tg. The overshoots of enthalpy below the equilibrium value are moved from a lower temperature to a higher one if the initial volume is decreased below a critical value.

The clusters in the rubbery material above Tg behave similarly to the time dependent process found in or below the glass transition. The kinetic effects in the breakdown of cooperative regions are rate and temperature dependent and can apparently exist between the glass and liquid-liquid transition as introduced by Boyer (11) years ago.

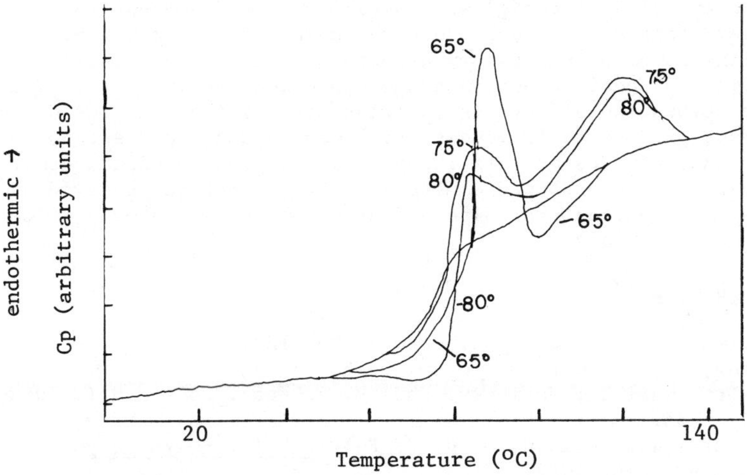

Fig. 8 DSC curves for PVC samples annealed for 1000 hours at different temperatures.

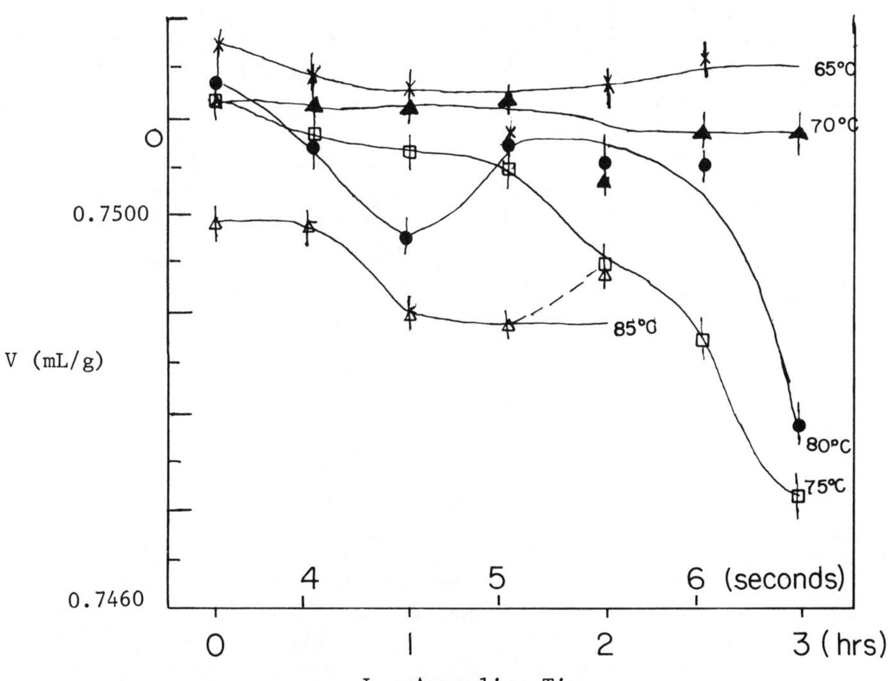

Fig. 9 Specific volume of PVC samples at 25°C for sample annealed at different temperatures as function of annealing time.

The theories of the glass transition should account for observed effects below and above the main glass transition. Annealing or other prehistory affects sub-Tg, Tg and post-Tg anomalies in specific heat and the expansion coefficients. Non-linear relaxation effects in the glass transition kinetics depend on different parameters in models which can be used to describe the data. An extension of this theory must include the data reported in this paper. The proposed clusters which can not be observed by X-rays are important for alterations of the ultimate properties of glassy materials.

Literature Cited

1. Illiers, K. H. Makromol. Chem. 1969, 127, 1.
2. Gray, A.; Gilbert, M. Polym. 1976, 17, 44.
3. Foltz, Carl R.; McKinney, Paul V. J. Appl. Polym. Sci. 1969, 13, 2235.
4. Straff, R.; Uhlmann, D. R. J. Polym. Sci., Polym. Phys. Ed. 1976, 14, 1087.
5. Berens, Alan R.; Hodges, Ian M. Macromolecules 1982, 15, 756.
6. Philips, R.; Cox, R. L.; Heiberger, C. A. Proc. 26th SPEANTEC Meeting New York. 1968, p. 216.
7. Hyndman, John R. Polym. Eng. Sci. 1966, 6, 169.
8. Matsuoka, S. Polym. Eng. Sci. 1974, 14, 162.
9. Ibar, J. P. J. Macromol. Sci., Phys. Ed. 1982, B21, 481.
10. Kovacs, A. J. Fortschr. Hochpolym. Forsch. 1963, 394, 3.
11. Boyer, R. F.; Gillham, J. K. J. Macromol. Sci., Phys. Ed. 1977, B13, 494.

RECEIVED January 30, 1984

STATE OF THE ART IN FIBER TECHNOLOGY

23

Modeling the Melt Spinning Process

HENRY H. GEORGE

Celanese Research Company, Summit NJ 07901

Melt spinning is, today, a major commercial process for the formation of synthetic fibers. Two of the major commercial fibers, nylon and polyester (PET), are produced by melt spinning. During the late 1950's and early 1960's, the mathematical modeling of melt spinning began to develop. This early activity is summarized by Ziabicki(1). Since then, there have been numerous models published until today we have models that can predict spun yarn physical properties from a knowledge of the process conditions and polymer properties, dynamic models that predict spin-line stability and simulate the effects of process disturbances.

A typical melt spinning process is shown in Figure 1. The key features are a source of molten polymer, in this case, from an extruder, a metering pump, and a spin pack that contains the spinneret and filter media. The spinneret is, in effect, a plate with a number of holes, one for each filament. The number can range from one to several thousand, and their diameter is typically of the order of tens of thousandths of an inch.There is a considerable body of technology extant dealing with the design and construction of spinnerets. When the polymeric filament leaves the spinneret, it enters a gaseous environment that is usually air at a temperature below the melting point of the polymer. The spin line generally contacts some guide surface before going onto the godet rolls which control the linear velocity of the process. The temperature of the spin line must be below its melting point before touching any surface. The linear density or denier (grams/9000 meters) is positively controlled by the metering pump and the godet rolls. On the other hand, the denier of the individual filaments is determined by the hydraulic split between the individual holes in the spinneret, which points up the need for precise control of the hole diameter and entry shape.

There are many objectives in modeling the melt spinning process; the three most pertinent are:

- Relating spun yarn properties to process variables.
- Evaluating process uniformity and robustness.
- Process stability.

0097-6156/84/0260-0355$06.00/0

The triangle shown in Figure 2 shows the interrelation between the process, the structure of the resulting product, and physical properties. In a modeling situation, one generally starts with a knowledge of the process conditions from which it is desired to determine structural and physical properties. A study[2] of spinning Poly(ethylene terephthalate) (PET) will be discussed where it was possible to use a model to calculate the stress at the solidification point, and then relate this to spun yarn tensile properties.

In a typical commercial spinning situation, there can be hundreds of filaments which experience slightly different histories as they proceed from the spinneret to the godet. A model can be useful in determining the product differences that result. A successful process should be insensitive to minor fluctuations in the operating conditions to avoid excessive defective product. A model can be used to estimate this robustness of the process.

In any of the situations mentioned above, the model can be used iteratively to help determine an optimum set of operating conditions.

The question of spin-line stability is an important one, and there is considerable literature in this area, but it is beyond the scope of this presentation.

In most models, the calculations deal with the filament from the moment it leaves the spinneret until it reaches a guide surface. Figure 3 shows schematically in some detail what is involved in the model of a single filament. Polymer exits from the spinneret at X = 0 at a temperature above its melting point, and immediately swells to some extent as it undergoes elastic recovery from the recent deformation history experienced in the spinneret hole and entry. It also is now exposed to a quenching or cooling gas, typically air, which is usually blown perpendicular to the spin line. In the past, when spinning speeds were relatively slow (500-1000 meters/min), the cross-velocity component increased the overall heat transfer coefficient. At the higher take-up velocities in current use, the effect on the heat transfer is fairly small, and the most important effect of the cross-flow of air is to provide aerodynamic stability to the spin line. Another result of the cross-flow velocity is to bend spin line at a slope, S. The spin line is accelerating all the way down to the freeze point where it reaches its solidification temperature and the velocity of the godet or other take-up device.

Published spin-line models are generally phenomenological, relying on correlations only for such engineering parameters as heat transfer and aerodynamic drag coefficients. In general, they all make recourse to equations of continuity; conservation of mass, force, and energy; and a constitutive equation for the polymer. In order to have a tractable set of equations, some assumptions are usually made, either of mathematical necessity or for convenience and conservation of computing resources.

The assumptions shown in Table 1 are from George's[2] work, and are fairly typical. The first two reduce the equations from second-order partial differential to first-order ordinary differen-

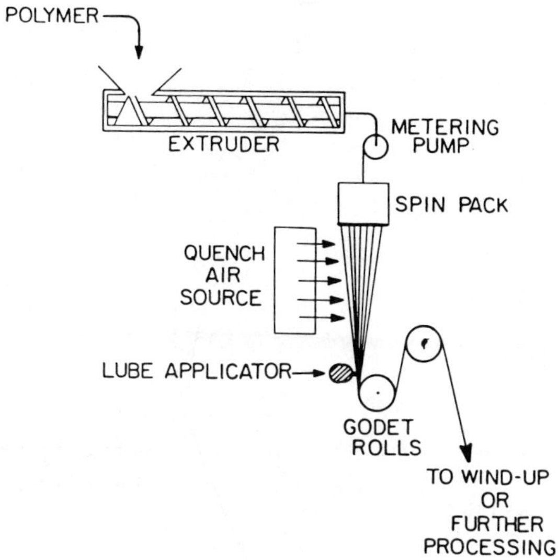

Figure 1 Schematic View of Melt Spinning Process
Reproduced with permission from Ref. 2. Copyright 1982.

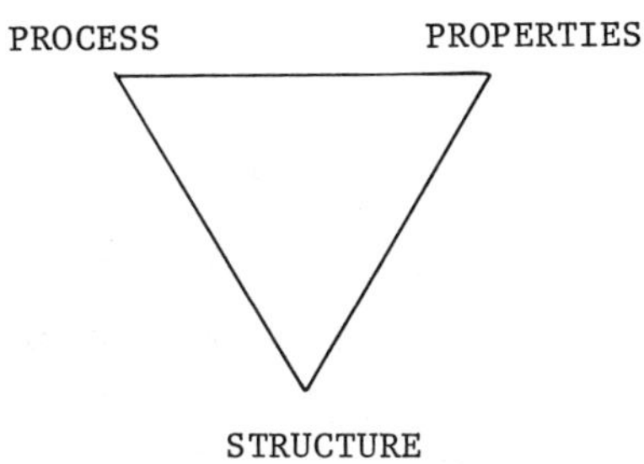

Figure 2 Process/Property/Structure Interrelation
Reproduced with permission from Ref. 2. Copyright 1982.

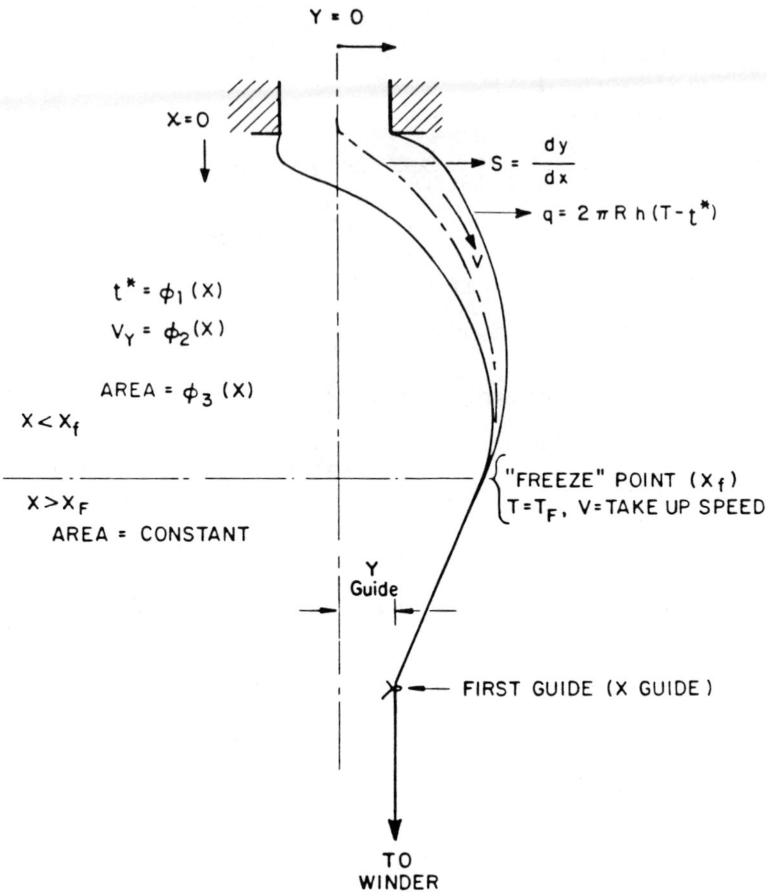

Figure 3 Schematic Diagram of Spinline

Table 1

STEADY STATE MODEL

MAJOR ASSUMPTIONS

- NO RADIAL TEMPERATURE GRADIENT
- NO AXIAL CONDUCTION
- EXTENSIONAL VISCOSITY = 3 x SHEAR VISCOSITY
- ELASTIC EFFECTS ARE NEGLIGIBLE
- SURFACE TENSION EFFECTS ARE NEGLIGIBLE
- CRYSTALLIZATION EFFECTS (THERMAL & RHEOLOGICAL) ARE NEGLECTED

tial ones. The radial gradient assumption is valid, except at very high speeds, as has been shown by Knox and Frankfort[3]. The assumption of a Troutonian type extensional viscosity was made in the absence of any data and the generally recognized difficulty in measuring it. Assuming no elastic effects allows the use of a powerlaw type constitutive equation, which again simplifies the mathematics; and, based on the shear viscosity data available[4], is quite reasonable for PET. The inclusion of a Barris effect at the top of the spin line takes into account the elastic effects experienced in the spinneret where the deformation rates are about an order of magnitude larger than experienced in the draw-down region of the spin line. Surface tension effects are negligible for polymers with their high viscosity. Surface tension must be considered in glass spinning. The crystallization assumption is made because we don't know how to deal with it in a non-isothermal, transient stress situation. Fortunately, for many commercially important polymers, crystallization is not encountered at normal process conditions. This is true for PET, through it will crystallize at high speeds[5,6].

The system of equations shown in Figure 4 are those of George[2], except for the addition of the spin-line deflection terms, S, the slope of the spin line with respect to the vertical, and Y, the distance from the vertical axis through the spinneret hole. Both these terms are illustrated in Figure 3.

An important step in any model development process is validation or confidence building. As a first step, one attempts to predict some of the primal properties; in this case, velocity and temperature profiles with respect to position. Figure 5 compares experimental and calculated velocity profiles. The agreement is good in general, though the 3000 meter/minute profiles do show a substantial difference. These data can be brought into line by increasing the heat-transfer coefficient by about 25% at the expense of the low-velocity data. Gagan and Denn[7] have analyzed the same data using more complex constitutive equations to reflect some elasticity. They show that, with the more complex constitutive equations, they can achieve a better fit. However, considering the imprecise nature of the heat transfer and elasticity data, it is not clear which situation is correct. The temperature profiles shown in Figure 6 show excellent agreement and illustrate the fact that, at a constant flow rate, the take-up velocity has virtually no effect on the temperature profile. The model predicts a very weak dependence of temperature with velocity. This observation was found to be most useful in some studies of high-speed spinning[5].

Historically, temperature and velocity profiles have been shown as functions of position, as in Figures 5 and 6. It is instructive to plot these profiles as functions of time, as in Figures 7 and 8. From this type of plot, we gain a sense of the time involved and the magnitude of the rates of change. The time plotted in these two figures starts at the instant an element of polymer leaves the spinneret hole and ends when the freeze point is reached. Typically, this transit time is a few hundred milliseconds but, as can be seen

$$\frac{dt}{dx} = \frac{-2\sqrt{1+S^2}\sqrt{\pi}\,h}{G C_P}\sqrt{A}\;(t-t^*)$$

$$h = 2.7804 \times 10^{-4}\left(\frac{G}{\rho A^2}\right)^{0.216}\left[\frac{G}{\rho A} + (8V_y)^2\right]^{0.167}$$

$$\frac{dA}{dx} = \frac{-\rho\sqrt{1+S^2}}{G}\;A^2\left[\frac{F}{\beta_o A}\right]^{1/N}$$

$$\beta_o = 2.787 \times 10^{-4}\,(IV)^{5.056}\left[\exp\left(\frac{6802}{t+273}\right) + \frac{100}{t-t_f}\right]$$

$$\frac{df}{dx} = \left[\frac{-G^2}{g_c \rho A^2}\right]\frac{dA}{dx} + \rho A + C\sqrt{1+S^2}\;G^{-1.19}A^{-0.69}\,(NFIL^{-0.31})$$

$$c = \rho^{-1.19}\left[\frac{K\pi}{2g_c}\,\rho_{AIR}^{0.19}\;\mu_{AIR}^{0.81}\left(\frac{4}{\pi}\right)^{0.095}\right]$$

$$\frac{ds}{dx} = \frac{-(1+S^2)^2}{F}\left[\left(\frac{S}{1+S^2}\right)\frac{dF}{dx} + \frac{\rho_{AIR}C_D V_y^2}{g_c}\sqrt{\frac{A}{4\pi}}\right]$$

$$\frac{dy}{dx} = S$$

Figure 4 Steady State Spinline Model Equations

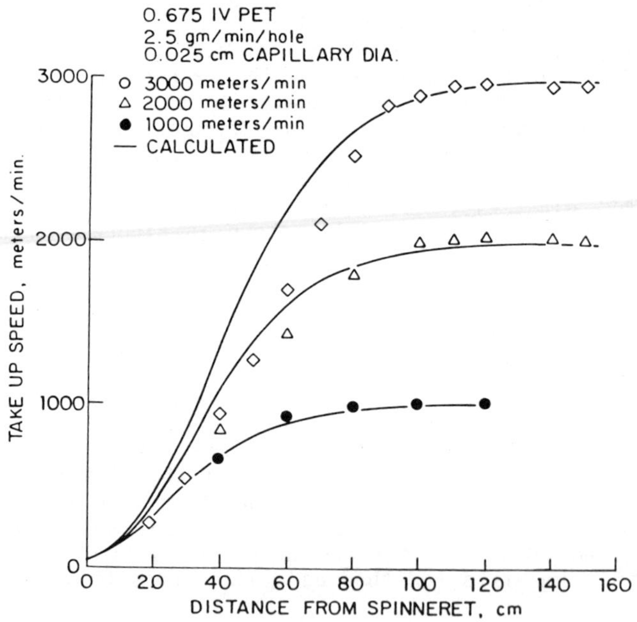

Figure 5 Comparison of Calculated and Measured Spinline Velocity Profiles
Reproduced with permission from Ref. 2. Copyright 1982.

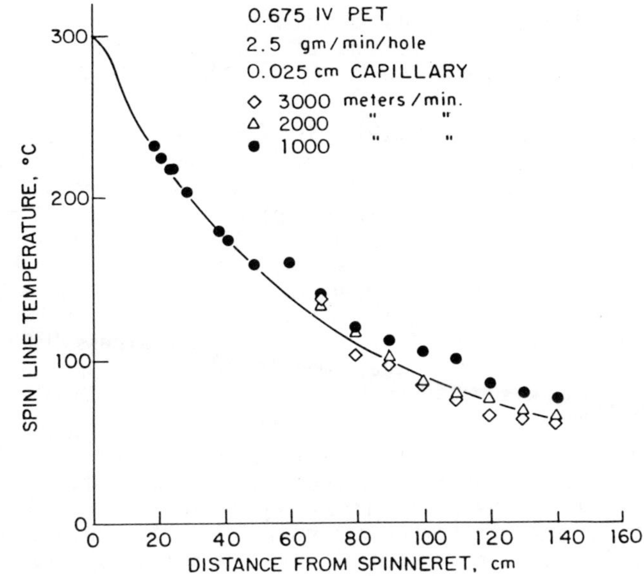

Figure 6 Comparison of Calculated and Measured Spinline Temperature Profiles Reproduced with permission from Ref. 2. Copyright 1982.

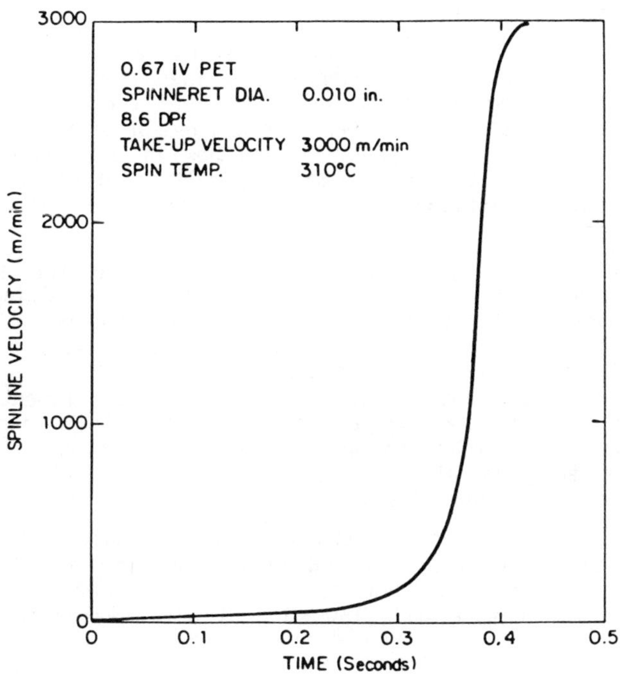

Figure 7 Calculated Spinline Velocity as a Function of Residence Time

here, most of the deformation and cooling occur over a much shorter period of time and near the end of the time interval. In other words, the polymer spends most of the time in a hot, slow-moving state.

In some early models[8,9], it was further assumed that the spin-line tension was constant along its length. This is approximately correct at low speeds, but is seriously in error at the higher speeds in use today. For example, spinning PET at 3000 meters/min. is illustrated in Figures 9 and 10. In the first figure, the velocity and temperature profiles are compared and, while different, do not suggest any profound differences. However, in Figure 10, a comparison of the stress profiles shows a difference in the stress at the freeze point of about 70%. This terminal stress is one of the most important properties that models of this type can compute, because it is the key property that enables us to estimate structural and, finally, physical properties of the spun yarn.

In Figure 11, the calculated stress at the freeze point is plotted versus the measured birefringence of as-spun fibers for several IV levels of PET. The three lines represent various other workers' measurements of the stress-optical coefficient by various techniques, and are all in excellent agreement with the model calculations. This plot demonstrates the bridge between process conditions and polymer properties and developed structure as measured by birefringence. It is interesting to note that this curve is essentially linear over the entire range of experimental conditions in spite of the fact that spin-line crystallization begins to occur at a spin-line stress of about 0.08 grams per denier[5].

The final link in getting from predeterminate operating conditions to spun yarn physical properties is presented in Figures 12, 13, and 14, which show the correlations between tenacity, elongation to break, and initial modulus as a function of birefringence for PET. With this link established, we are now in a position to deal with more realistic situations involving multifilament spinning and trying to minimize the filament to filament differences that occur in a spinning process.

Looking back to Figure 1, it can be seen that the quench air blows across the spin line in such a way that some filaments are upstream of others. The upstream filaments are quenched by fresh air that is heated as it cools the filament. Consequently, the next filament is cooled by slightly warmer air and is, therefore, produced by a slightly different process. In a case where the filament count is high, the air can pass through many rows of filaments. In multifilament spinning, one also has to become concerned with the tendency of the filaments to pump air down the spinway which alters the air velocity and convects heat downward.

Matsuo, et. al[14] have published a methodology for dealing with multifilament problems, including air heating and pumping, in which they solve for the conditions of the upstream filament and calculate the temperature and flow rate of air leaving the filament

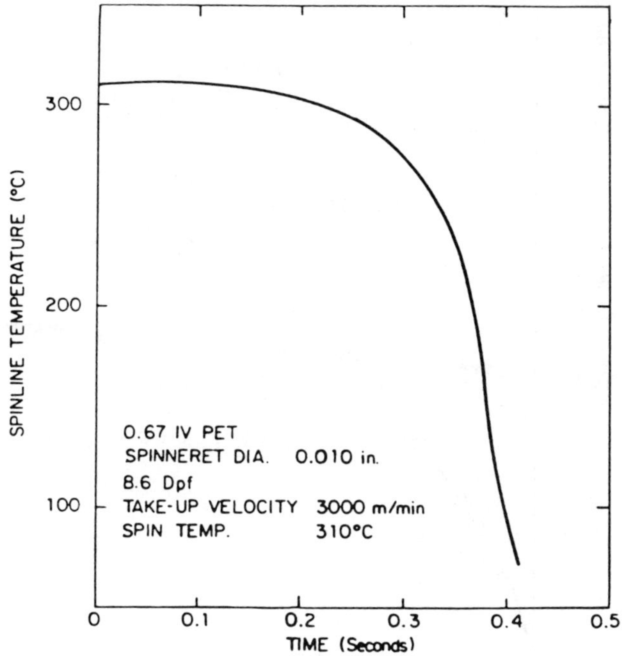

Figure 8 Calculated Spinline Temperature as a Function of Residence Time

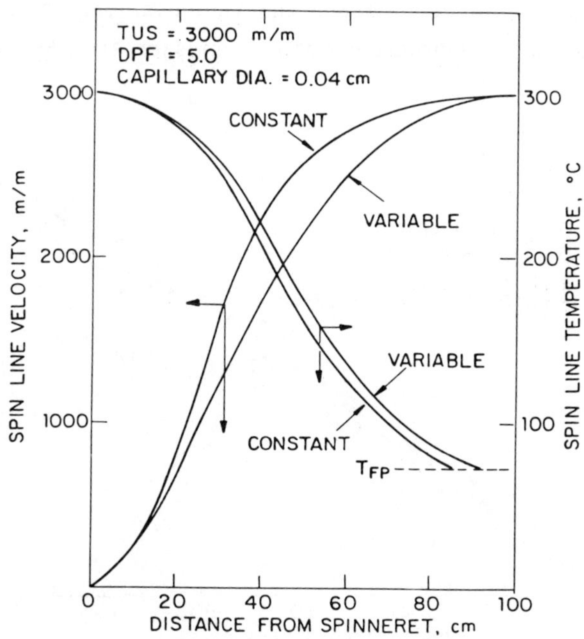

Figure 9 Comparison of Constant and Variable Tension Models; Velocity and Temperature

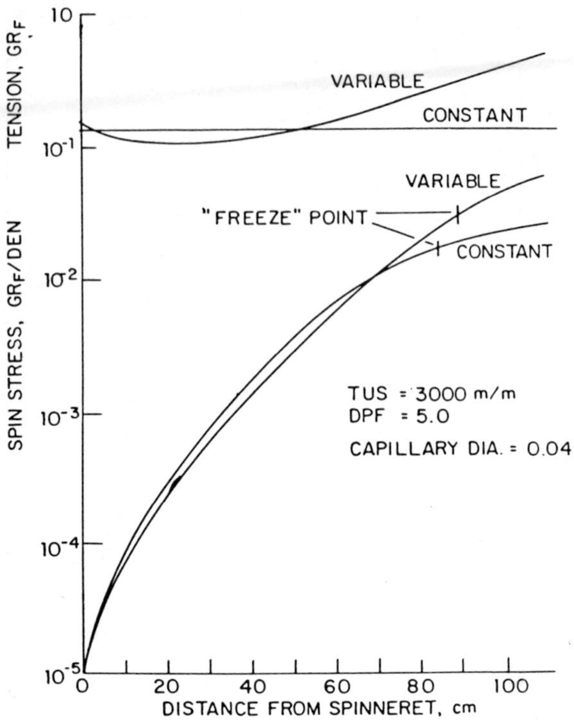

Figure 10 Comparison of Constant and Variable Tension Models; Tension and Stress Reproduced with permission from Ref. 2. Copyright 1982.

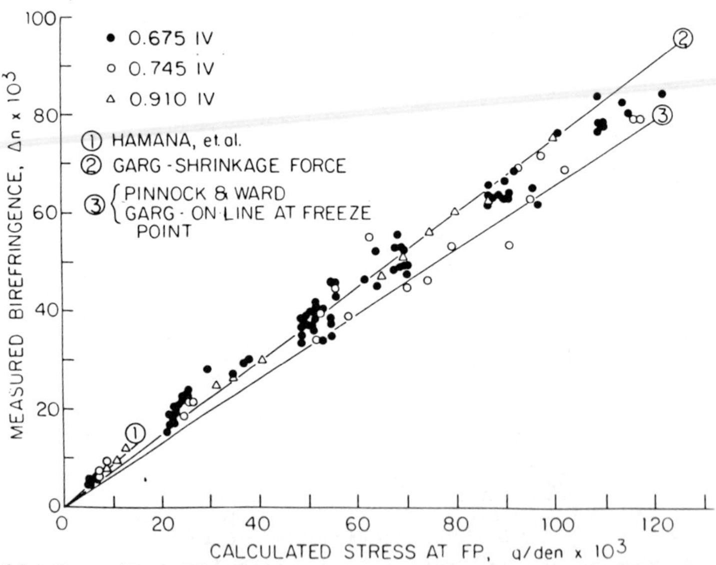

Figure 11 Spun Yarn Birefringence as a Function of Calculated Stress at Freeze Point Reproduced with permission from Ref. 2. Copyright 1982.

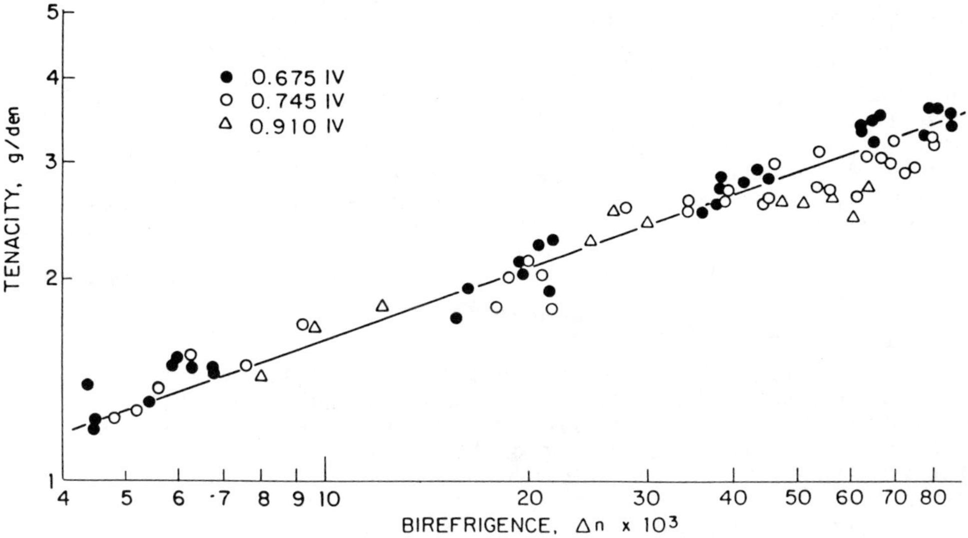

Figure 12 Spun Yarn Tenacity as a Function of Spun Yarn Birefringence Reproduced with permission from Ref. 2. Copyright 1982.

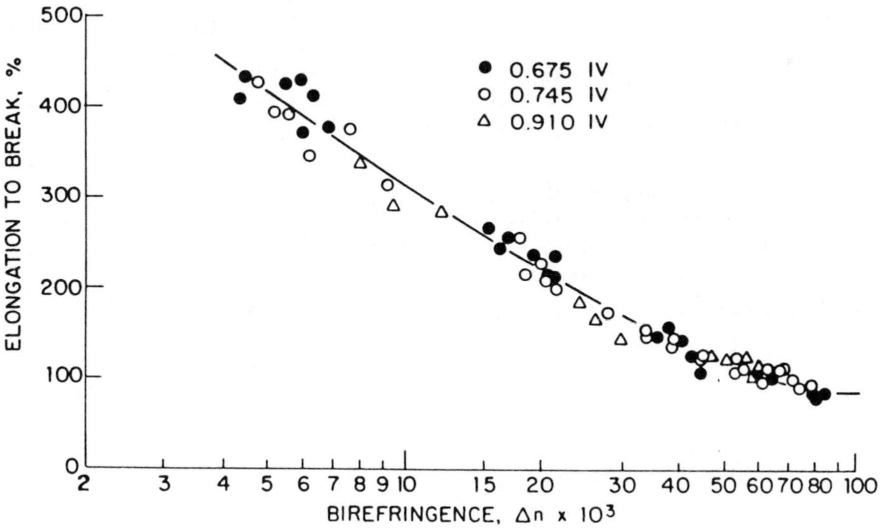

Figure 13 Spun Yarn Elongation as a Function of Spun Yarn Birefringence Reproduced with permission from Ref. 2. Copyright 1982.

which, in turn, become process conditions for the next filament. Their approach is shown schematically in Figure 15. The extent of this upstream/downstream effect is shown graphically in Figure 16, which deals with a case where there were 7 rows of filaments. The cross-sectional area profile shows some deviation along its length, but finally reaches the same value as would be expected from the material balance. The tension (or stress) at the freeze point is different; consequently, the properties will be different. The real utility of the model in this situation is that a number of approaches to reducing the spread in stress can be evaluated quickly and cheaply, and only the most promising pursued in the pilot plant. Also, the use of these models allows one to try solutions that would not be evaluated in the pilot plant due to cost or low likelihood of success.

Matsuo, et. al[14] also published an interesting table (Figure 17) that relates several commonly observed spinning instabilities which should be useful to any spinner. In their paper, they are not clear how this table was developed, but it is interesting to note that they show the relevant calculable variables. It is not unreasonable that one can obtain stability information from a steady-state model. For example, Kase[9] develops a linearized transient model which can be expressed in terms of time and position-dependent deviation variables and position-dependent but time-independent parameters derived from the steady-state solution. Thus, the stability of the system of deviation equations is determined solely by the values of functions of the steady-state solution only.

The objective of this review has been to provide an insight into the considerations involved in constructing spin-line models and some of the ways in which they can be used. Like any area, the task is never complete and the development is continuing. There are many areas in which the models can be improved and made more useful. Some of the main areas that need attention:

- Development models that handle radial temperature gradients. With the ever-increasing availability of computing resources, this should not be too formidable a challenge.
- Continue work on transient models.
- Develop models that deal with spin-line crystallization.

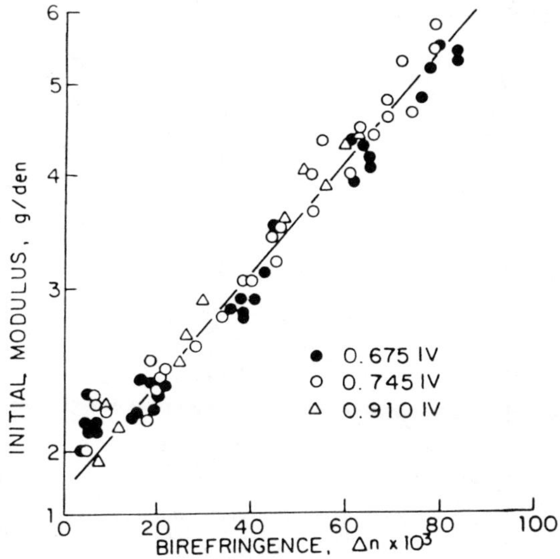

Figure 14 Spun Yarn Initial Modulus as a Function of Spun Yarn Birefringence Reproduced with permission from Ref. 2. Copyright 1982.

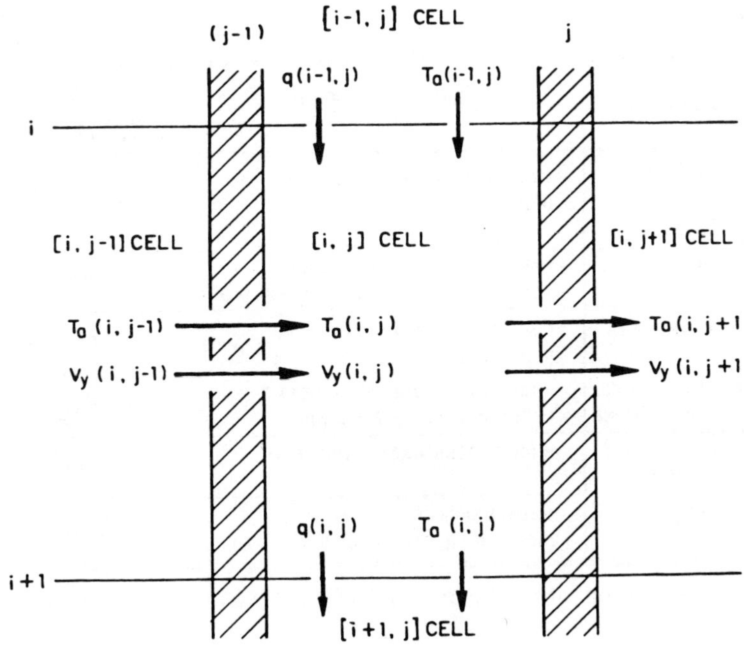

Figure 15 Multifilament Solution Methodology (14)

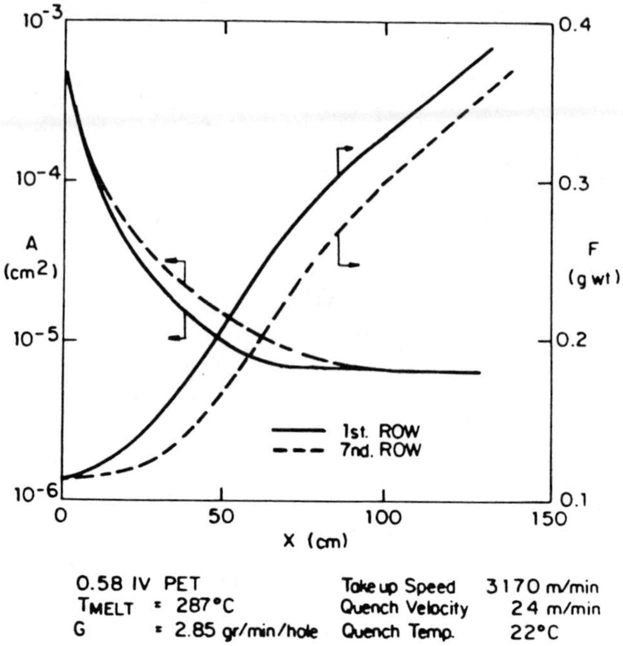

Figure 16 Interfilament Variation Between Rows of Filaments (14)

INSTABILITY	CONDITIONS LEADING TO INSTABILITY	CALCULABLE VARIABLES RELEVENT TO INSTABILITY
TWITCHING	- LOW SPINLINE TENSION - HIGH DRAW RATIO	F_0, V_0/V_{TU}
KNEEING AT SPINNERET	- LOW SPINLINE TENSION - PARTIALLY FOULED SPINNERET - TEMPERATURE GRADIENT ACROSS HOLE	F_0
DRAW RESONANCE	- HIGH DRAW DOWN RATIO - INSUFFICIENT COOLING	S_T, T^*, V_0/V_{TU}
FILAMENT BREAKS IN LOWER SPINLINE	- HIGH STRESS DUE TO AIR DRAG - EXCESSIVE GUIDE FRICTION	σ_{TU}
FILAMENT BREAKS IN DRAW ZONE	- HIGH STRESS ALONG SPIN LINE	$\sigma(X)$
ROUGH EXTRUDATE	- MELT FRACTURE - HIGH SHEAR STRESS IN SPINNERET	—

Figure 17 Relation of Spinning Instabilities to Process Parameters (14)

Literature Cited

1. Ziabicki, A. "Fundamentals of Fiber Formation"", J. Wiley & Sons, London (1976).

2. George, H. H., Polym. Eng. Sci., **22,** 292 (1982).

3. Frankfort, H. R. E., Knox, B. H., U. S. Patent No. 4,195,051.

4. Gregory, D. R., Watson, M. T., J. Polym. Sci., Part C, 399 (1970).

5. George, H. H., Holt, A., Buckley, A., Polym. Eng. Sci., **23,** 95 (1983).

6. Heuvel, H. M., Huisman, R., J. Appl. Polym. Sci., **22,** 2229 (1978).

7. Gagon, D. K., Denn, M. M., Polym. Eng. Sci., **21,** 844 (1981).

8. Kase, S., Matsuo, T., J. Polym. Sci., Part A, **3,** 2541 (1965).

9. Ibid, J. Polym. Sci., **11,** 251 (1967).

10. Humana, I., et al., Melliand Text., **50,** 382 (1969).

11. Garg, S., in press.

12. Pinnock, P. R., Ward, I. M., Trans. Faraday Soc., **62,** 1308 (1966).

13. Garg, S., Celanese Research Co., unpublished data.

14. Matsuo, T., et al., Prepr. Mezhdunar. Simp. Khim. Voloknam, 2nd, **2,** 206 (1977).

RECEIVED May 31, 1984

Design of Tire Cords

Optimization of Viscoelastic Properties

D. C. PREVORSEK, C. W. BERINGER, and Y. D. KWON

Corporate Research and Development, Allied Corporation, Morristown, NJ 07960

Pneumatic tire is a viscoelastic composite structure which undergoes large periodic deformation during the service. This leads to the mechanical loss from tire components, cords and rubber, causing the tire temperature to rise and resulting in tire rolling resistance. A higher temperature in service also reduces tire durability via degradation of rubber in tire shoulder. In order to improve the fuel economy and tire durability, it is therefore desirable to reduce the mechanical loss of the cord and rubber to the practical minimum. In this paper, we review the dynamic viscoelastic properties of various tire cords and discuss the approaches to minimizing the mechanical loss. In high speed driving, however, a tire constructed with materials of lower mechanical loss is more susceptible to the formation of standing wave which is detrimental to the safety. Therefore, the optimal loss for tires is usually not at their minimum but at an intermediate level. This situation necessitates determination of a most effective compromise regarding the optimal level of mechanical loss of tire components. An example is presented showing an approach to minimize the probability of forming standing wave at high speed driving without affecting the tire rolling resistance adversely.

Tire cords, as the reinforcing element of tire wall, must meet numerous criteria such as strength, dynamic modulus and loss, dimensional stability at ambient as well as elevated temperature during manufacturing and end use, chemical stability with respect to the chemicals present in rubber stock, resistance to fatigue and abrasion.

0097-6156/84/0260-0371$07.50/0

Consequently, an optimization for a specific tire is a prohibitively complex task. Improvements in performance are, therefore, achieved by separate studies focusing on a particular combination of properties.

This study is concerned exclusively with the role and optimization of viscoelastic properties of cords, an area where tire mechanics made very significant progress during the recent few years.

When a viscoelastic material such as tire cord or rubber is subjected to a small amplitude sinusoidal straining, the resulting stress-strain curve is an ellipse and the material properties are characterized by the real and imaginary moduli E' and E" or the ratio E"/E' (= tan δ). These properties are dependent on temperature and frequency in the experiment but independent of stain amplitude.

When the strain amplitude is relatively large as in the case of tire cord in a running tire, the viscoelastic behavior is no longer linear. The stress-strain loop is not elliptic but distorted (Figure 1). The material properties in the nonlinear regime can not be represented with the real and imaginary moduli. In the present study, we characterize the viscoelastic properties in nonlinear regime by the effective dynamic modulus and mechanical loss.(1)

Referring to Figure 1 (b), the dynamic modulus is given by $\Delta F/\Delta \varepsilon$ and mechanical loss is the area of loop. It should be noted that, in addition to the frequency and temperature dependence encountered with E' and E", the effective dynamic modulus and mechanical loss depend also on the strain amplitude as well as on the strain and temperature history of the sample.

In tire technology, the mechanical loss per cycle is frequently referred to as the heat generation per cycle. The importance of mechanical loss comes from the fact that it causes tire temperature to rise which in turn affects adversely many aspects of tire performance. Among the factors affecting mechanical loss of tire cords, the chemical structure of polymer must be considered as a dominant one, but mechanical loss of tire cord is affected by many other variables. For a given polymer, very significant and technologically critical effects are achievable also by altering fiber morphology. These effects are related to the degree of crystallinity, crystallite dimensions, the mode of coupling between the phases, structure of the amorphous domains and orientation functions. A great deal of research has been dedicated to the study of the role of these morphological parameters on the dynamic properties of fibers.

Mechanical loss also varies with the cord construction variables such as denier per yarn strand, number of strands and number of twists per unit length (2). Significant changes in mechanical loss are also observed during the post treatments of greige tire cords. These include the steps of cord tensilization, application of the adhesives and curing of tires.

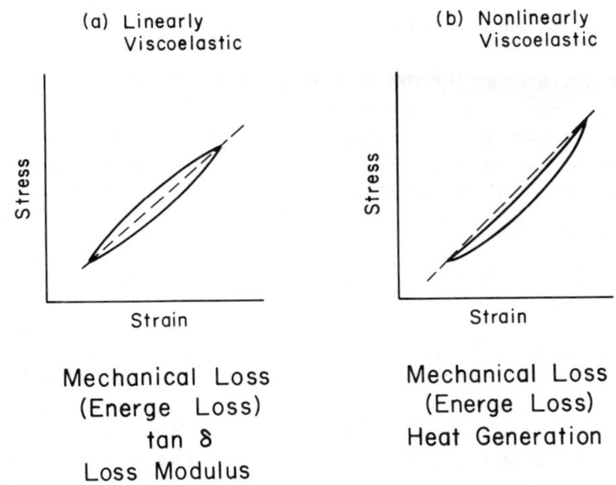

Figure 1. Stress-strain curves in cyclic tensile straining (a) linear viscoelastic (b) nonlinear viscoelastic.

Since the amplitude of cord deformation in tire decreases with the increasing cord modulus, it is important that, in a quantitative analysis of cord heat generation, both factors, mechanical loss and dynamic modulus of tire, be taken into consideration. A systematic study to determine the effect of cord modulus on strain amplitude yielded the following relationship (2).

$$\frac{\Delta\varepsilon_1}{\Delta\varepsilon_2} = \left(\frac{E_2}{E_1}\right)^{1/2}$$

Here, $\Delta\varepsilon$ is the strain amplitude, E is the dynamic modulus and the subscripts refer to the two samples being compared.

Consequences of Mechanical Loss

Tire temperature and mechanical properties of cords. The heat generation from cord and rubber under cyclic straining in rolling naturally causes the tire temperature to rise. Dynamics of this temperature rise can be expressed by the following heat transfer equation with the heat generation rate terms for cord and rubber. (3)

$$\rho C_p \frac{\partial T}{\partial t} = K \nabla^2 T + Q_c v_c + Q_r v_r \tag{1}$$

In equation (1), ρ, C_p, T, t, K, ∇^2, Q_c, Q_r, v_c, v_r are density, heat capacity, temperature, time, thermal conductivity, Laplacian operator, heat generation rates of cord and rubber, volume fractions of cord and rubber, respectively. The equation can be solved numerically by use of either the finite difference approximation or the finite element method. The solution of the equation with related boundary conditions provide the temperature profile in the tire wall cross section. Examples of such solutions are shown in Figure 2. The procedure of obtaining such solutions is outlined in Appendix I.

For the purpose of this study, it is desirable to consider two tires having an identical design and same rubber stock but are reinforced with cords which exhibit different heat generation characteristics. We want to determine how this difference would affect the resulting tire temperature profile.

Until very recently, it has been widely believed that the cords do not contribute significantly to the total amount of heat generated in a rolling tire (4). These views were disproved. Recent studies and experimental results presented below show that the cords contribute significantly to the heat generation. Moreover, the results show that the contribution from cord is even more critical because it affects the hot spot temperature in the tire shoulder, an area where tire failure is initiated frequently.

Figure 2(a) compares the heat generation rate of nylon 6 and nylon 66 cords which were used in reinforcing the bias truck

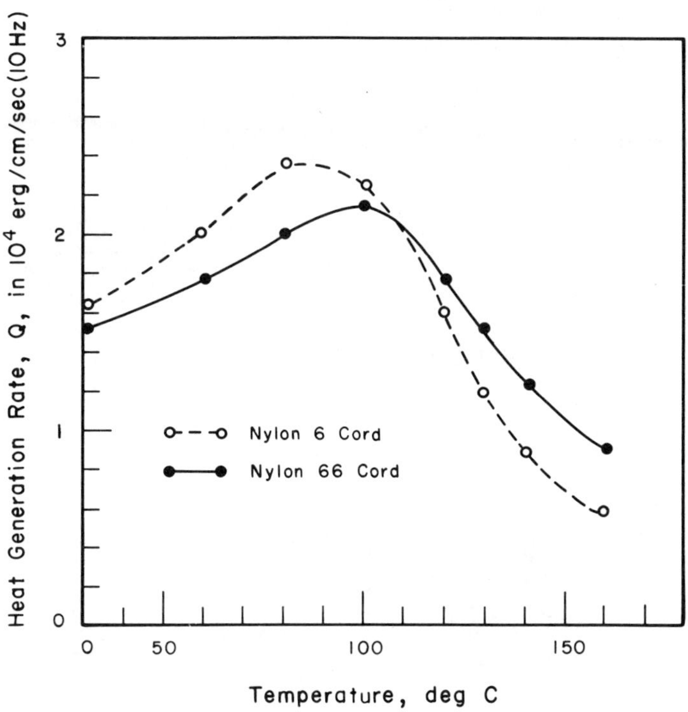

Figure 2a. Effect of the cord heat generation on the tire temperature profile (Bias truck tires, 65 MPH); heat generation rate.

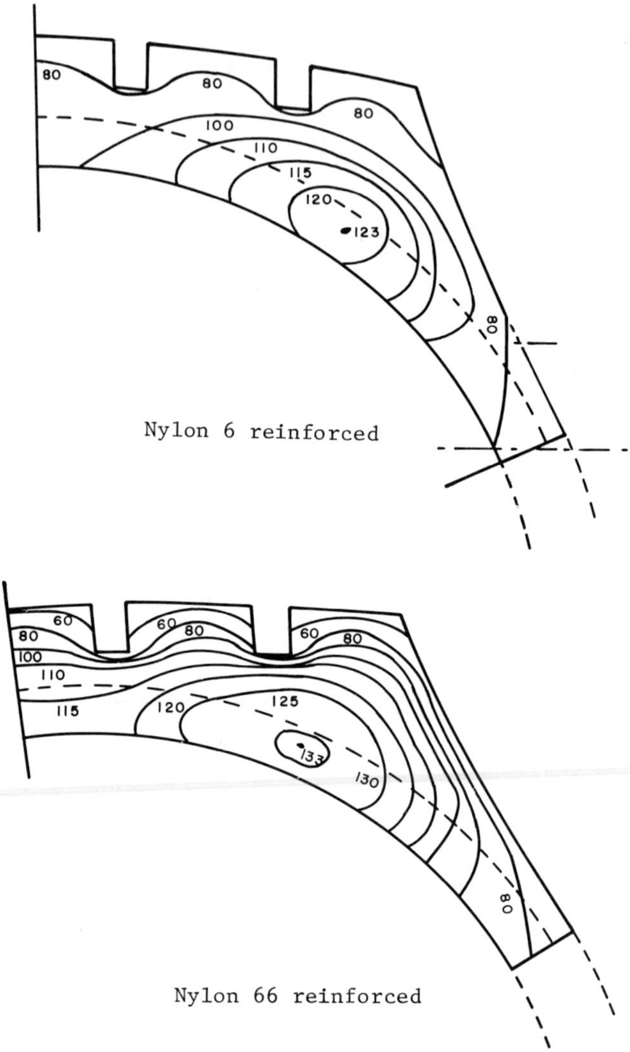

Figure 2b. Effect of the cord heat generation on the tire temperature profile (Bias truck tires, 65 MPH); temperature profile.

tires of identical design. Note that, when the temperature is below about 110 deg. C, nylon 6 cord generates more heat than nylon 66 cord while, at temperatures above 110 deg. C, nylon 66 cord generates more heat. Effect of this difference is manifested in the level of hot spot temperature in the shoulder (Fib. 2(b)). The hot spot temperature in nylon 66 tire is about 10 deg. C higher than that of nylon 6 tire. We shall discuss the effect of this temperature difference on the tire fatigue endurance later. According to our study (5), contribution from the cord to the total heat generated can be as much as 40 - 50% under certain conditions.

Having seen that the tire temperature is affected significantly by the characteristics of cord heat generation, we would like to know how the temperature affects the fatigue endurance of cords. We took nylon 6 and nylon 66 cords which were used in constructing these tires and subjected them to cyclic stressing under various stress amplitudes to measure the number of cycles to failure. The experiments were conducted at the room temperature and 85 deg. C. Results in Figure 3 show that, when the temperature rose from the room temperature to 85 deg. C, the number of cycles to failure decreased by 10 fold. Nylon 6 and nylon 66 showed similar level of endurance although, at room temperature, nylon 6 seems to show slightly better endurance. In the course of our studies, we examined numerous failures of tires reinforced with nylon 6 and nylon 66 tires. Since we have never encountered a case where the failure in straight rolling could be attributed to cord fatigue, we consider the minor differences in the mechanical properties of cords inconsequential to the tire performance.

Searching for some explanations regarding the subtle but important differences between the characteristics of nylon 6 and nylon 66 cords, we carried out additional viscoelastic studies including the analysis of changes in viscoelastic properties of tire cords during the cyclic straining under a large strain amplitude which is encountered in a rolling tire (6). This was done by superimposing a small amplitude, high frequency straining on the fundamental, large amplitude straining of the cord. Such an experiment allows the determination of viscoelastic properties at each instant during the straining.

Studies of nylon 6 and nylon 66 cords by this method showed that, in addition to the strain dependence of modulus and mechanical loss, these materials also undergo reversible, structural changes during the cyclic straining. Surprisingly, the types of these reversible structural changes were found to be different between nylon 6 and nylon 66. When a nylon 6 cord is being stretched during the cyclic straining, the sample is observed to release energy and this could be related to some improvement in order or degree of crystallinity. In contraction, the energy is absorbed back, indicating the reduction in order to the level of unstrained state. With nylon 66, on the other hand, the situation is reversed. On stretching, the sample is observed to absorb energy (7). Decrease in the fiber density, cavitation, reduction in order or degree of crystallinity are the effects which are consistent with this observation.

The causes of this phenomenon have not been elucidated. Based on the available x-ray diffraction data of fibers under strain, we speculate, however, that crystallites of nylon 6 adjust more readily to stress, undergoing a larger crystallographic change than the crystallites of nylon 66. Thus, the crystallites of nylon 66 act more like a rigid filler. We believe that this inherent difference between nylon 6 and nylon 66 could be a factor in the superior impact resistance of nylon 6.

Tire rolling resistance. Once the tire temperature profile is established, the tire rolling resistance due to the viscoelastic hysteresis of the components (i.e. mechanical loss) can be estimated by taking the volume integral of the heat generation rate. The relationship between the rolling resistance, rolling speed and the heat generation rate of the tire components in a rolling tire is as follows.

$$R \cdot V = \int_v (Q_c v_c + Q_r v_r)\, dv \qquad (2)$$

In equation (2), R is the rolling resistance, V is the rolling speed and v refers to the volume of tire wall. The rolling resistance estimated in this manner has been found to make up more than 90% of the total rolling resistance observed. Table 1 illustrates an example of relevant data.

Table 1. Comparison of the measured rolling resistance and the rolling resistance which was calculated from the total heat generation rate by use of the relationship given by equation (2)

	Tire 1. Truck tire reinforced with nylon 6 cord	Tire 2. Truck tire reinforced with PET cord.
Speed	50 MPH	50 MPH
Total heat generation rate per tire	313.6 cal/sec	333.4 cal/sec
Rolling resistance calculated from the heat generation rate (Rc)	13.2 lbf	14.0 lbf
Rolling resistance measured (Rd)	13.7 lbf	14.9 lbf
Rc/Rd	0.96	0.94

Mechanism of tire failure and fatigue of rubber. Tires examined in this study were first subjected to laboratory wheel testing conducted under progressively increasing load until the tires failed. Inspection of the failed tires revealed complex failure modes in the shoulder area the cause of which could not be determined by inspecting the failed tires. In order to determine the mechanism of failure and its locus, we subjected a few sets of tires to same testing and in this case the tire surface temperature was continuously monitored. All tires developed after some time a modest but discernible overheating localized in the shoulder area. At this point, the test was interrupted and the tire shoulder disected at the hot spot area. This revealed a small crack in the rubber area close to the reinforcing layer of the cords but still outside the adhesive interface. (Figure 4)

These analyses demonstrated that the failure in these tires was initiated in the rubber. Several tires were then submitted to similar investigation but they were held longer in order to let the crack grow. This procedure which yielded the actual initiation and propagation of the crack until the catastrophic failure will be described in detail in a separate report. Locus of the crack initiation and its propagation is illustrated in Figure 11. For the purpose of present discussion, it is necessary to take into consideration that the failure is initiated in rubber and that this method establishes quite precisely the location of crack initiation and the direction of crack propagation.

For further analysis of the performance of these tires in wheel testing, it was then necessary to determine the fatigue resistance of rubber in the areas of crack initiation under a condition which is as close as possible to that in rolling tires. For this purpose, we devised a fatigue test which uses the specimen cut out from the shoulder region as shown in Figure 5. The specimen is given notches as shown and then the specimen is held between a pair of clamps and placed in a temperature-controlled chamber. Then, the specimen is subjected to a cyclic straining to cause the crack propagate toward failure. The frequency is set to the rotational frequency of a tire and the strain amplitude is set to such a level that the failure would occur in a reasonable length of time. The test was performed by use of Allied's High Strain Dynamic Viscoelastometer (6,8) which has the features of cyclic straining, elapsed time measurement and creep measurement.

Figure 6 shows the fatigue endurance of tire wall specimens from different tires as the functions of temperature. First, we observe that the fatigue endurance is sensitively dependent on the temperature. At about 130 deg. C which is the temperature range in the shoulder of these truck tires at 65 MPH, it is seen that a temperature increase of 10 deg. C reduces the number of cycles to failure by a factor of two. These experiments show clearly that the tire shoulder temperature is a critically important factor affecting the tire durability and that a lower

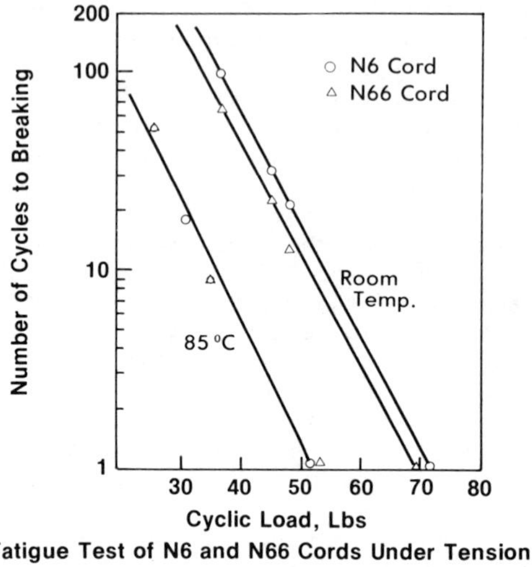

Figure 3. Effect of temperature on the cord fatigue endurance under cyclic tensile straining.

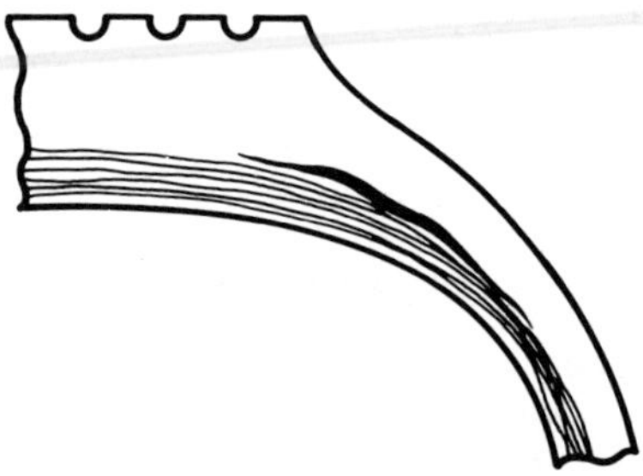

Figure 4. Locus of crack formation in the tire shoulder.

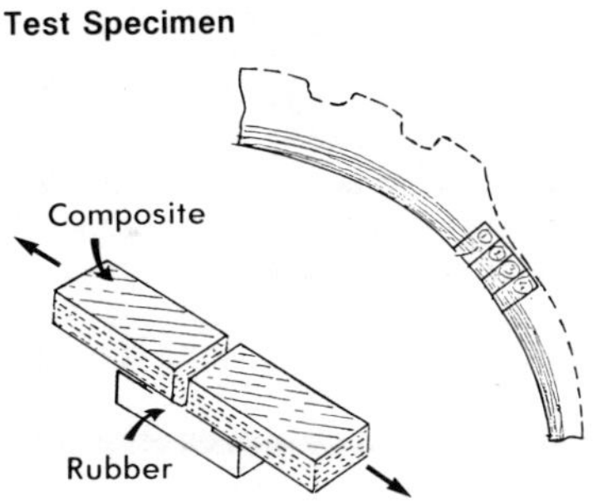

Figure 5. Effect of the tire shoulder temperature on fatigue endurance: test specimen.

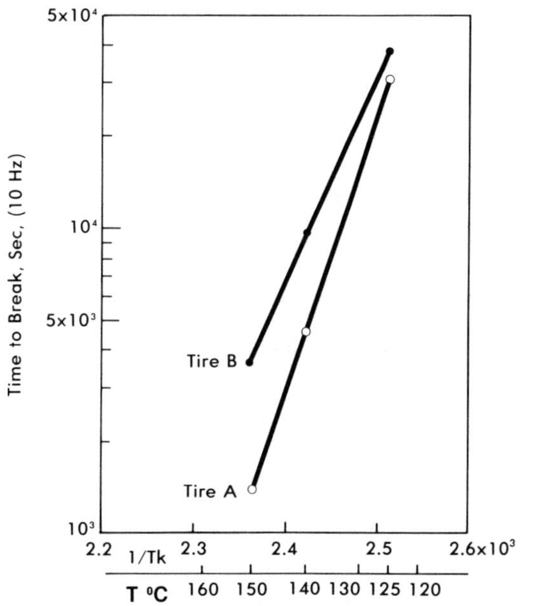

Figure 6. Effect of tire shoulder temperature on fatigue endurance: test results.

temperature is desirable for improved durability. This can be achieved, of course, by lowering the rate of heat generation of cord and rubber via the optimizaton of their composition, construction variables and fabrication procedures.

Minimization of Mechanical Loss from the Cord in Running Tires

We have shown that the mechanical loss of the cord contributes to the tire rolling resistance and to the acceleration of fatigue process. In order to minimize the effect of this contribution, therefore, one has to minimize the mechanical loss of the cord. This can be achieved in many ways. Of particular interest are:

Optimization of the cord tensilization process

High stress spinning of the fibers

Optimization of the cord construction variables

Choice of the materials of fibers.

Optimization of the tensilization process. In order to minimize the mechanical loss of nylon 6 and nylon 66 cords, we carried out the following experiments to optimize the cord tensilization process. The variables studied were:

Cord twist (9 - 11/inch)

% Strecth (8 - 10%)

Temperature (390 - 455 deg. F)

% Shrinkage (0 - 2%)

The tensilization was done in two stages. In the first stage, the cord was stretched at a temperature and, in the second stage, the cord was allowed to shrink at a temperature. The cord twist was varied to two levels, stretch % to two levels, temperature of the first and second stages to three combinations and the shrinkage % to two levels. The experiments of tensilization were designed factorially. In Figure 7, we show the heat generation rate of nylon 6 and nylon 66 cords which resulted from the experiments. The figure shows the maximum and minimum levels of heat generation rates among the nylon 6 and nylon 66 cords which were tensilized in this study.

The following conclusions can be drawn on the basis of these data.

1. The difference between the maximum and minimum values of the heat generation rate is significant and, therefore, the tensilization should be done at a condition where the heat generation rate of the resulting cord is at the minimum.

2. The minimum heat generation rates of nylon 6 and nylon 66 cords are at about the same level. But we observe a very important difference between the two types of cords. In the temperature range of 80 - 110 deg. C, nylon 6 generates more heat than nylon 66. However, outside this range of temperature, nylon 66 cord generates more heat. Therefore, depending on the operating temperature of a tire section, either nylon 6 or nylon 66 cord will be generating more heat during service. For example, in a truck tire operating at a speed of 60 MPH, the shoulder temperature is in the neighborhood of 130 deg. C while the sidewall temperature is about 80 deg. C. Consequently, in the shoulder region, nylon 66 would generate more heat and, hence, contribute more to the temperature rise and fuel consumption. But the opposite would be the case for tire sidewall. Recently, it was confirmed (9) in a laboratory study that the bias truck tire reinforced with nylon 66 cord exhibits higher shoulder temperature and shorter lifetime than the one reinforced with nylon 6 cord under a same design.

High stress spinning of the fibers. Recently, it became known that, when the yarn is spun under a high stress, the resulting cord generates less heat (10). Figure 8 shows a comparison of the heat generation rate of PET cords which were prepared by the conventional spinning and by the high stress spinning. It is seen that the cord prepared by the high stress spinning generates considerably less heat than the cord prepared by the conventional spinning process at all temperatures.

Optimization of the cord construction variables. In a recent study in this laboratory (2), we investigated the effect of the degree of cord twisting, number of cord ends per inch in the carcass, number of plies and crown angle of the cord in bias passenger tires on tire temperature rise in rolling. The results indicated that the degrees of cord twisting affects the heat generation rate of the cord significantly.

Figure 9 illustrates the effect of twisting on the heat generation rate and dynamic modulus of PET cords. It is seen that an increase in the number of twist per unit length caused decrease both in the heat generation rate and dynamic modulus. The lower dynamic modulus would result in a larger strain amplitude under a same cyclic loading and the increased strain amplitude causes an increase in the heat generation rate. Thus, a part of the advantage of lower heat generation rate with a higher degree of twisting seen in Figure 8 would be lost. But the initial advantage is large enough so that a part of it would remain even after discounting the effect of lower modulus.

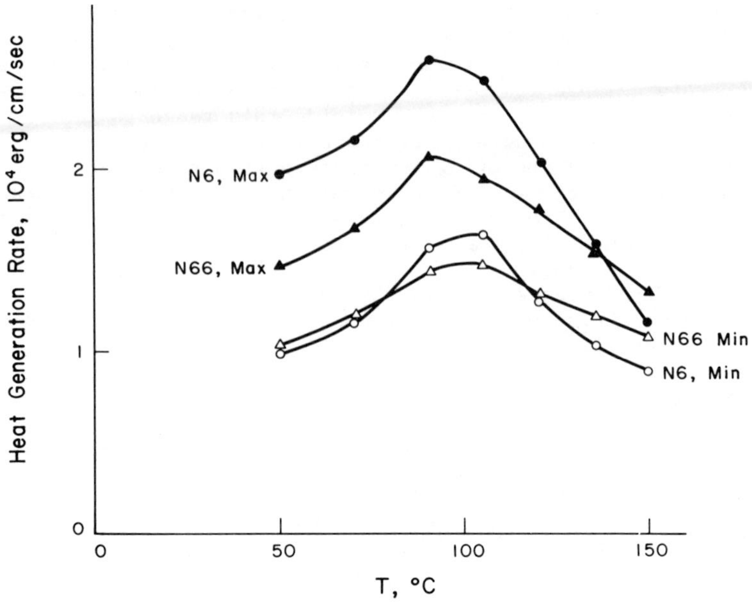

Figure 7. Results of the optimization of tensilization process (strain amplitude 1.0%, frequency 10 Hz).

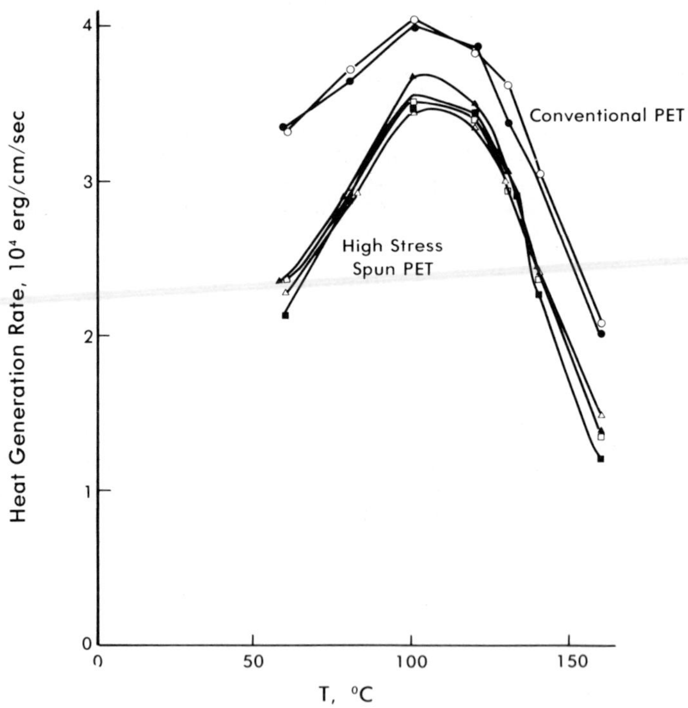

Figure 8. Heat generation rate of high stress spun PET cords.

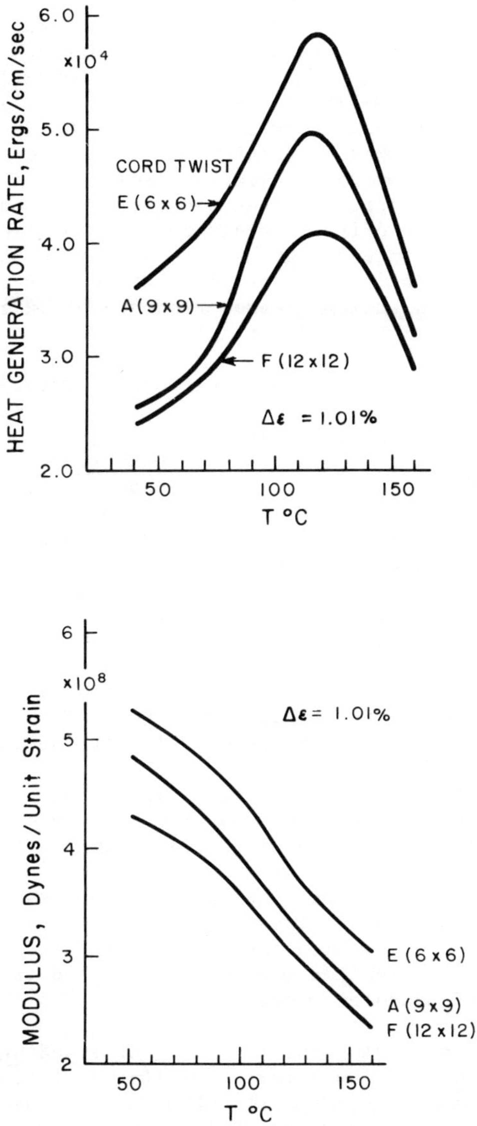

Figure 9. Effect of twist on viscoelastic properties of PET cords. Reproduced with permission from Ref. 2. Copyright 1979 ASTM,

Choice of the Materials of Fibers

As different materials have different heat generation characteristics, it is obvious that selection of materials is another approach to the minimization of mechanical loss of cord.

An immediate example can be seen in Figure 2(b). In a bias truck tire, the operating temperature of the shoulder region (where the hot spot is) is in the order of 120 - 140 deg. C when the tire is reinforced with nylon 6 or nylon 66 cord. Because of the heat generation characteristics of the two cords in this temperature range, use of nylon 6 would result in a lower temperature at the hot spot and thus reduce the probability of heat failure under severe driving conditions. In an extensive study of truck tires reinforced with nylon 6 and nylon 66 cords which we carried out recently, we observed that the tires reinforced with nylon 66 cords show a greater tendency of undergoing a shoulder failure which is heat-initiated.

Effect of Excessively Low Mechanical Loss of Cords on Tire Vibration

Formation of standing wave at high speed driving. So far, we have considered how to minimize the mechanical loss of the cord so that fuel economy and fatigue endurance can be improved. However, if the mechanical loss of the tire components is reduced too much, there arises the problem of standing wave which would be formed because of the insufficient damping effect (11,12). The relationship between the mechanical loss and formation of standing wave is outlined in Appendix II.

Figure 10 shows the relationship between the damping coefficient (which is equivalent to the mechanical loss) and the critical speed at which the standing wave would form. Thus, the critical speed inceases with the increase of the damping coefficient. This means that a tire would be more free of standing wave hazard if its rolling resistance is higher. The figure also shows that a higher density of the tire wall material would push the critical speed down because of its inertia effect during the contour recovery phase in tire rotation. Therefore, a steel-reinforced tire has a lower critical speed unless its rolling resistance is increased to provide the necessary damping effect.

From this figure, we see that the rolling resistance of a tire can not be reduced too much without regard to the effect on the formation of standing wave.

An option for suppressing the formation of standing wave. Suppose we achieved a low level of rolling resistance within the limit of preventing the formation of standing wave at a reasonably high tire speed (e.g. 80 MPH). Under this circumstance, is there a way of lowering the level of the hot spot temperature in the shoulder without affecting the damping effect i.e. the mechanical loss? If this is possible, it would mean that the tire fatigue endurance can be improved while the formation of standing wave is suppressed.

We have studied the effect of shifting some of the heat generation from the cord to the rubber in the tire shoulder region. Figure 11 illustrates the result of shifting half of the cord heat generation to the rubber in the shoulder. Thus, the total rolling resistance remained same but the temperature profile shows that the hot spot temperature at the shoulder dropped by 2.4 deg. C. This drop in the hot spot temperature would help improve the fatigue endurance according to the fatigue-temperature relationship which was shown earlier in Figure 6.

Implementation of this concept would involve using a cord of lower mechanical loss and rubber of correspondingly higher mechanical loss.

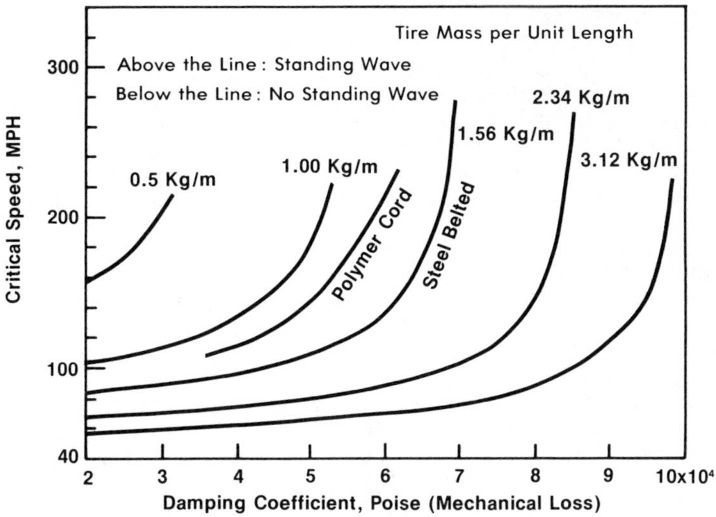

Figure 10. Effect of mechanical loss (damping) on formation of standing wave.

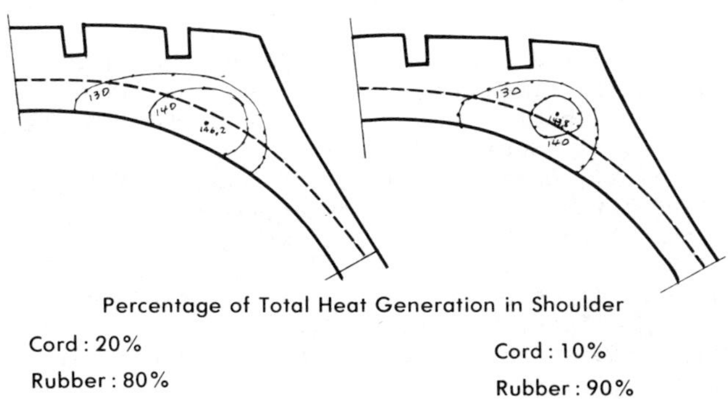

Figure 11. An option of increasing the mechanical loss of rubber to improve fatigue endurance while preventing standing wave.

Conclusions

Recognizing the role of the mechanical loss of tire cords in the tire temperature rise, rolling resistance and fatigue endurance, we have considered some ways of minimizing the mechanical loss of tire cords. However, an excessive reduction of the mechanical loss would make the tires more susceptible to the formation of standing wave. Therefore, a compromise would be necessary in the optimization of tire cord properties. An option for lowering the hot spot temperature in the shoulder without reducing the mechanical loss while suppressing the formation of standing wave was also considered.

Literature Cited

1. Kwon, Y. D.; Sharma, R. K. ; Prevorsek, D. C. ASTM SPECIAL TECHNICAL PUBLICATION 694, ASTM, 1979; p. 239.
2. Prevorsek, D. C.; Beringer, C. W.; Kwon, Y. D.; Sharma, R. K. ASTM SPECIAL TECHNICAL PUBLICATION 694, ASTM, 1979; p. 298.
3. Prevorsek, D. C.; Kwon, Y. D.; Sharma, R. K. SAE CONFERENCE PROCEEDING P-74 ON TIRE ROLLING AND FUEL ECONOMY - AN R & D PLANNING WORKSHIP, 1977; p. 75.
4. Schuring, D. J. Rubber Chemistry and Technology, 1980, 53, p. 600.
5. Sharma, R. K.; Kwon, Y. D.; Prevorsek, D. C. ASTM SPECIAL TECHNICAL PUBLICATION 694, ASTM, 1979; p. 263.
6. Kwon, Y. D.; Sharma, R. K.; Prevorsek, D. C. In "Probing Polymer Structures", Koenig, J. L. Ed.; ADVANCES IN CHEMISTRY SERIES No. 174, American Chemical Society; Washington, D. C., 1979; p. 35.
7. Prevorsek, D. C.; Kwon, Y. D.; Sharma, R. K. J. Applied Polymer Science, 1980, 25, 2063.
8. "RJS Model 167 Viscoelastometer, Instruction Manual", RJS Corporation, Akron, Ohio, 1981.
9. Prevorsek, D. C.; Beringer, C. W.; Palley, I.; and Kwon, Y. D. "Application of Fracture Mechanics in Tire Endurance Analysis", submitted for presentation at the Tire Society meeting, March 28 - 29, 1984, Akron, Ohio.
10. Unpublished Data; Corporate R & D, Allied Corporation.
11. Prevorsek, D. C.; Kwon, Y. D. "Fuel Economy Combined with Safe High Speed Driving: A Challenge for Future Tire Design", The Fifth Annual National Conference of the Plastics and Rubber Institute -TYRES-, 14 - 16 November 1979 at the Hilton Hotel, Stratford-upon-Avon.
12. Kwon, Y. D.; Prevorsek, D. C. "Formation of Standing Wave in Rolling Tires: Radial Passenger Tires", Tire Society Meeting, March 25 - 26, 1982, Akron, Ohio.
13. Pacejka, H. B. "Standing Waves", p. 743 of "Mechanics of Pneumatic Tires", National Bureau of Standard Monograph 122, ed. S. K. Clark, November 1971.
14. Kwon, Y. D. and Prevorsek, D. C., SAE Technical Paper Series No. 800242, 1980.

Appendix I

Procedure of Obtaining the Solutions to Heat Transfer Equations Representing the Dynamics of Temperature Change in a Rolling Tire

In order to determine the temperature profile in the tire cross section shown in Figure 2, the heat transfer equation (Equation 1) has to be solved with proper boundary conditions, material properties and the heat generation rate of cord and rubber appearing in Equation (1). In this appendix, we give a brief description of the procedures involved in this.

Determination of the Heat Generation Rate of Cord and Rubber under Cyclic Straining

Starting from the initial condition of Equation (1) which is usually a uniform temperature in the tire cross section (at environmental temperature), the tire temperature rises because of the heat generated by the cord and rubber under cyclic straining. Q_c and Q_r in Equation (1) are the heat generation rate of cord and rubber, respectively. Q_c and Q_r are functions of temperature and strain amplitude, i.e.

$$Q_c = f(T, e_c, n)$$
$$Q_r = f(T, e_r, n) \qquad \text{(A-1)}$$

where e_c and e_r are the strain amplitudes of cord and rubber, respectively and n is the frequency.

Q_c and Q_r are experimentally determined from the cord and rubber samples by use of Allied High Strain Dynamic Viscoelastometer (6,8) at various temperatures in the temperature range of room temperature - 160 deg. C and under various strain amplitudes. The resulting data are subjected to regression analysis to store the data in a computer-retrievable form. Thus, by giving T and e for cord or rubber, Q_c and Q_r can be produced for the sample in the computation.

Measurement of Tire Temperature at Strategic Positions in the Tire while Tire is Running on a Test Wheel

We developed a technique of inserting thin thermocouples into the tire wall at strategic positions (see Fig. A-1) so that the temperature of the tire can be monitored while the tire is running. Results of this temperature measurement provide the basis for obtaining the complete temperature profile as well as the effective strain amplitudes.

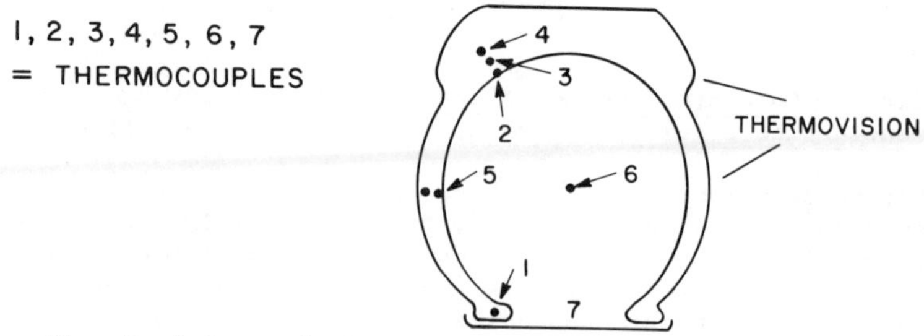

Figure A1. Positions of thermocouples in tire for measuring tire temperature. Reproduced with permission from Ref. 2. Copyright 1979 ASTM, 1916 Race Street, Philadelphia, PA 19103.

Iterative Solution of the Heat Transfer Equation by Finite Difference Approximations

Geometrical simplification of the tire wall and the corresponding heat transfer equations. In order to facilitate the numerical solutions of the heat transfer equations, the tire cross section was represented as a combination of simpler geometrical regions as shown in Fig. A-2. Thus, the side wall region is looked at as a one dimensional system and the crown and shoulder region is divided into three zones (inner zone, crown zone, shoulder zone) each of which is represented by two dimensional geometrical systems. The bead region is not included in this analysis. Note that some of the boundaries of these zones are overlapping.

For the side wall region, Equation (1) assumes the form of Equation (A-2). For the inner zone of crown-shoulder region, Equation (A-3) is used and, for the crown zone and shoulder zone, Equation (A-4) is used.

$$\rho C_p \frac{\partial T}{\partial t} = k \frac{\partial^2 T}{\partial x^2} + Q_c V_c + Q_r V_r \qquad \text{(A-2)}$$

$$\rho C_p \frac{\partial T}{\partial t} = \frac{k}{r} \frac{\partial}{\partial r}\left(r \frac{\partial T}{\partial r}\right) + \frac{k}{r^2} \frac{\partial^2 T}{\partial \theta^2} + Q_c V_c + Q_r V_r \qquad \text{(A-3)}$$

$$\rho C_p \frac{\partial T}{\partial t} = K \frac{\partial^2 T}{\partial x^2} + K \frac{\partial^2 T}{\partial y^2} + Q_c V_c + Q_r V_r \qquad \text{(A-4)}$$

As for the boundary conditions, the following types were applied:

1) Surfaces in contact with air : Neumann condition

2) Overlapping boundary lines : Dirichlet condition

3) Boundary line in the rubber : No heat flux

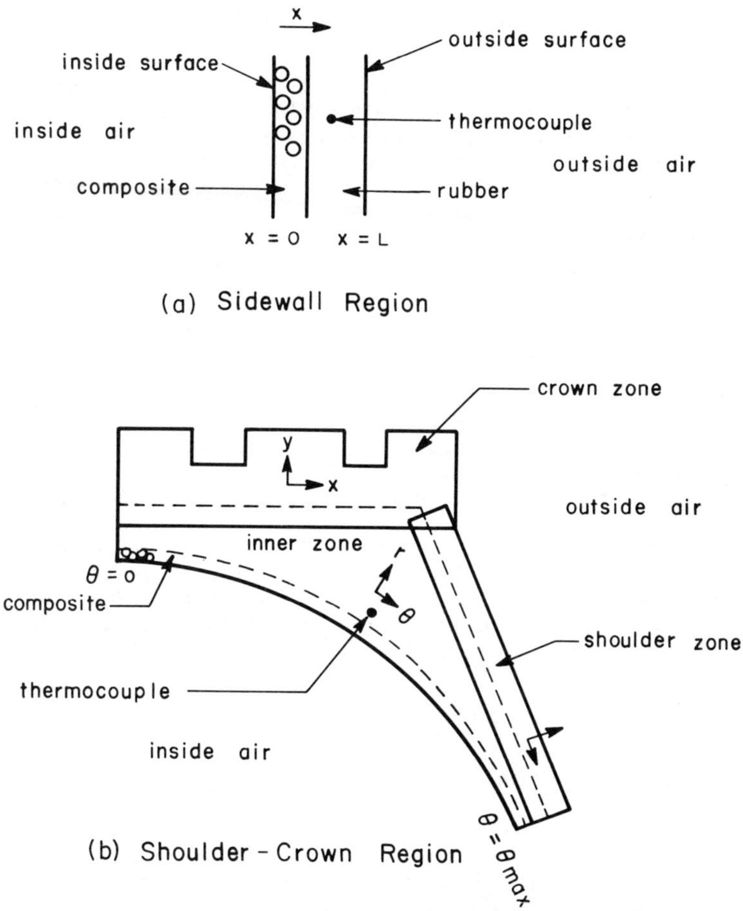

Figure A2. Geometry of the tire cross section. Reproduced with permission from Ref. 2. Copyright 1979 ASTM.

Numerical solutions. Numerical solution was obtained by the implicit alternating method for the above shown equations. As the strain amplitudes are not known a priori, solutions were obtained with several combinations of e_c and e_r and the solution which would match the measured temperatures best was selected as the final solution (2). Equations (A-2) - (A-4) represent unsteady state changes in temperature. By allowing sufficient number of time steps, the steady state solution could be obtained.

In the application of the Dirichlet type boundary conditions at the overlapping boundary lines between the inner zone, crown zone and shoulder zones, the temperature in the neighboring region at the previous time step was used as the known temperature at first and this was repeated a few times to refine the values of temperature. The heat transfer coefficients needed in the application of Neumann type boundary conditions were determined by trial and error by repeating the solutions several times.

Results of the Solution

From the results of solution, the temperature profile can be plotted as illustrated in Figure 2a and b. It shows that, in bias truck tires, the temperature at the hottest spot can reach as high as 130 - 140 deg. C. The hot spot is at the shoulder section where the tire wall is thickest, providing higher resistance to heat flow.

The solution is also useful in following the change of temperature as the function of time. In the simulation of tire temperature changes in varying speeds, the solution of Equation (1) will provide the necessary solutions as long as the varying speed (n) is put into the computation to reflect the actual change in speed.

Appendix II

Relationship between the Mechanical Loss and Standing Wave Formation

If a tire is built with a purely elastic and deformable wall material, the tire wall will obviously vibrate by the natural frequency when the contact deformation undergoes the recovery process unless the tire rotates very slowly. The mechanical loss from the viscoelastic tire wall material provides the damping force which prevents the vibration (standing wave) up to a certain speed depending on the level of damping force. In this appendix, we describe briefly how the results of Figure 9 (Effect of mechanical loss (damping) on the formation of standing wave) were obtained.

Mathematical Model (Circular Membrane Model)(13)

Dynamics of the formation of standing wave in a rolling tire can be approximated by the following equation (12).

$$\left(\eta\rho\Omega^2 - \frac{S_o}{r^2}\right)\frac{d^2w}{d\phi^2} - k_r\Omega\frac{dw}{d\phi} + (c' - \rho\Omega^2)w = 0 \qquad \text{(A-5)}$$

where

η : fraction of centrifugal force restored by radial force

ρ : mass per unit length, kg/m

Ω : angular velocity of rotation, rad/sec

S_o : tension force on the nonrotating tire, N

r : tire radius, m

w : outward displacement of the rotating membrane, m

ϕ : angular position of a tread element with respect to the vertical through the wheel axis, the vertical being fixed in the space

k_r : damping coefficient, Pa-sec

c' : $= c_r + c_s/r^2 \ -2bp_i/r$

c_r : radical stiffness of foundation per unit length, N/m^2

c_s : tensile carcass stiffness of unit length (average elastic modulus times tread band cross section), N

$2b$: effective tread-band width, m

p_i : inflation pressure, N/m^2

Critical Speed of Standing Wave Formation

General solution of Equation (A-5) is

$$w = c_1 e^{\lambda_1\phi} + c_2 e^{\lambda_2\phi} \qquad \text{(A-6)}$$

with $$\lambda_{1,2} = \frac{K_r\Omega \pm \sqrt{\zeta}}{2(\eta\rho\Omega^2 - S_o/r^2)} \tag{A-7}$$

$$\zeta = K_r^2\Omega^2 - 4\left(\eta\rho\Omega^2 - \frac{S_o}{r^2}\right)(c' - \rho\Omega^2) \tag{A-8}$$

Standing wave would occur if

$$\zeta < 0 \tag{A-9}$$

Therefore, the critical speed of standing wave formation is where

$$\zeta = 0 \tag{A-10}$$

Denoting the critical speed by V_c,

$$V_c = r\,\Omega_c \tag{A-11}$$

where the subscript refers to the critical speed. Equating ζ of (A-8) to zero and denoting

$$X = V_c^2 \tag{A-12}$$

x can be obtained by solving the quadratic equation

$$X^2 + BX + C = 0 \tag{A-13}$$

with

$$B = \frac{r^2}{4\eta}\cdot\frac{K_r^2}{\rho^2} - \frac{r^2c'}{\rho} - \frac{S_o}{\eta\rho} \tag{A-14}$$

$$C = \frac{S_o c'}{\eta\rho^2} \tag{A-15}$$

By solving Equation (A-13) for X with various values of the parameters including k_r, one can obtain the relationship between the critical speed where the standing wave forms and the parameters. Fig. 9 was obtained by varying k_r while keeping other parameters at the standard values (12).

Relation Between the Damping Coefficient k_r and Mechanical Loss of Tire

The damping coefficient k_r can be measured from the mechanical loss of tire wall specimen under cyclic flexing (12). It was also shown that the mechanical loss of tire, i.e. the rolling resistance, can be related to the mechanical loss in cyclic flexing (14). Thus, k_r represents a measure of the tire mechanical loss.

RECEIVED March 17, 1984

25

Deformation Profiles

Effect on the Properties of High Density Polyethylene Drawn by Solid-State Extrusion

TETSUO KANAMOTO, MASAYOSHI ITO, KATSUYUKI OGURA, and KOJI TANAKA

Department of Applied Chemistry, Science University of Tokyo, Kagurazaka, Shinjuku-ku, Tokyo 162, Japan

ROGER S. PORTER

Materials Research Laboratory, Polymer Science and Engineering Department, University of Massachusetts, Amherst, MA 01003

The deformation profiles on drawing of the spherulitic, semicrystalline-state of high density polyethylene (HDPE) by solid-state extrusion are markedly affected by the draw ratio and the choice of molecular weight (MW). At higher draw ratios ($\geq$ 12), a typical parabolic shear flow profile is observed on draw of a low MW, but a complex W-shape pattern for a high MW HDPE. The shear, extensional, and complex chevron pattern in these deformation profiles have been studied with respect to their effect on the efficiency of draw in achieving chain orientation and extension. DSC melting behavior, elastic recovery, and small- (SAXS) and wide-angle x-ray diffraction (WAXS) are among the techniques that have been used to characterize the deformation. The results show that the macroscopically different deformation modes within a drawn extrudate, reflecting the complex stress-strain field in an extrusion die designed for extensional flow, have no measurable effect on the resultant morphology and on the efficiency of extrusion draw at $\leq$ 120°C and at least for draw ratios and pressures up to 30 and 230 MPa. The differences and similarities in the deformation process at low draw ratio ($\leq$ 6) by solid-state extrusion and by tensile drawing are discussed on the basis of the WAXS and SAXS data.

The crystalline (solid) state extrusion of semicrystalline polymers has successfully been applied for the preparation of ultradrawn, highly-oriented, high modulus and high strength polymer morphologies (1,2). Recently (3), we have demonstrated that the deformation mode in crystalline state extrusion of high density polyethylene (HDPE) is more variable and complex than that in the extensional flow expected in conventional drawing. The deformation profile on crystalline state extrusion of melt-crystallized spherulitic HDPE changed markedly from a shear (parabolic) shape for low molecular weight (MW) to a W-shaped pattern for high MW ($\geq$ 200,000) HDPE. Other extrusion variables such as temperature, pressure, die angle, and use of lubricant affected only slightly the deformation patterns. The parabolic flow is

0097–6156/84/0260–0397$06.00/0

apparently composed of both a shear (a sheath and wall-adhesion region) and extensional (core region) deformation component, whereas the complex W-shaped pattern, shear (sheath) and chevron (core) components, reflecting a more complex stress-strain field in a die designed for extensional flow, i.e. smooth, conical surfaces of low, ~20°, angle.

An earlier electron microscopic study (4) of HDPE extrudates prepared at higher temperatures in the crystalline state of 134-137°C (4-7°C above the melting temperature of the isotropic HDPE under atmospheric pressure) reports that the fibril size is markedly different across the radius of the extrudated fibers. This observation was interpreted in terms of the macroscopic shear vs. extensional deformation components which were expected to affect the resultant morphology. However, many experimental results (5-10) such as modulus, long period, crystalline core length, amorphous and crystalline orientation, density, and thermal properties are significantly affected by the deformation temperature which may independently influence the reorganization of polyethylene morphology during deformation and annealing. Thus, the effect of apparent deformation components is considered important in understanding the morphological reorganization in crystalline-state extrusion.

In this paper we report the effect for each of the two major deformation components on the resultant morphological properties of low and high MW HDPE. The drawn extrudates exhibit the typical shear parabolic and W-shape deformation pattern, respectively. The extrusion was performed at low temperatures of 90-120°C to minimize the annealing which might occur during and after deformation in an extrusion die (9).

Experimental. Two HDPE's, duPont Alathon 7050 (M_w=58,000 and M_w/M_n=3.0), and Philips Marlex 6003 (M_w=200,000, M_w/M_n=7-13) were extruded in an Instron capillary rheometer by the solid-state coextrusion technique (10) at three extrusion temperatures (T_e) of 90, 110, and 120°C. Solid billets were prepared by melting HDPE pellets at 160°C in vacuo in a special apparatus followed by cooling (~8°C/min) to <50°C. A slight pressure was maintained during cooling to avoid the formation of voids. The billet was split longitudinally into two halves with smooth surface. Melt-crystallized spherulitic films of ~1 mm thick were compression molded at 160°C. The crystalline mass fraction of the films estimated from the heats of fusion was 70 and 68% for the low and the high MW HDPE, respectively. A ribbon of 1 cm wide cut out from the films was imprinted with lateral ink marks to follow the deformation flow profile and measure the extent of draw. The ribbon was placed between two split billet halves of the same polyethylene, and the assembly coextruded through conical brass dies which had an included entrance angle of 20° and nominal extrusion draw ratios, EDR, of 12, 25, and 30. No lubricant was used. The EDR calculated from the displacement of the line mark was in good agreement with the nominal EDR defined as the ratio of entrance to exit cross-sectional area of a die. The extruded films were used only for the evaluation of the effect of deformation components on the resultant morphology and properties. The semiperipheral coextrudates obtained simultaneously from the extruded assembly were used for the x-ray study of the deformation mechanism for extrusion drawing.

Perkin-Elmer DSC I-B and II, differential scanning calorimeters, were used for measurements of melting temperature (T_m) and heats of fusion (ΔH_f). The sample size was varied from 0.05 to 3 mg depending on the DSC heating rate to minimize the time lag effect related to the thermal conductivity and melting rate. Melting temperatures and heats of fusion were calibrated by the melting transition of Indium measured at the corresponding heating rates.

Wide-angle x-ray diffraction (WAXS) profiles were recorded on a Rigaku RAD-III diffractometer equipped with a pulse-hight discriminator. The intensity was collected by step-scan at 0.02° intervals in 2 θ, and the counting time was adjusted to accumulate the peak intensity of about 10^4 counts. The crystallite sizes; D_{200}, D_{020}, and D_{002}, along the three principal axes of the unit cell were estimated from the integral breadths of the (200), (020), and (002) reflections by the use of Sherrer's formula. The observed width were corrected for $K_{\alpha 1}$ and $K_{\alpha 2}$ radiations and for instrumental broadening by using the reflections from hexamethylenetetramine, and assuming Gaussian intensity distributions for both extrudates and the standard sample. The azimuthal scans for the (200), (020), and (002) planes were made using a Rigaku fiber specimen holder. Crystalline chain orientation was evaluated by the Herman-Stein orientation functions and from the split angles of the reflection maxima from the equator. Small-angle x-ray scattering (SAXS) and WAXS from the sample portion were simultaneously recorded photographically with a modified JEOL point collimation small-angle vacuum camera. In all x-ray work, Ni-filtered CuKα radiation was used.

Thermally induced elastic shrinkage was measured by quickly immersing a sample of ~1 cm long, ~0.2 mm thick, and ~0.5 mm wide in silicon oil bath kept at 160°C for 30 sec (11,12).

Results and Discussion. In a previous study (3), the deformation flow profile in crystalline-state extrusion of HDPE was observed for the MW range of 6-30 x 10^4 by the deformation of a lateral line pattern preimprinted on the flat surface of a split billet half (Figure 1-a), or by inserting a colored HDPE strand of the same HDPE as the billet to preformed split billets (Figure 1-b) or unsplit billets (Figure 1-c) normal to the extrusion direction. The deformation pattern changed dramatically depending on choice of MW, as shown in Figure 2 for typical samples of a low (Alathon 7050) and a high (Marlex 6003) MW HDPE.

In the present extrusion of HDPE ribbons, the deformation patterns were examined by the deformation of parallel ink marks preimprinted on the surface of a HDPE (Figure 1-d). At EDR $\geq$12, the low and high MW HDPE exhibited a typical shear parabola and a W-shape deformation profile, respectively, with both characteristics enhanced at higher EDR as shown in Figure 2. These characteristics of the deformation patterns are in well agreement with our previous observations (3) and further confirm the previous conclusion that there is no significant effect of cutting a billet into two halves and/or coextrusion of a film with the split billet halves on the deformation flow patterns.

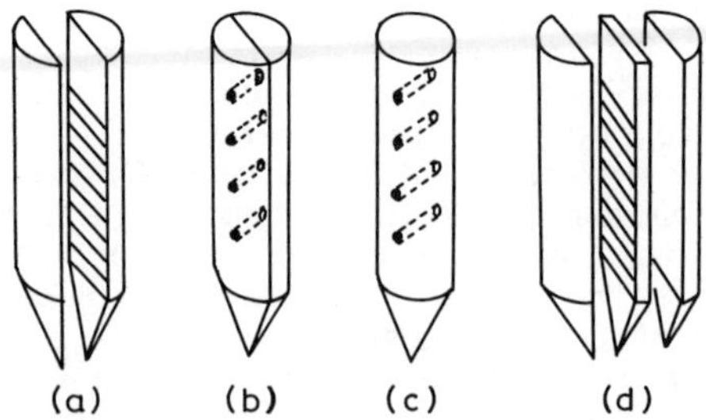

Figure 1. Observations of deformation profiles by different methods: (a); parallel ink marks on the flat surface of a split billet half, (b); inserting colored strands of the same HDPE as the split billet halves perpendicular to the flat surface and the extrusion direction, (c); the same as in (b) except an unsplit billet was used, and (d); a ribbon with parallel ink marks on the surface coextruded with two split billet halves of the same HDPE.

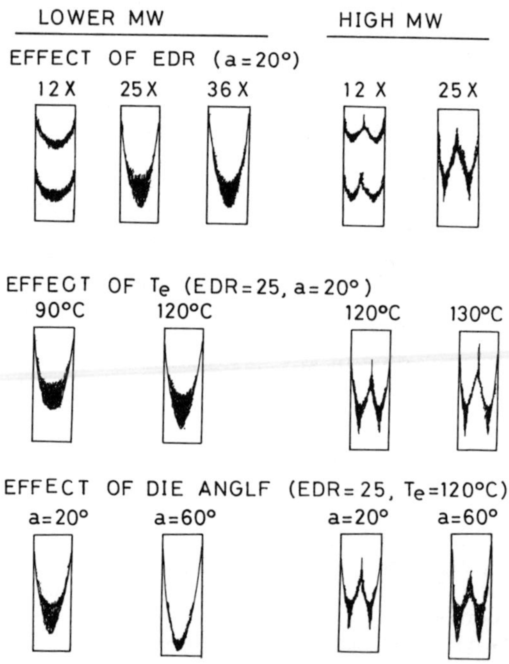

Figure 2. Schematic drawing of the effects of EDR (top), extrusion temperature (middle), and die angle (bottom) on the deformation flow profiles for two typical HDPE's having a low and a high MW.

DSC Melting Behavior

To examine the efficiency of draw for the apparently different deformation flow components, the extruded films were separated into shear (portion c in both low and high MW HDPE), extensional (a and b in the low and high MW, respectively) and concave (a in high MW) deformation components as shown in Figure 3. The samples containing all components were also prepared by cutting out the specimens perpendicular to the fiber axis (A-sample). The effect of EDR on the DSC melting endotherms are shown in Figure 4 for these A-samples from the low MW HDPE extrudates. The T_m became sharper and shifted to higher temperature with increasing EDR. The T_m and ΔH_f values for extrudates, corresponding to shear parabola (low MW) and W-shape profiles (high MW), are listed in Table 1. The DSC heating rate was 8°C/min. Both T_m and ΔH_f increase with EDR and reach a limit at EDR $\geq$ 25. The influence of T_e is not significant in the range 90 $\leq$ T_e $\leq$ 120°C studied in this work. These DSC results are consistent with our earlier work (8,13).

Table I. Melting Peak Temperatures (Tm) and Heats of Fusion (ΔH_f) of Extruded Films Measured for Samples Cut Out Perpendicular to the Extrusion Axis (Sample A). Heating Rate = 8°C/min.

		Lower MW HDPE				High MW HDPE		
Te (°C)		1#	12	25	30	1#	12	25
90	Tm (°C)	131.0	134.5	136.0		132.2	136.6	
	ΔH_f(cal/g)	48.2	52.3	57.0		47.0	47.7	
110	Tm (°C)	135.0	137.0	137.1		137.0		
	ΔH_f(cal/g)		53.2	58.8			52.7	
120	Tm (°C)				137.3			138.3
	ΔH_f(cal/g)				59.4			55.3

#; Same starting films.

Figure 5 shows the heating rate dependence of T_m for the a, b, and c portions of a W-shape profile in the high MW HDPE extrudates of EDR 12 and 25. Although the T_m is higher at higher EDR (8,13), there is no substantial difference in the T_m values for different portions deformed under different flow components. The lower MW HDPE extrudates, which showed the typical shear parabola, also showed similar melting behavior. To evaluate the effect of deformation components further, the T_m and ΔH_f for different portions of the two typical deformation patterns were measured at 10°C/min, and summarized in Table II. The extrudates were prepared at T_e=110°C, except the EDR 25 of high MW HDPE (T_e=120°C), and EDR=12, 25, and 30. Each portion of the deformation profile exhibits substantially the same T_m and ΔH_f which change depending primarily on the extent of draw.

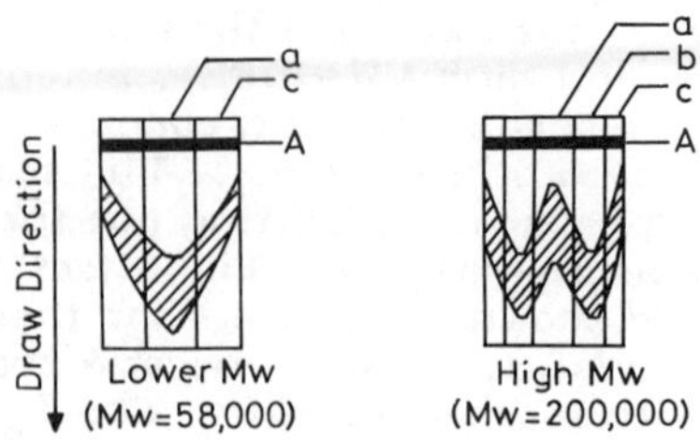

Figure 3. Separation of the sample portions, a-c, formed under different deformation components from the extruded films of a lower and a high MW HDPE.

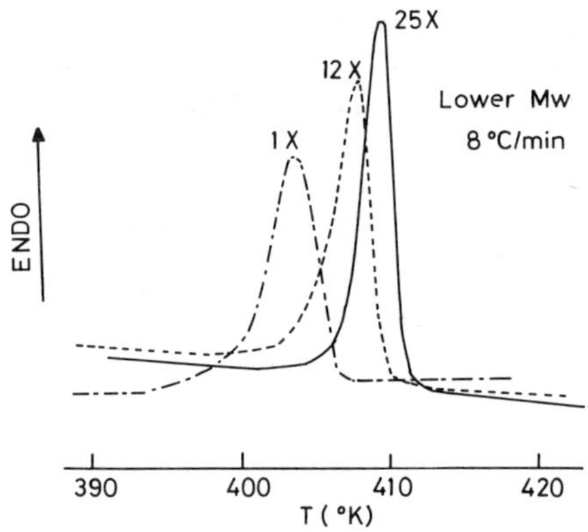

Figure 4. Effects of EDR on the DSC thermogram for the lower MW HDPE extrudates prepared at 90°C.

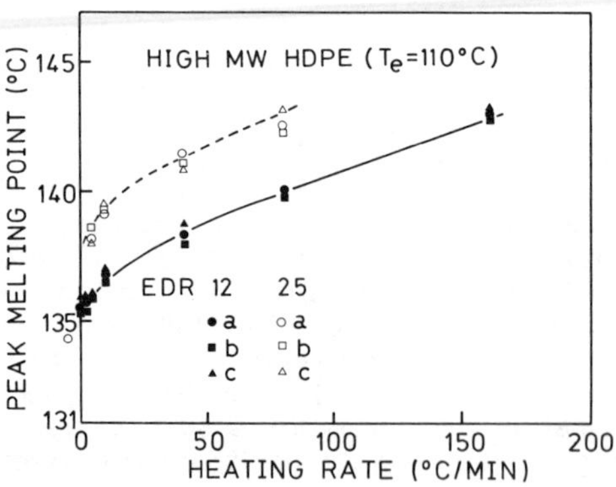

Figure 5. Melting peak temperature as a function of heating rate for the sample portions, a-c, of a high MW HDPE extruded at 110°C to EDR 12 and 25.

Table II. Melting Peak Temperatures and Heats of Fusion at Different Deformation Components for HDPE Extrudates Prepared at 110°C. (Heating rate=10°C/min).

Lower MW HDPE						
EDR	12		25		30	
Sample portion	a	c	a	c	a	c
Tm(°C)	136.5	136.4	137.0	137.1	137.9	137.7
ΔH_f(cal/g)	53.7	54.0	59.8	58.2	59.3	59.5

High MW HDPE						
EDR	12			25#		
Sample portion	a	b	c	a	b	c
Tm(°C)	136.8	136.3	137.1	139.1	139.3	139.6
ΔH_f(cal/g)	53.2	53.1	53.4	55.5	55.4	55.8

#, Prepared at 120°C.

It has been reported (8,13) and confirmed here, that the DSC T_m and ΔH_f of HDPE extrudates became insensitive to draw at higher EDR. However, as shown in Table I and II, and Figures 4 and 5 there are definitely detectable differences in T_m and ΔH_f between EDR 12 and 25 samples at heating rate of 10°C/min, indicating that there is still some insensitivity for DSC melting behavior to some morphological variations within an extrudate. This observation suggests that the shear, extensional, and concave deformation components in the complex stress and strain field in an extrusion die do not significantly affect the resultant morphology and hence the efficiency of draw as evaluated by the DSC melting behavior.

Elastic Shrinkage

It has been shown that the elastic recovery in drawn HDPE can be a sensitive measure for the efficiency of draw (11,12). The correlations between this property and morphology (14) and tensile modulus (11) have been discussed. Thus, we measured thermally induced elastic recovery for each sample portion separated from the characteristic deformation components in the low and the high MW HDPE extruded at 90 and 110°C to EDR 12 and 25. When a drawn specimen was immersed and freely floated in a silicon oil bath kept at 160°C, the elastic shrinkage quickly occurred and completed within a few seconds. The shrinkage was evaluated in terms of % recovery (R) and molecular draw ratio (MDR) defined previously (11,12).

$$R = \frac{L_t - L_s}{L_t - L_o} \times 100$$

$$MDR = \frac{L_t - L_s}{L_o} + 1$$

where, L_o is the sample length prior to extrusion, L_t the total length after extrusion, and L_s the shrunken length after elastic recovery. The average values of four measurements for each portion separated from the extrudates prepared at 110°C are shown in Table III. The samples extruded at 90°C also showed the same shrinkage behavior.

Table III. Elastic Recovery Data for Different Deformation Components in HDPE Extrudates Prepared at 110°C.

	EDR	Sample portion	Recovery (%)	MDR
Lower MW	12	a	72.9	9.0
		b	72.3	9.0
Lower MW	25	a	74.4	18.9
		b	73.6	18.7
High MW	12	a	97.6	11.7
		b	97.6	11.7
		c	97.8	11.8
High MW	25	a	97.3	24.4
		b	98.6	24.7
		c	97.3	24.4

Independent of sample portions, and EDR, the recovery is about 73 and 98% for the low and the high MW HDPE extrudates, respectively. The efficiency of draw (EFF) was evaluated by the ratio of MDR to macroscopic deformation, EDR (L_t/L_o).

$$EFF = 1 - \frac{(L_s - L_o)}{L_t}$$

The EFF depends on MW and increased from ~0.75 for the low MW (58,000) to near unity for the high MW (200,000), independent of the extrusion variables. The pronounced effect of MW on the shrinkage properties was previously observed (11) but not by the other workers (15). This is likely due to the difference in the MW ranges studied by two groups and possibly to the time heated samples at a high temperature. A prolonged heating sometimes causes an additional shrinkage due to the surface tension of a molten sample (12). The present results show that the deformation components as well as other extrusion variables have actually no important influence on the efficiency of draw as evaluated by the elastic recovery. The shrinkage is highly dependent on the MW of polyethylene.

X-Ray Analysis

The morphology of sample portions, a, b, and c formed under the characteristic deformation flow components were studied by SAXS and WAXS. The EDR 12 extrudates prepared at 90°C were used for the x-ray analysis. The SAXS and WAXS patterns of each portion showed well-developed fibrous morphology with a high degree of chain orientation. This indicates that the transformation of the initial lamellae to a fibrous morphology has been completed at EDR $\leq$ 12. More quantitative evaluation of the WAXS was made by diffractometer scans. Table IV shows the lamellar long period (ℓ) and crystallite sizes along the a (D_{200}), b (D_{020}), and c (D_{002}) unit cell axes for each deformation component in the low and the high MW HDPE. Each of these structural parameters shows a specific dependence on T_e and EDR (7,9). The D_{200} and D_{020} which likely correspond to the lateral sizes of a microfibril in extrudates are about 100 Å for all the portions. The thickness (ℓ) of folded-chain crystal blocks within microfibrils is ~205 Å, again independent of the deformation components and polyethylene MW. The average crystal stem length, D_{002}, sensitively depends on the extent of deformation (7,9,16), and has been shown to be an important morphological parameter which is well correlated with the mechanical properties of drawn HDPE. The D_{002} of each portion shows substantially the same value of 170 ± 5 Å, indicating a similar efficiency of draw for different deformation components.

Table IV. X-Ray Characterization Along the Radius of Extruded HDPE Films. (Te=90°C, EDR=12)

Sample portion		D_{200}(Å)#	D_{020}(Å)#	D_{002}(Å)#	L (Å)#
Lower MW	a	87	113	173	209
	c	95	114	171	205
High MW	a	86	110	165	207
	b	91	120	169	206
	c	90	119	175	210

#; Estimated error ± 5 Å.

+; Sample portions, a-c, as designated in the text and Figure 3.

All the results of DSC, elastic recovery, and x-ray measurements on the sample portions formed under characteristic deformation flow components indicate that the complex stress strain field in an extrusion die produces markedly different deformation flow profiles on crystalline-state extrusion of HDPE, but results in no significant effect on the morphology and properties of the extrudates prepared under the conditions used in this work. It is remembered, however, that an earlier electron microscopy study found marked morphological variations across the radius of HDPE extrudates prepared at higher temperatures (134 to 137°C) under the combined effects of chain orientation and pressure (4). The T_m of the undeformed isotropic sample was about 130°C. Therefore, at such high temperatures, the morphology of extrudates might be exposed to severe annealing during and after deformation in the complex stress field in an extrusion die, as suggested in our previous study (9). In the present experiments, extrusion was performed at temperatures ($T_e \geq 120°C$) well below the melting point (~130°C), where the effect of annealing can be suppressed minimal.

Morphological Change During Extrusion

The deformation mechanism for tensile drawing of semicrystalline polymers (17-20) has been extensively discussed based on the detailed data of the morphology and properties of drawn materials. Although the mechanism for drawing by solid-state extrusion has not been explicitly studied, a similar process of deformation is assumed (20). The results described above for highly drawn samples of HDPE (EDR ≥ 12) seem to be consistent with such an assumption. However, in some cases, the important role of the shear vs. extensional deformation components (4) in extrusion of spherulitic HDPE and the compressive force in extrusion of powders of polyethylene single crystals (21) and ultrahigh MW polyethylene (22) has been suggested. Indeed, we have shown (23) recently that single crystal mats of various polyethylenes, which could not be extended by tensile force below ~90°C can be effectively drawn even at room temperature up to EDR 20-26 by the solid-state coextrusion (10). Although some prior papers (7,24,25) report on the morphological change during extrusion of HDPE, we think it worthwhile to re-examine the process of lamellar deformation and transformation into well oriented fibrous structure. These changes in morphology will be discussed based on the SAXS and WAXS data, and compared with those previously reported for tensile drawing. For this, a semiperipheral coextrudate of a low MW HDPE (Alathon 7050) prepared simultaneously with a film coextrudate at 90°C and 230 MPa was used. It should be noted that the initial billet had a thermal history slightly different from the film used in the coextrusion. In this sample, EDR increased along the extrusion length from EDR 1 at the beginning to a constant EDR of 25 at a steady state extrusion. The EDR at a given extrudate length L, before reaching the steady state, was estimated assuming a constant volume by the following equation (2),

$$EDR(L) = \left(1 + \frac{3L\tan\theta}{r}\right)^{2/3}$$

where, θ is the half cone angle of an extrusion die and r the radius of the die exit.

Figure 6 shows some typical WAXS and SAXS photographs at different EDR taken along the extrudate length. The SAXS patterns were also schematically traced for the clarity. The uniform rings in WAXS of an isotropic sample transform to arcs which vary depending on the reflecting planes and EDR. At EDR 1.8, the WAXS intensity maxima of the (hk0) reflections are split into two arcs on both sides of the equator showing a preferential orientation for each of the a- and b-axis at an angle to the equator. The split angle decreases rapidly with increasing EDR. At EDR 3.2, the intensity maximum of the (200) reflection occurs on the equator, whereas those of the (020) are

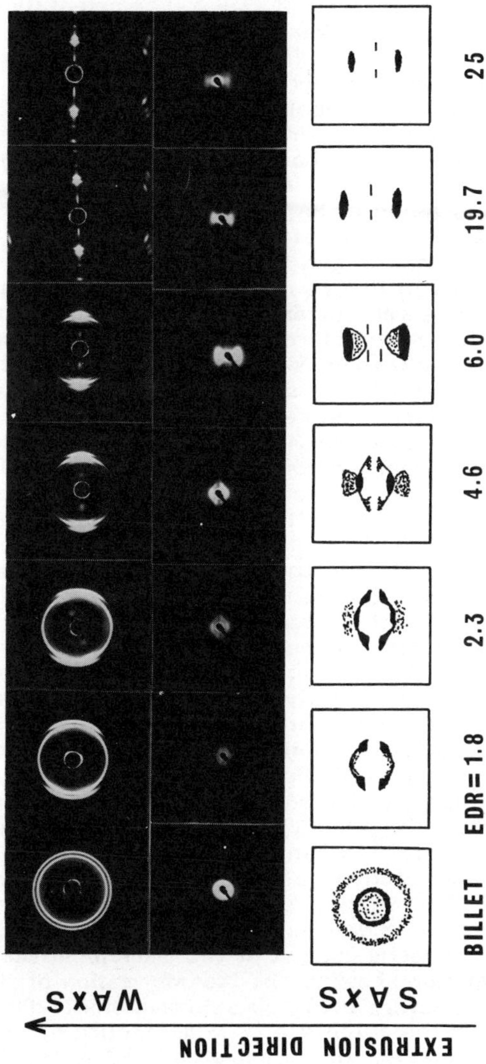

Figure 6. Typical WAXS (top) and SAXS (middle) patterns at different EDR taken along the extrudate length of a lower MW HDPE extruded at 90°C and 230 MPa. Schematic drawings of the SAXS patterns are also shown (bottom).

still slightly off equator, suggesting more rapid orientation for the a-axis than for the b-axis. Thereafter (EDR $\geq$ 4), all the (hk0) reflection maxima stay on the equator and the azimuthal intensity distribution becomes narrower with increasing EDR. The quantitative evaluation of these features were made by diffractometer scans for the (200), (020), and (002) reflections along the azimuthal direction. In Figure 7, Herman-Stein orientation functions of a a- (f_a), b- (f_b), and c- (f_c) axis are shown as a function of EDR. The orientation of the unit cell axes proceeds rapidly at lower EDR, and f_a, f_b, and f_c approach -0.5, -0.5, and 1, respectively, at EDR > 6. This means that the molecular chain align nearly perfectly along the extrusion direction and the a- and b-axes perpendicular to that at EDR > 6 for extrusion at 90°C. The high degree of chain orientation is manifested by the f_c=0.992 $\pm$ 0.001 at EDR 25. Although the orientation functions f_a and f_b clearly show that the a-axis orients faster than the b-axis at low EDR, no difference is found at higher EDR. These results are consistent with our previous study of solid-state extrusion (7) and those of tensile drawing by Kakudo and Kasai (17).

The SAXS patterns in Figure 6 show more complex changes with increasing EDR. In the small strain range (EDR $\leq$ 1.8), the initially uniform SAXS ring for an undeformed sample portion looses the intensity first on the equator without significant changes on the meridian. Concurrently, four diagonal intensity maxima grow on the diagonals and move to a slightly larger angle (from ℓ_α=305 A to 295 A), whereas the meridional maxima stay at the initial angle (ℓ_m=305 A). Thus, the SAXS intensity from partially deformed lamellae periodically varies along an ellipsoid which has the longer axis on the equator. With increasing EDR further, the intensity of the diagonal scatterings increases up to EDR ~3.4, and then decreases gradually until it disappears around EDR 6. During these changes, the scattering angle rapidly increases toward a constant angle (ℓ_m=205 A) at higher EDR (> 6). The intensity of rather sharp meridional scatterings on the ellipsoid increases with EDR in the EDR range of 2.1 to 4.6. Around EDR 4.6, the meridional scattering intensity is markedly stronger than the diagonal one which has already lost most of the intensity. Thereafter, the intensity of the meridian also diminishes rapidly. During these changes, the scattering maxima on the meridian also move toward the higher angle (ℓ_m=205 A) at higher EDR. Concurrent with these changes in the meridional and the diagonal scattering on the ellipsoid, new diffuse scatterings (ℓ_m=205 A) due to a fibrous structure appear at around EDR 2.1 on the meridian just outside the sharper scatterings (ℓ_m=305 A). The intensity of this scattering increases steadily with increasing EDR (2.1 to 6) first at the expense of the diagonal scatterings and later the equatorial scatterings on the ellipsois. At EDR > 6, the SAXS patterns consist of two meridional scatterings extended along the equator showing the transformation of the initial lamellae to fibrous structure being almost completed. The four very weak scattering along the equator at EDR 6 (Figure 6) are likely due to a small amount of the residual lamellae inclined to the fiber axis. The characteristics of the fibrous pattern shows no significant change at higher EDR ($\geq$ 10), but the SAXS intensity decreases steadily with increasing EDR.

The inclination of the lamellar (ω) and the molecular chains (ϕ) with respect to the fiber axis are schematically shown in Figure 8.

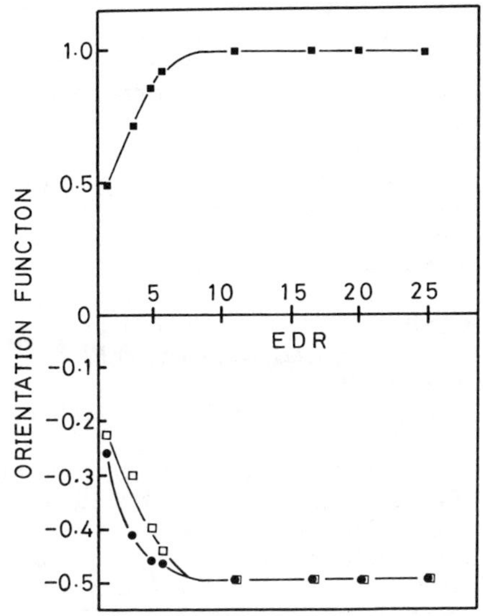

Figure 7. Orientation functions f_a (●), f_b (□), and f_c (■), as a function of EDR measured along the extrudate length of a lower MW HDPE extruded at 90°C and 230 MPa.

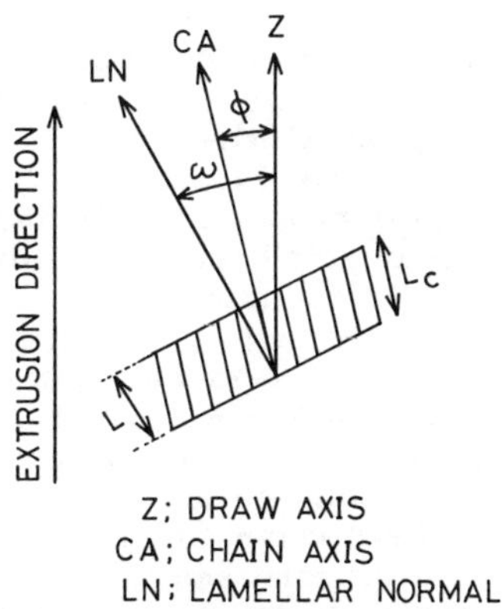

Figure 8. Orientation of the lamellae and the molecular chains with respect to the extrusion direction. The angles ω and ϕ represent tilting of the lamellar normal and the chains from the extrusion direction, respectively. L_c and L show the crystalline stem length and the lamellar long period, respectively.

The angle ω between the SAXS maxima on the diagonals and the equator determined from the photographs is plotted in Figure 9 as a function of EDR. The angle ω corresponds to the tilt of the lamellar normal to the fiber axis. The chain tilt was evaluated both from the split angles of the (200) and (020) reflection maxima and from the orientation function f_c in Figure 7. The former corresponds to the most probable angle and the latter to the weight average value.

The variations in WAXS and SAXS with increasing EDR show that the morphological change during extrusion drawing of HDPE can be divided into three major stages of deformation as has been extensively discussed for tensile drawing of semicrystalline polymers (18,20). It should be remembered that at the first stage of extrusion draw (EDR ≤ 1.8), the macroscopic deformation flow profile was nearly linear. The SAXS patterns in this range of draw indicate that the lamellae within spherulites initially oriented with the fold surfaces parallel to the extrusion direction (parallel lamellae with ω of ~90°), either distort to form inclined lamellar stacks with the lamellar normal at ~63° (ω) to the extrusion direction or transform to a fibrous structure. At this stage, however, no fibrillar scattering appears on the meridian inconsistent with the former process. The WAXS at EDR 1.8 reveals that the molecular chains orient at ~32° to the extrusion direction, indicating a tilting of chains within lamellae by $(\omega - \phi) = 31°$ to the lamellar normal (Figure 8). Another significant change is the de- in the lamellar long period from the initial $\ell_o = 305$ Å for the undeformed sample to $\ell_d = 295$ Å for the inclined lamellae at EDR 1.8. This thinning of lamellae is likely caused by the chain inclination within lamellae. The diagonal long period ℓ_d may be estimated by $\ell_d = \ell_c \cos(\omega - \phi)$, where ℓ_c is the crystalline core length. If the chains in the undeformed lamellae are initially oriented perpendicular to the fold surface, $\ell_c = \ell_o$. The ℓ_d at EDR 1.8 is thus expected to be 261 Å, significantly smaller than the observed ℓ_d of 295 Å. The ℓ_d value for the inclined lamellae (ω=63°) with chains tilted at ϕ=32° to the fiber axis can be explained when the molecular chains within the initial lamellae are inclined by 27.5° to the lamellar normal. Indeed, Keller and Sawada (26) report that the molecular chains within a lamella isolated from melt-crystallized spherulitic HDPE are tilted by 15-36° to the lamellar normal depending on the crystallization conditions. The chain tilt within a lamella plays an important role in the morphological transformation at the following stage.

At the second stage (EDR = 2.1-6), which involves disruption and unraveling of the crystalline lamellae, the macroscopic flow pattern gradually bears a shear parabolic nature with increasing EDR. The SAXS patterns show that the number of inclined lamellar increases up to EDR ~3.4 and then rapidly decreases until it almost disappears around EDR 6. During these changes, the lamellae slightly rotate toward the extrusion direction (Figure 9) and the long period ℓ_d decreases from 295 Å at EDR 1.8 to 240 Å at EDR 4.6 due to further chain tilting within lamellae. These results suggest that the inclined lamellae transform to a fibrous structure with keeping the lamellar normal tilted at 62-72° and the molecular chains at 30-17° to the extrusion direction (Figure 8).

The lamellae oriented with the fold surfaces perpendicular to the extrusion direction (perpendicular lamellae) also increases in number up to EDR ~4.6, and then rapidly disappear between EDR 4.6 and 6.

The meridional long period of these lamellae, ℓ_m, which stayed almost invariant at lower EDR (ℓ_m = 305 Å at EDR < 1.8), slightly decreases with increasing EDR (ℓ_m = 270 Å at EDR 4.6). This lamellar thinning is likely caused by the further tilting of the chains within lamellae. The rapid disappearance of these lamellae at EDR > 4.6 suggests that when the chain tilt reaches at an angle, the lamellae suddenly broke into smaller blocks due to shear slip along the chain direction and are reorganized in microfibrils oriented along the draw direction.

Concurrent with these changes, new fibrous structure appears around EDR 2.1 and steadily grows with increasing EDR. The observation that the fibrous scattering grows first at the expense of the diagonal and soon later the equatorial scatterings indicates that the transformation of lamellae to fibrils proceeds first for the inclined lamellae and then for the perpendicular lamellae. At EDR 6, thus, the transformation is almost completed. The molecular chains rapidly align to the extrusion direction but still are tilted by ~10° at this stage (Figure 7). These processes of lamellar rotation, tilting, orientation, thinning, and transformation are schematically shown in Figure 10.

As discussed above, the predominany mechanism for the morphological transformation of lamellae is shear slip along the chain direction, as has been established for tensile drawing (20,27,28), whether they are oriented at an angle or perpendicular to the extrusion direction. At lower draw ratio, however, there is a distinct difference between extrusion drawing and tensile drawing in the effect of initial lamellar orientation on the rotation or inclination of lamellae. When HDPE films were drawn by tensile force, the perpendicular lamellae first and then the parallel lamellae rotate to form inclined lamellar stacks (19). A similar behavior has been more clearly shown by Tsvankin et al. (29) in tensile drawing of low density polyethylene films with preoriented lamellar structure. A reverse effect of the initial lamellar orientation was observed on extrusion drawing of HDPE. At the initial stage of extrusion, the parallel lamellae rotated or tilted to form inclined lamellar stacks and the perpendicular lamellae tended to increase in number. The perpendicular lamellae increased up to around EDR 4.6 in the second stage where the inclined lamellae started to transform steadily into fibrils. This perpendicular lamellar orientation characteristic for extrusion drawing is likely caused by the compressive force applied for extrusion. Such an effect of the compressive force has been more clearly observed in our previous study of crystalline-state extrusion of polyethylene single crystal powder (21).

At the final stage (EDR ≥ 6), the deformation flow profiles exhibit more prominent parabolic nature. The SAXS pattern characteristic for a fibrous morphology showed only minor changes. The orientation of microfibrils as well as molecular chains increased slightly from EDR 6 to ~10 and the SAXS intensity steadily decreased with increasing EDR. The WAXS and SAXS patterns indicated a morphological change similar to tensile drawing. The deformation in this range of draw proceeds by interfibrillar slip induced by the tilt of chains and fibrils and by a shear component coming from inhomogenieties in the sample. The shear deformation component in an extrusion die may also accelerate such deformation. The steady decrease in SAXS intensity has previously been ascribed to the combined effects of the

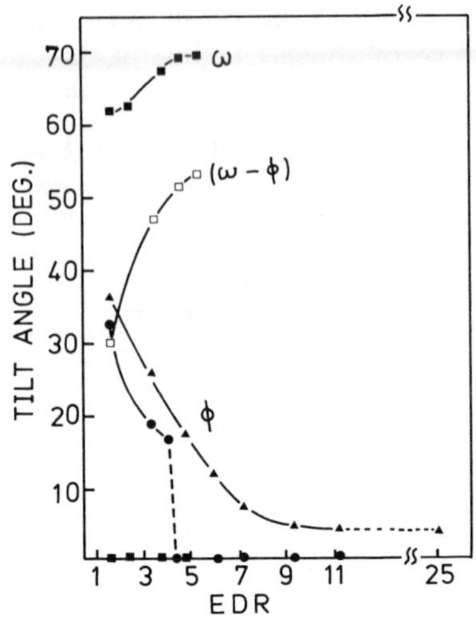

Figure 9. The angles between the extrusion direction and the lamellar normal (ω; ■) and the chains (ϕ; ▲, and ●), and the chain tilt within a lamella (ω−ϕ; □) as a function of EDR. The chain tilt is expressed by the most probable angle (●) and weight average value (▲).

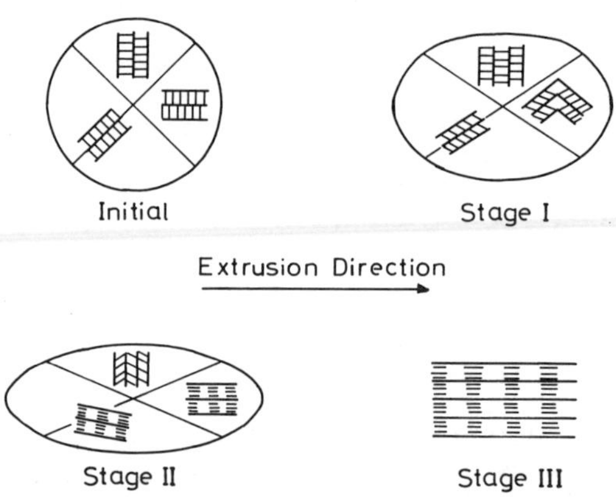

Figure 10. Models for the deformation of lamellae within a spherulite during drawing by solid-state extrusion based on the WAXS and SAXS patterns.

increase in the amorphous density and the decrease in the crystal density (30,31) as well as the decrease in the amount of chain folds in microfibrils (9).

Although some specific differences in the lamellar deformation process have been found between extrusion drawing and tensile drawing at low draw ratio, the present x-ray study shows that no significant effect of draw techniques on the basic deformation mode and the resultant morphology at higher EDR. According to Peterlin (20,27,28), the plastic deformation of fibrous materials proceeds by mutual displacement of fibrils which produces interfibrillar tie molecules. The frictional forces on the lateral surface of each microfibril during slippage yield a shear stress on each fibril which causes a displacement of chains inside the crystal blocks. Such slip produces new crystalline bridges in the interior of microfibrils (27). These all suggest that the macroscopically extensional deformation in tensile drawing is basically shear internally, and the important morphological elements of inter- and intra-fibrillar tie molecules are formed during the shear deformation. This is consistent with our conclusion that macroscopically shear and extensional deformation components in crystalline-state extrusion of HDPE have no significant influence of the efficiency of draw as discussed above.

Conclusions. We confirm that the deformation profiles on crystalline-state extrusion of melt-crystallized HDPE dramatically changed with choice of molecular weight, MW. A low MW HDPE exhibited a typical shear parabora, whereas a high MW gave a complex W-shape profile with this tendency increasing with EDR, irrespective of the extrusion conditions, as reported previously. The small- and wide-angle x-ray diffraction show some differences between crystalline-state extrusion and tensile drawing in the lamellar deformation process only at a low deformation ratio. The DSC melting behavior, elastic recovery, and x-ray diffraction indicate that the macroscopically-distinct deformation components, i.e. shear versus extensional and combined chevron components, do not significantly affect the resultant morphology and hence also does not influence the efficiency of draw for extrusion performed at temperatures well below the melting point of the isotropic samples.

Literature Cited

1. Zachariades, A.E.; Mead, W.T.; Porter, R.S. In "Ultra-High Modulus Polymers"; Cifferi, A.; Ward, I.M., Ed.; Appl. Sci. Pub.; London, 1979.
2. Zachariades, A.E.; Porter, R.S.; Eds.; "Strength and Stiffness of Polymers"; Plastic Engineering Series, 1983; Vol. 4, Marcel Dekker.
3. Kanamoto, T.; Zachariades, A.E.; Porter, R.S. Polym. J. 1979, 11, 307.
4. Crystal, R.G.; Southern, J.H. J. Polym. Sci. 1971, A-2, 9, 1641.
5. Perkins, W.G.; Porter, R.S. J. Mater. Sci. 1977, 12, 2355.
6. Mead, W.T.; Desper, C.R.; Porter, R.S. J. Polym. Sci., Polym. Phys. Ed. 1979, 17, 859.
7. Kanamoto, T.; Fujimatsu, S.; Tanaka, K.; Porter, R.S. Repts. Progr. Polym. Phys. Japan 1981, 24, 185.

8. Kojima, S.; Porter, R.S. J. Appl. Polym. Sci., Appl. Polym. Symp. 1978, 33, 129.
9. Tsuruta, A.; Kanamoto, T.; Tanaka, K.; Porter, R.S. Polym. Eng. Sci. 1983, 23, 521.
10. Griswold, P.D.; Zachariades, A.E.; Porter, R.S. Polym. Eng. Sci. 1978, 18, 861.
11. Watts, M.P.C.; Zachariades, A.E.; Porter, R.S. J. Mater. Sci. 1980, 15, 426.
12. Porter, R.S.; Daniels, M.; Watts, M.P.C.; Perira, J.C.; DeTeresa, S.J.; Zachariades, A.E. J. Mater. Sci. 1981, 16, 1134.
13. Kojima, S.; Desper, C.R.; Porter, R.S. J. Polym. Sci., Polym. Phys. Ed. 1978, 16, 1721.
14. Barham, P.; Keller, A. J. Polym. Sci., Polym. Letters Ed. 1975, 13, 197.
15. Capaccio, G.; Ward, I.M. Colloid & Polym. Sci. 1982, 260, 46.
16. Gibson, A.G.; Ward, I.M. Polymer 1978, 19, 683.
17. Kasai, N.; Kakudo, M. J. Polym. Sci. 1964, A-2, 2, 1955.
18. Hay, I.L.; Keller, A. Kolloid-Z & Z. Polym. 1965, 204, 43.
19. Meinel, G.; Morosoff, N.; Peterlin, A. J. Polym. Sci. 1970, A-2, 8, 1723.
20. Peterlin, A. Polym. Eng. Sci. 1977, 17, 183.
21. Kanamoto, T.; Sherman, E.S.; Porter, R.S. Polym. J. 1979, 11, 497.
22. Zachariades, A.E.; Watts, M.P.C.; Kanamoto, T.; Porter, R.S. J. Polym. Sci., Polym. Letters Ed. 1979, 17, 485.
23. Kanamoto, T.; Porter, R.S. Presented at the Annual Meeting, Polym. Soc., Japan, Polym. Prep. Japan 1983, Vol. 32, No. 4, 743.
24. Imada, K.; Yamamoto, T.; Shigematsu, K.; Takayanagi, M. J. Mater. Sci. 1971, 6, 537.
25. Sherman, E.S.; Porter, R.S.; Thomas, E.L. Polymer 1982, 23, 1069.
26. Keller, A.; Sawada, S. Makromol. Chem. 1964, 74, 190.
27. Peterlin, A. J. Appl. Phys. 1977, 48, 4099.
28. Peterlin, A. Polym. Eng. Sci. 1979, 19, 118.
29. Gerasimov, V.I.; Genin, Y.V.; Tsvankin, D.Y. J. Polym. Sci., Polym. Phys. Ed. 1974, 12, 2035.
30. Peterlin, A.; Corneliussen, R. J. Polym. Sci. 1968, A-2, 6, 1273.
31. Adams, W.W.; Thomas, E.L.; Lin, J.S. Polym. Prep. 1983, 24, No. 2, 372.

RECEIVED May 10, 1984

26

Tailoring Polymer Molecules for Specific Fiber and Ribbon Properties

THADDEUS E. HELMINIAK and ROBERT C. EVERS

Air Force Wright Aeronautical Laboratories, Materials Laboratory, Wright-Patterson Air Force Base, OH 45433

Based on the unique film forming characteristics of certain rigid-rod aromatic-heterocyclic polymers capable of forming liquid crystalline solutions, a research and development effort to explore the potential of these "ordered polymers" as structural materials was initiated. Both poly (p-phenylenebenzobisoxazole) (PBO) and poly(p-phenylenebenzobisthiazole) (PBT) exhibited outstanding thermooxidative stability and formed nematic solutions in methanesulfonic acid or polyphosphoric acid. Attempts to modify polymer structure to permit polymer solubility and processing from noncorrosive solvents were unsuccessful. PBT and PBO were processed into fiber specimens which exhibited a high degree of order. In the case of PBT, modulus and tenacity values as high as 2600gpd (53.3Msi) and 30gpd (615Ksi), respectively, were obtained after appropriate heat treatment. Uniaxially oriented ribbons of PBT were also extruded which after heat treatment exhibited a tensile modulus of 1800gpd (36.9Msi) and a tenacity of 25gpd (512Ksi). Efforts were directed toward an improved understanding of the morphology of these "ordered polymers," PBO and PBT, either neat or in "composite" form with flexible matrix polymers.

The Nonmetallic Materials Division of the Materials Laboratory has long been interested in aromatic-heterocyclic polymers as potential structural materials because of their high thermooxidative stability and environmental resistance. More recently, certain of these polymers have been intensely investigated because of their unique film forming characteristics and ability to form liquid crystalline solutions in strong acids. Their resultant potential as high performance structural materials has consequently been demonstrated in the fabrication of high strength-high modulus fibers, ribbons, and molecular composites. This paper describes early research leading to the discovery of these "ordered polymers" and the joint research and development program by Air Force Wright Aeronautical Laboratories (AFWAL) and Air Force Office of Scientific Research (AFOSR) to exploit the potential of rigid-rod aromatic-heterocyclic polymers as high strength fibers and ribbons through modification of polymer structure and morphology.

Background

Extensive research carried out within and under the sponsorship of the Materials Laboratory during the last decade sought improvement in the thermooxidative stability and environmental resistance of polymeric materials. These efforts focused on aromatic-heterocyclic structures and included consideration of ladder polymer structures which ultimately led to the observation of a unique film forming phenomena. The first observation of this was for the ladder polymer BBL.(1)

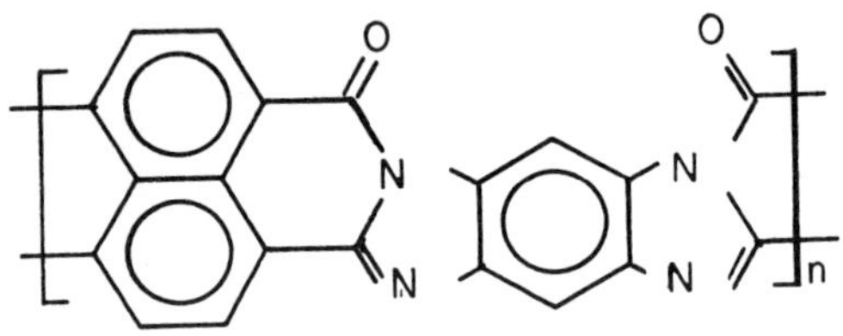

BBL

When BBL was dried from a nonsolvent precipitated slurry, a precipitated film, possessing excellent mechanical properties, was formed by the coalescence of discrete particles of solid matter without going through a melt phase. Although further research considering the relationship between the macromolecular structure and this film forming phenomenon showed that the molecular geometry of the polymer chain was the critical factor,(2) the film forming phenomenon was only observed for aromatic-heterocyclic ladder polymers. Concurrent work within the Materials Laboratory attempting to improve the molecular ordering of rigid-rod aromatic-heterocyclic polymers led to the observation that certain rigid-rod, aromatic-heterocyclic polymers capable of forming liquid crystalline solutions also formed precipitated films comparable to those previously observed only for highly extended ladder polymer molecules. Furthermore, it was proposed that the high strength of these unique precipitated films stemmed from their "composite" character.(3,4) They were considered "composites" because the precipitated films were formed by the aggregation and coalescence of individual microscopic sheets of precipitated polymer. The high strength was attributed to the inherent strength of the microscopic sheets due to their high degree of molecular order. These observations and other considerations led to the hypothesis that a nonreinforced composite with useful mechanical properties might be achieved with appropriate polymer chain geometry and suitable processing. The key factor for success would be the ability to achieve a high degree of molecular order. Further, the tendencies of those rigid-rod polymers to form liquid-crystalline solutions strongly suggested the possibility of highly anisotropic fiber and ribbon formation. The Materials Laboratory with the additional support of AFOSR undertook the task of exploring the potential of this "ordered polymer" approach to structural materials for aerospace applications. The principal efforts of this program initially have been concerned with the choice of the macromolecular chemical structure while concurrent studies, theoretical and experimental, have been considering the problems of characterization, evaluation and processing into high strength fibers and ribbons.

Discussion

Materials. On the basis of the observations made above, the choice of molecular design has centered on that of an extended chain, rigid-rod molecule for reasons of molecular ordering and an aromatic-heterocyclic structure for thermal and oxidative stability. Three such polymers have been synthesized, PDIAB,(5) PBO,(3,6,7) and PBT.(8)

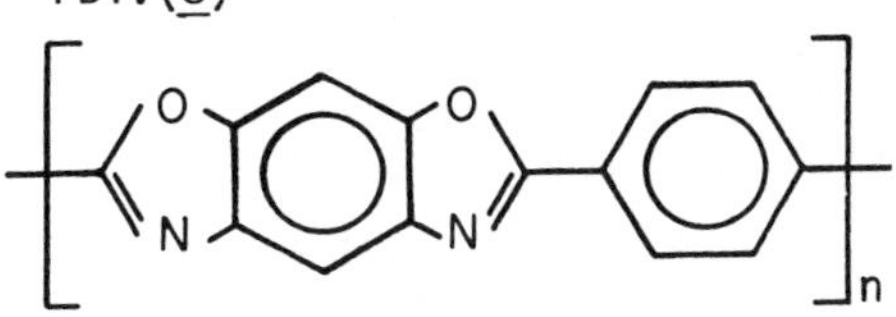

PBO

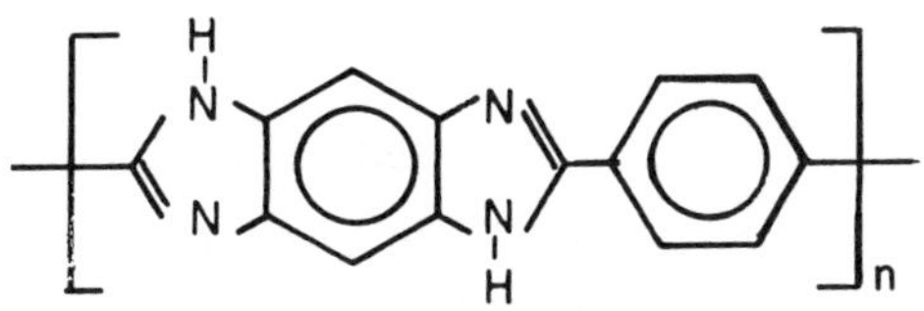

PDIAB

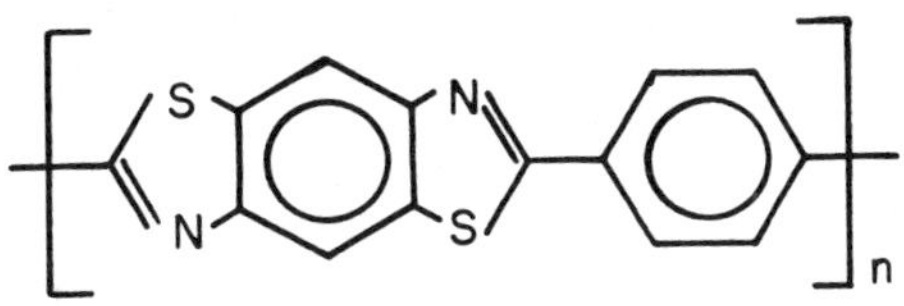

PBT

The PBO polymer system offers improved thermooxidative stability over PDIAB and does not pick up the moisture that the benzimidazole structure does, whereas PBT provides the best thermal and oxidative stability of the three systems (weight retention of $>$ 50% after 200 hours at 371°C in circulating air).

Although these three materials have been successfully synthesized, they have presented special processing problems because of the extended chain, rigid-rod structural character of the molecules. Present processing requires strong mineral or organic acid solvents such as methanesulfonic acid or polyphosphoric acid. This problem of solubility has during the course of the research manifest itself in several ways; the limited number and type of solvents, relatively low solubility and difficulty in polymerization to high molecular weight due to apparent insolubility. This problem has been addressed with two approaches. The first attack of the problem has been an extensive solubility study to discover noncorrosive organic solvents or to determine organic chemical structures that could be expected to dissolve these polymers.(9) The results have not provided a pure organic single solvent nor do the results hold significant promise for a new single processing solvent.(10) A second approach to the improvement of solubility consists of polymer chain structural modifications either in the form of pendant additions or the introduction of swivel joints into the main chain forming an articulated molecular structure. The former has been accomplished for PBO(6) and PBT(8) to give the modified PBO and PBT structures shown below.

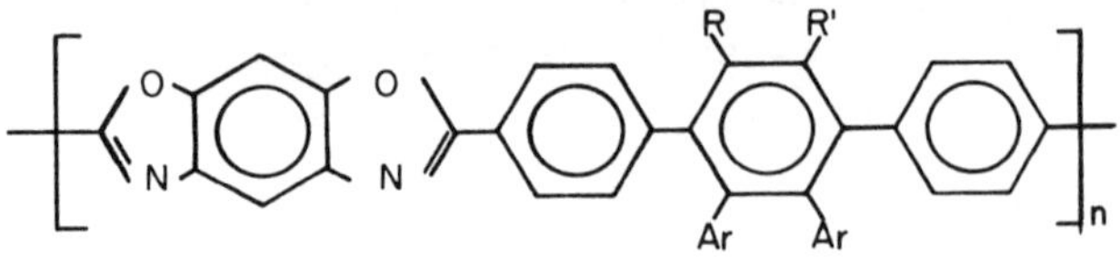

Modified PBO where R and R' can be H or Ar

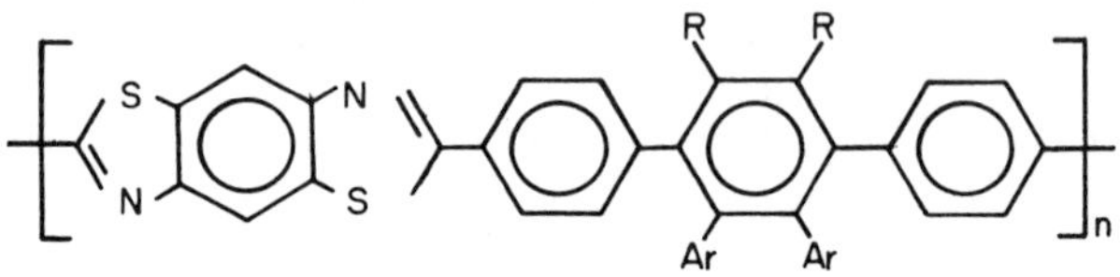

Modified PBT where R and R' can be H or Ar

In neither case were pendant phenyl groups of sufficient influence to significantly improve the solubility or offer processing solvent alternatives of practical value. The synthesis of articulated molecular structures was based on theoretical considerations of the phase equilibria of rodlike polymers.(11) Articulated PBO, PBT, and PDIAB polymers which contained diphenoxybenzene "swivels" were synthesized(12) as were PBT and PBO structures with thermooxidatively stable biphenyl and bipyridyl "swivels".(13) Although differences from PBO, PBT and PDIAB in film-forming properties and solution behavior were observed,(14) the solubilities were not sufficiently altered to afford processing alternatives.

Characterization. Extensive characterization has been carried out on PBO and PBT establishing that the polymers exhibit rod-like behavior in solution and can exist in either optically isotropic or anisotropic states depending on temperature, concentration and molecular weight.(15-17) These data show that PBT is rodlike, protonated in strong acids used as its solvent and is of a fairly high degree of polymerization. A weight average degree of polymerization of approximately seventy is readily attainable and intrinsic viscosity values indicating much higher molecular weights have been recorded.(18) Phase equilibria data indicate PBT can form stable solutions in the nematic state as can PBO. However, PBO and PDIAB, as measured by intrinsic viscosity, have never been polymerized to as high a molecular weight as PBT ($[\eta]$=47dl/gm).(18) Concerns regarding the kinetics and mechanism for the polymerization of extended chain rigid-rod macromolecules to high molecular weight have prompted a study to address this problem.(19)

Processing. Processing of the "ordered polymers" into fiber specimens was carried out in order to assess the potential of these polymers for applications requiring high modulus and strength. Through this approach, it was possible to minimize the sample quantities required and take advantage of developed technology to provide the requisite specimens for evaluation of mechanical properties and morphology. Initially, PBO of moderate molecular weight ($[\eta]$=3.88dl/g) was wet spun from methanesulfonic acid dopes into fiber specimens exhibiting tensile modulii of approximately 700gpd (14 Msi) after heat treatment.(7) Low tenacity values of 5gpd (102Ksi) were recorded, attributable to the limited polymer molecular weight and/or processing defects. The processing defects arose from incursions into the fiber surface which occurred during coagulation, resulting in a reduction of cross-sectional area having well oriented molecules and, thus, a lower tenacity. With the subsequent availability of higher concentration dopes of high molecular weight PBT, dry-jet wet spinning of these methanesulfonic acid and polyphosphoric acid dopes led to PBT fibers with as-spun modulus and tenacity as high as 2050gpd (42Msi) and 20gpd (400Ksi), respectively.(20,21) Through appropriate heat treatment of the PBT fiber, these values were increased to 2600gpd (53Msi) and 30gpd (615Ksi), respectively, with an elongation of 1.1 percent being recorded. Efforts are continuing to improve on the spinning and coagulation process to eliminate or minimize incursion defects and to optimize fiber properties.

Efforts have also been made to process both PBO and PBT into a continuous ribbon. While initial attempts to extrude PBO dopes were not particularly promising,(7) subsequent efforts with PBT dopes have yielded uniaxially oriented ribbons (1.0in. x 0.5mil) with as spun properties of 650gpd (13Msi) modulus, 13gpd (250Ksi) tenacity, and 3.6 percent elongation.(22) Appropriate heat treatment of the ribbons increased the modulus and tenacity values to 1800gpd (37Msi) and 25gpd (512Ksi), respectively, and decreased the elongation to one percent. As with the PBO and PBT fibers, investigations of the film morphology indicated a high degree of orientation. Efforts are currently underway to obtain wider ribbons as well as biaxially oriented specimens.

Concurrent to these efforts to process the neat polymers, the rigid-rod molecules were used as reinforcement in flexible aromatic-heterocyclic polymer and thermoplastic polymer matrices to provide composites at the molecular level which are analogous to chopped fiber composites.(23) Initial investigations revealed that these molecular composites exhibited significant increases in tensile modulus and strength with only ten percent of rigid-rod polymer in the matrix.(24) Subsequent morphology studies for vacuum cast and shear quenched films showed that maximum dispersion and orientation of rod-like PBT molecules in a poly-2,5(6)-benzimidazole matrix is achieved from processing from solution at or near its critical concentration.(25) These studies are continuing.

Future efforts are being directed toward an improved understanding of the morphology of the "ordered polymers" PBO and PBT, neat or in "composites" form with flexible matrix polymers. Emphasis will continue to be placed on learning to process the extended chain, rigid-rod aromatic-heterocyclic polymers into useful forms for structural applications.

Literature Cited

1. Arnold, F. E.; Van Deusen, R. L. J. Appl. Polymer Sci. 1971, 15, 2035.
2. Sicree, A. J.; Arnold, F. E.; Van Deusen, R. L. J. Polymer Sci., Polymer Chem. Ed. 1974, 12, 265.
3. Helminiak, T. E.; Arnold, F. E.; Benner, C. L. Am. Chem. Soc., Prepr. Polymer Div. 1975, 16 (2), 659.
4. Helminiak, T. E.; Arnold, F. E. U.S. Patent 4 051 108, 1977.
5. Kovar, R. F.: Arnold, F. E. J. Polymer Sci., Polymer Chem. Ed. 1976, 14, 2807.
6. Wolfe, J. F.; Arnold, F. E. Macromolecules 1981, 14, 909.
7. Choe, E. W.; Kim, S. N. Macromolecules 1981, 14, 920.
8. Wolfe, J. F.; Loo, B. H.; Arnold, F. E. Macromolecules 1981, 14, 915.
9. Bonner, D. C. Air Force Materials Laboratory Technical Report AFML-TR-76-91 1976.
10. Bonner, D. C. Air Force Materials Laboratory Technical Report AFML-TR-77-73 1977.
11. Flory, P. J. Macromolecules 1978, 11, 1141.
12. Evers, R. C.; Arnold, F. E.; Helminiak, T. E. Macromolecules 1981, 14, 925.
13. Evers, R. C. U.S. Patent 4 359 567, 1982.
14. Berry, G. C., private communications, Carnegie-Mellon University (Jan 1983).
15. Berry, G. C.; Casassa, E. F.; Metzger, P.; Venkatraman, S. Air Force Materials Laboratory Technical Report AFML-TR-78-164 (Part II, 1980.
16. Berry, G. C.; Casassa, E. F.; Lee, C. C.; Furukawa, R.; King, R. S.; Veukatraman, S. Air Force Wright Aeronautical Laboratories Technical Report AFWAL-TR-80-4099, 1980.
17. Chu, S. G.; Vemkatraman, S.; Berry, G. C.; Einaga, Y. Macromolecules 1981, 14, 939.
18. Wolfe, J. F., private communications, SRI International (Jan 1983).
19. Cotts, D. B.; Berry, G. C. Macromolecules 1981, 14, 930.
20. Hwang, W-F.; Wiff, D. F., private communications, University of Dayton Research Institute (Jan 1983).
21. Allen, S. R.; Filippov, A. G.; Farris, R. J.; Thomas, E. L.; Wong, C. P.; Berry, G. C.; Chenevey, E. C. Macromolecules 1981, 14 1135.
22. Chenevey, E. C., private communications, Celanese Research Company (Jan 1983).
23. Helminiak, T. E.; Benner, C. L.; Arnold, F. E.; Husman, G. U.S. Patent 4 207 407, 1980.
24. Husman, G.; Helminiak, T. E.; Adams, W. W.; Wiff, D. R.; Benner, C. L. Amer. Chem. Soc., Prepr. Org. Coat. Plast. Div. 1979, 40, 797.
25. Hwang, W-F.; Wiff, D. R.; Helminiak, T. E. Amer. Chem. Soc., Prepr. Org. Coat. Plast. Div., 1981, 44, 32.

RECEIVED May 11, 1984

INDEXES

Author Index

Subject Index

D

E

M

N

O

P

Production and indexing by Karen McCeney
Jacket design by Pamela Lewis

Elements typeset by Hot Type Ltd., Washington, D.C.
Printed and bound by Maple Press Co., York, Pa.

RECENT ACS BOOKS

"Treatment and Disposal of Pesticide Wastes"
Edited by James N. Seiber and Raymond F. Krueger
ACS SYMPOSIUM SERIES 259; 384 pp.; ISBN 0-8412-0858-1

"Stable Isotopes in Nutrition"
Edited by Judith R. Turnlund and Phyllis E. Johnson
ACS SYMPOSIUM SERIES 258; 240 pp.; ISBN 0-8412-0855-7

"Bioregulators: Chemistry and Uses"
Edited by Robert L. Ory and Falk R. Rittig
ACS SYMPOSIUM SERIES 257; 286 pp.; ISBN 0-8412-0853-0

"Polymeric Materials and Artificial Organs"
Edited by Charles G. Gebelein
ACS SYMPOSIUM SERIES 256; 208 pp.; ISBN 0-8412-0854-9

"Pesticide Synthesis Through Rational Approaches"
Edited by Philip S. Magee, Gustave K. Kohn, and Julius J. Menn
ACS SYMPOSIUM SERIES 255; 351 pp.; ISBN 0-8412-0852-2

"Advances in Pesticide Formulation Technology"
Edited by Herbert B. Scher
ACS SYMPOSIUM SERIES 254; 264 pp.; ISBN 0-8412-0840-9

"Structure/Performance Relationships in Surfactants"
Edited by Milton J. Rosen
ACS SYMPOSIUM SERIES 253; 356 pp.; ISBN 0-8412-0839-5

"Chemistry and Characterization of Coal Macerals"
Edited by Randall E. Winans and John C. Crelling
ACS SYMPOSIUM SERIES 252; 192 pp.; ISBN 0-8412-0838-7

"Conformationally Directed Drug Design:
Peptides and Nucleic Acids as Templates or Targets"
Edited by Julius A. Vida and Maxwell Gordon
ACS SYMPOSIUM SERIES 251; 288 pp.; ISBN 0-8412-0836-0

"Ultrahigh Resolution Chromatography"
Edited by S. Ahuja
ACS SYMPOSIUM SERIES 250; 240 pp.; ISBN 0-8412-0835-2

"Chemistry of Combustion Processes"
Edited by Thompson M. Sloane
ACS SYMPOSIUM SERIES 249; 286 pp.; ISBN 0-8412-0834-4

"Polymer Blends and Composites in Multiphase Systems"
Edited by C. D. Han
ADVANCES IN CHEMISTRY SERIES 206; 400 pp.; ISBN 0-8412-0783-6